SONIC WIND

SONIC WIND

THE STORY OF JOHN PAUL STAPP AND HOW A RENEGADE DOCTOR BECAME THE FASTEST MAN ON EARTH

CRAIG RYAN

LIVERIGHT PUBLISHING CORPORATION

A Division of W. W. Norton & Company

Independent Publishers Since 1923

New York · London

Copyright © 2015 by Craig Ryan

For information about permission to reproduce selections from this book,
write to Permissions, Liveright Publishing Corporation, a division of
W. W. Norton & Company, Inc., 500 Fifth Avenue, New York, NY 10110

For information about special discounts for bulk purchases, please contact
W. W. Norton Special Sales at specialsales@wwnorton.com or 800-233-4830

Manufacturing by RR Donnelley, Harrisonburg, VA
Book design by Dana Sloan
Production manager: Anna Oler

Library of Congress Cataloging-in-Publication Data

Ryan, Craig, 1953–
Sonic wind : the story of John Paul Stapp and how a renegade doctor became the
fastest man on Earth / Craig Ryan.
    pages cm
Includes bibliographical references and index.
ISBN 978-0-87140-677-4 (hardcover)
1. Stapp, John P. (John Paul), 1910–1999. 2. Biophysicists—United States—
Biography. 3. United States. Air Force—Officers—Biography. 4. Motor vehicles—
Safety appliances—Research—United States—History—20th century.
5. Aeronautics—Safety measures—Research—United States—History—20th
century. I. Title.
TL140.S688R93 2015
571.4092—dc23
[B]
                            2015013245

ISBN 978-0-63149-191-0 pbk.

Liveright Publishing Corporation, 500 Fifth Avenue, New York, N.Y. 10110
www.wwnorton.com

W. W. Norton & Company Ltd., Castle House, 75/76 Wells Street,
London W1T 3QT

1 2 3 4 5 6 7 8 9 0

For the Alaskans
Autumn, Cedar, Ava, Molly and Porter

CONTENTS

PART III: ABOVE THE ARMSTRONG LINE

PART IV: THE GHOSTS THAT NEVER HAPPENED

AUTHOR'S NOTE: All quotes from written sources are rendered verbatim, including spelling and punctuation.

A life's work in the agony and sweat of the human spirit.

—William Faulkner

SONIC WIND

PROLOGUE

O N THE AMBULANCE RIDE to the Holloman Air Force Base hospital following the most spectacular experiment in human tolerance to dynamic force ever attempted, John Paul Stapp was already looking ahead to Hurricane Mesa. In the exhilaration that always hit him when he survived one of these things, his heart still banging away, Stapp was onto the next challenge, the next piece of the puzzle—though the medical people didn't yet know how badly he'd hurt himself on this one.

Months earlier, he had been invited to inspect Hurricane Mesa as engineers kicked off construction of the 12,000-foot test track at the top-secret Hurricane Supersonic Research Site in southern Utah. They'd built the facility to prove out new designs of ejection seat systems for high-speed jets. The seats were extremely sophisticated machines created by some of the most talented aeronautical engineers in the world. Still, in action, one out of every four pilots using ejection seats was being killed in the process. They needed better, more reliable escape systems, and they needed a way to qualify them for use by the Air Force's most valuable assets: its trained fighter pilots.

Once completed, the track would slice across the plateau of a dramatic tabletop mesa rising nearly half a mile above the Virgin River valley, the heavy-duty rails bolted to flat bedrock and extending right to the edge of the canyon wall. As Stapp had stood there and looked out across the magnificent high desert, he'd been able to imagine it all in perfect detail, strapped to the sleek new test sled, in a very bad mood as he always was in the moments before the launch, waiting for technicians to light the rockets.

Once the telemetry was dialed in and everyone had scampered into the concrete control bunker, they would fire him down the track so fast he'd lose all sensation of time. He would cover the two-plus miles across the mesa top in a blur, his cheeks flapping like rubber and the windblast grinding into his chest like the heel of a giant hand, and would be propelled by the power of twelve rockets right off the lip of the cliff and out across the desert valley. As he pierced the sound barrier and finally hit 1,000 miles per hour, the g-forces would smash him against the back of the seat so hard it would squeeze the last bit of air from his lungs.

The ejection sequence: A cartridge activates a catapult propelling Stapp's seat up a vertical set of guide rails. As he clears the sled, a small rocket fires him straight up, the rocket's thrust fighting the forces trying to flip him like a coin. The seat executes a slow counterclockwise quarter revolution. A couple of seconds later the parachute deploys, triggering a crank that pulls the pins from the harness mechanism, releasing Stapp and parachute from the seat. He falls free and glides to a one-somersault landing on the banks of the Virgin. Gentlemen, Stapp would say to the press, like a magician revealing that his beautiful assistant hadn't been sawed in half after all, we've now proven this technology can be trusted with the lives of the best pilots in the world.

Back at Holloman, shortly after his arrival at the hospital, the base public information officer informed him that, under the circumstances, they were going to wait at least two days before bringing in the reporters and photographers. One of the nurses handed him a mirror, but he couldn't see anything other than a gray fuzz. They told him he looked hideous. Two massive shiners and eyeballs like raw meat. The newspaper guys would love that. They'd want to know how fast he'd gone, and what the purpose of it had been. They'd want to know what it *felt* like. Most of all, they'd want to know about his injuries. Had he been afraid he was going to die?

He never talked about fear. He approached the whole business with a nihilist mind-set. Afterward came the euphoria; he had made it through one more time, and proved the doubters and the "Stamp Out Stapp" boys wrong. In a couple of days, when he'd healed up enough to be presentable and they let the press in to see him, he'd give them a show and reveal his plans for the 1,000-miles-per-hour run. Stapp loved an audience, and he'd ham it up. He'd get them laughing. He'd explain that, hell yes, they'd just managed to *simulate* a high-speed ejection on the track at Holloman ... but Hurricane Mesa! Hurricane Mesa would give him a chance to conduct controlled airborne tests with a live volunteer. He wouldn't tell them everything, of course. He never did. He needed to tread carefully with the Air Force brass in order to secure the approvals. It was like Stapp had told the movie producer in L.A.: he had the master plan and they would all be told what they needed to know in good time.

Meanwhile, after his medical examination and some bandaging, and a quick call patched through to Systems Command headquarters to inform General Flickinger that it had all gone according to script and to assure him they hadn't lost anybody, Stapp's thoughts drifted back to Hurricane Mesa as he tried to work out the math for the windblast at Mach 1.3. Never mind the g-forces that wanted to suck the eyes out of your skull: could you really expose a man to supersonic friction and keep from killing him?

Now *that* was a problem worth solving.

PART I

THE GLORIOUS REBELLION AGAINST DISSOLUTION

1

WHERE ARE MY CHILDREN TO-NIGHT?

> My only diversion is to dream all night about one or another of you . . . We might even worry about you if we were not worked to the limit each day and hour of our existence.
>
> —*Mary Louise Stapp, letter to her sons from Brazil*

IN JANUARY OF 1910, when Charles Franklin Stapp left behind the frontier town of Burnet in the Texas Hill Country, the place of his birth and a place, in his words, "as beautiful as fairly land was ever imagined to be," he was twenty-eight years old and burning with the Cause. His wife and fellow Baylor University graduate, Mary Louise, was a year and a half his senior and newly pregnant with their first child. The journey, which would take them from the only place they'd ever really known and loved, was not merely their assignment and lot as missionary Baptists; it was the grand adventure they expected would define both their time on earth and their destiny. As a result, they embraced it with a fervor so powerful they sometimes literally trembled in each other's arms.

The trip took nearly three weeks: by train to New York and from there by ship to South America. The first leg, however, on the back roads from the Burnet cotton fields to Austin, was by means of horse-drawn wagon. Gasoline-powered motorcars had been running on American roads since the 1890s, but few Hill Country residents could afford even the $650 for one of Ransom Olds' one-cylinder Oldsmobiles. In the cities, cars were already

common—common menaces, some were saying. In the year the Stapps left
Texas, American motorcar accidents surpassed horse-and-buggy accidents
for the first time, numbering about 4,000 annually. It was as good an indication as any that the modern world, for better and worse, had arrived.

Charles and Mary Louise, though, were still very much of the world
familiar to their parents and grandparents, and were about to join a historic connection between Texas Baptists and the nation of Brazil. Dr. William Bagby, a Texan, had gone there in 1881 and cofounded the first Baptist
church in Salvador da Bahia, the oldest city in Brazil. When the Stapps
disembarked in Bahia that January, twenty-nine years after Bagby had
arrived, it was 85 degrees, and they found the coastal plains and flowering
jungles exotic. The couple gorged on the local fruits—mangos, strawberry
guavas, and delicious bananas—and introduced themselves to the mostly
friendly but equally exotic inhabitants. Bahia had once been the capital
of the Portuguese slave trade in the New World and had become, by the
early twentieth century, the center of a rich Afro-Brazilian culture. It was
a lush, extravagant land bursting with millions of souls seemingly ripe for
the picking.

The Stapps had never known luxury and did not expect to find it in
South America. Yet with Charles's appointment as president and director
of the American Baptist College, they found themselves ensconced in the
forty-seven-room compound of a former Portuguese nobleman. The property overlooked tamarind- and bamboo-covered hills above Bahia de Todos
os Santos (All Saints' Bay). Still, the Stapps understood from the first that
Brazil would also be a brutal place. There were drenching rains followed by
drought, biting insects, and tropical disease. Even worse were the heart-
breaking hunger and poverty that surrounded them. An economic bubble,
the Encilhamento, that had burst in Brazil two decades before the Stapps'
arrival had led to years of spiraling inflation, food shortages, and disease. On
more than one occasion the worst of the cases fell on their doorstep to die.

If their mission was a challenge for Reverend and Mrs. Stapp, it had
been doubly so for their predecessors. Missionaries in Brazil over the years
had died in all manner of mishaps and crimes—some of them horrific.
More than a few were drowned, kidnapped, or murdered. Maltreatment of
Protestants at the hands of Catholics, who outnumbered Baptists forty to

one, did not consist merely of snubs and harassment. Dr. Z. C. Taylor, a late nineteenth-century missionary who spent thirty years in Brazil, reported to his contacts in the U.S.: "Persecutions are pressing us on every side, now. The *padres* have dropped their pens to use the sword. Four soldiers who attended the Presbyterian church have been in prison for a month. All efforts to get them out have been in vain so far."

Nevertheless, Charles continued to preach his gospel and Mary Louise—"Dona Luiza," as she preferred to be known now, as they adapted to the local customs and language—managed the couple's affairs, taught classes to the local children, and learned to prepare staples such as *feijão*, the bean stew that formed the basis of most of their meals.

Those who knew and worked with Reverend Stapp described him as unfailingly warm and generous to a fault. He was five feet ten inches tall, rail-thin, almost brittle in appearance, with a full mustache, neatly trimmed goatee, and jug ears. He spoke with a slight stutter, and a childhood injury—he'd lost three toes on his right foot—had left him with a pronounced limp. Though he had shown early promise as a mathematician, he believed the spirit was in his blood. Mary Louise was genteel, but in some ways more imposing than her husband. She wore wire-frame spectacles and had the appearance of a studious, determined woman who intended to carry her weight and expected others to do the same.

She would certainly expect nothing less of her own offspring, and on July 11, 1910, just after sundown, she gave birth to the Stapps' first child, a brown-eyed, six-pound son. John Paul was named after his father's two favorite apostles and his birth was promptly recorded with the United States consulate to establish American citizenship. In spite of the fact that Charles and Mary Louise spoke English with each other, John Paul was forbidden to converse in any language but Portuguese until he reached the age of six. If he wanted a cup of juice: *"Eu quero um pouco de suco."* A year and a half after John Paul, Mary Louise gave birth to Robert Grady. Twenty-three months after that, to Carlos Celso. The boys grew up surrounded by music and teachers, but were confined to the school grounds in a near-captivity that was relieved only by the constant stream of students and visitors. It was a strict upbringing. Due to their parents' fears of communicable disease and disapproval of idle frivolity, the Stapp boys were seldom allowed to play with other children at the school. If any of them rebelled against the requirement of a daily afternoon nap, Mary Louise strapped him to his bed.

Then, in 1916—Celso was only two—the Stapps accepted a one-year furlough from their duties and returned to Texas. With a world war raging in Europe, the Brazilian government had adopted a policy of neutrality in order to protect overseas markets for its coffee and latex, a stance that seemed to please almost no one in Brazil. Anti-German, pro-war protesters battled nationalist antiwar protesters. A furlough after seven years was standard for Baptist missionaries stationed out of country, and for a year, at least, the Stapps left chaos and hardship behind.

Charles and Mary Louise spent much of their homecoming year traveling the Southern states lecturing and fundraising for the Foreign Mission Board. In the United States for the first time in their lives, the boys experienced their first snowfall, and—now that cars had found their way into even the cotton farm towns of central Texas—took their first automobile ride. John Paul found it thrilling. The most surprising thing about Texas from the perspective of the Stapp boys, however, was the huge doting family suddenly surrounding them and hanging on their every move—grandparents and uncles and aunts and cousins. As John Paul would remember it: "I was filled with wonder at the number and homeliness of my relatives."

When the furlough was up the following year, Charles took his family back to Bahia on a Brazilian steamer. John Paul enrolled in the second grade and quickly made the acquaintance of a missionary doctor who had taken up residence at the school. The doctor was a shambling, red-faced Missouri native who fed the boy's curiosity by sharing his stacks of medical journals—secretly and against the wishes of Charles and Mary Louise—and regaled him with lurid accounts of grisly medical procedures. The Stapps encouraged their children in the love of books, but mostly restricted them to religious fare such as Foxe's *Book of Martyrs*. John Paul would remark of that particular volume: "What I read frightened the hell out of me."

Doctors were sorely needed in Brazil in the early decades of the twentieth century. Contaminated water supplies, primitive sewers, and inadequate garbage systems spawned rampant disease: tuberculosis, smallpox, measles, typhus, and leprosy. Surrounded by these dangers, the boys required constant surveillance, but even that failed to protect them from becoming infected with all manner of parasites. Not long after their return to Bahia, the Stapp boys contracted tertian malaria, an infection that causes recurring fevers on a three-day cycle and is the most difficult form of malaria to cure. "We had malaria all the time," John Paul would recall. Charles gave them native herbal medicines and the Missouri doctor dosed them with quinine, which prevented the most acute attacks. All along, the doctor's calm authority and encyclopedic knowledge of sufferings and their cures made a considerable impression on John Paul.

Despite constant illnesses, family life revolved around the school and Reverend Stapp's mission. When his classroom and administrative duties allowed, Charles—sometimes with Dona Luiza alongside—traveled greater Brazil, mostly to missionary conferences. Riding and tromping his way through the jungles and highlands and along the rivers, he established a reputation as an honest and hardworking servant of the Cause. Once in Pernambuco, not long after the family's return from the States in the summer of 1917, Charles stopped to visit a mission in a jungle village. A local directive prohibited the construction of schools or churches in adobe because such structures never lasted more than a few years in the withering climate. Mission rules required brick—except there were no bricks in the area and there was no money to import them. Reverend Stapp began testing local soils and discovered a readily available clay from which excellent bricks could be made. With a work crew he recruited himself, half-crippled Charles built a kiln and began turning out bricks. As a result, the local men began to build: a church first, and then a school. Charles oversaw the building of roads and bridges all over northeastern Brazil. By one account he introduced parts of inland Brazil to dairy cattle and the mule. "He could have gone into mathematics," a colleague observed, "but he went the other way."

A few months after Wilford Lee, the fourth and youngest Stapp, was born, in 1918, Charles was transferred north to Aracaju in the state of Sergipe. There were no adequate schools there, so Dona Luiza taught the boys mostly by herself. She covered all the basic grammar school subjects, including French, German, and Latin. "My mother," John Paul would say, "was the most dedicated advocate of education that has ever lived, so far as I know." In preparation for future studies, he was required to learn twenty new English words a day to be selected by him from the family dictionary. Mary Louise introduced the boys to the dramatic arts by directing them in recitations and then in scenes from Shakespeare, and brought in tutors to teach them French literature and classical Portuguese. Charles had managed to procure a battered piano and Luiza drilled her sons in scales and harmonies. They all learned to play multiple instruments, the boys staging elaborate competitions that allowed them to demonstrate their prowess.

In Aracaju the boys, older and bolder now than when they'd been restricted to the school grounds in Bahia, had an easier time getting away and exploring the forests and waters of Brazil. One story the family never tired of telling involved John Paul's attempt to construct a secret treehouse in a coconut tree on the grounds of a Sergipe sugar plantation, and sawing off a limb that was supporting him. He fell 15 feet and landed flat on his back on the sandy ground. Don't pray for me, he later claimed to have said, congratulate gravity.

Meanwhile, living conditions across Brazil remained squalid. Malnutrition and disease were everywhere. Charles kept up with the outside world mostly through newspapers, and tried to keep his family up to date on current events. In 1922, a pair of Portuguese navy pilots made the first aerial crossing of the South Atlantic to mark the centennial of Brazil's

The missionary Stapps of Brazil. Clockwise from upper left: Reverend Charles Stapp, Robert, Mary Louise, Wilford, John Paul, and Celso. (Photo courtesy of Wilford Stapp)

independence, stirring the imagination of the Stapp boys just as the family was preparing for a second furlough home. John Paul, who had become expert at building bows and arrows and kites, and carving model boats, had begun designing little glider airplanes that he'd launch from the limbs of cashew trees.

The Stapps returned to Texas—by sea rather than by air, to the boys' great disappointment—and settled in the town of Brownwood, where Charles had arranged to attend postgraduate seminary school at Howard Paine College. John Paul enrolled in school there that fall. He was a good student—his mother's home academy had prepared him well; preliminary tests placed the twelve-year-old with the high school freshman class—and in spite of his slight stature and unremarkable appearance, he made an impression. Many years later, when a schoolmate from Brownwood saw John Paul Stapp's face on the cover of *Time* magazine, he tracked him down. "I was just thumbing through a copy of the high school annual for that year, *The Pecan*, and I see a picture of you, with a label underneath: 'A Tuff Customer.'"

Mary Louise knew that her lessons could only take the boys so far. So when their furlough was up early that summer and the family returned to Brazil, only three sons made the trip. John Paul, judged ready now for an American education, stayed behind with relatives in Burnet—named after the first president of the Republic of Texas and pronounced (fittingly, John Paul thought) "burn it." There he picked cotton for two cents a pound to earn a little spending money before going, that September, to one of the oldest boarding schools in the state of Texas: San Marcos Baptist Academy. This would be the way for the Stapps. Rather than send their children away to school, Charles and Mary Louise sent them home.

San Marcos was a well-regarded institution that had transformed itself into a military academy during the World War I years, and was a step up in John Paul's mind from the jungles of Brazil or the isolation of the dryland farm in Burnet. He felt like he was finally joining the modern world. The nation's first radio broadcasts, advertised in newspapers as "the craze of the age," were just coming online, and the students at San Marcos could occasionally pick up WRR-AM in Dallas beaming out crop reports and the news of the day on one of the school's crystal sets.

The discipline and regimentation at San Marcos, however, did not make for an easy transition. Cadet Stapp attended classes and drilled on the parade field with his unit by day—close order drills with heavy Springfield rifles—and spent his nights poring over his textbooks and any other reading material he could get his hands on. He understood math, but he loved books. He liked to play with words—in both Portuguese and English, occasionally in French and German—sometimes mixing them all together.

When John Paul joined the sophomore class at San Marcos he was thirteen years old, four feet nine inches tall, and weighed not quite eighty pounds. His size, coupled with an attitude he himself described as "eccentric," set him apart. While a few of his classmates found him witty, he became a target for bullies. He had to put up with ongoing abuse during his first few weeks on campus. Then, on a class hike and campout, he snapped and challenged an older boy who outweighed him by fifteen pounds. Stripped to the waist, the two of them whacked away at each other. Stapp got the worst of it. He knew he couldn't outmuscle his opponent, but he was pretty sure he could outlast him. Eventually, the older boy quit out of pity. John Paul, his face bloody and swollen, got the nickname "Demon Stapp" out of respect for his sheer cussedness. Later that same year, according to a classmate, the Demon put his reputation on the line by standing up for some Mexican boys at San Marcos who had been getting picked on by upperclassmen. It didn't make him particularly popular, but it gave him a certain notoriety that he found satisfying.

Over the next two years, Stapp spent his summers in Burnet, working in his uncle's cotton fields where he could earn $1.50 a day. He graduated from San Marcos at the age of sixteen, but Charles and Mary Louise, who'd kept in close touch through letters, insisted that he remain at the school for one additional year to complete a business course.

While he didn't find much to interest him in his business studies, John Paul Stapp found 1927 a mostly exciting—if dangerous—time to be alive. In January, a bus carrying the Baylor University basketball team traveling from Waco to Austin was hit broadside by a speeding train at a railroad crossing in Round Rock, killing ten of the students aboard. Nearly 25,000 Americans would die in vehicle crashes by the end of the year. The news that spring, however—at least, the news that caught Stapp's attention—was all about flying. On May 4, Army aeronaut Hawthorne Gray set a new lighter-than-air altitude record when he took a balloon filled with hydrogen gas to 42,240

Demon Stapp at the San Marcos Academy, a self-described "eccentric."
(Photo courtesy of Wilford Stapp)

feet, a full eight miles into the sky above southern Illinois. Then, a little more than two weeks later on May 20, a twenty-five-year-old mail pilot named Charles Lindbergh took off from Roosevelt Field in Long Island, New York, and flew a single-engine airplane solo across the Atlantic Ocean, 33.5 hours nonstop, to land near Paris, where he was greeted by 150,000 cheering fans. The audacity and courage of Gray and Lindbergh were the talk of the nation, and Stapp, along with some of the other boys at San Marcos, tracked their stories closely.

That fall John Paul followed in his parents' footsteps and enrolled at Baylor. He focused initially on music—studying theory and harmony, playing first-chair bassoon with the symphonic band and becoming proficient on both piano and cello—and joined the drama fraternity Alpha Psi Omega. He staged, directed, and acted in a play performed at the Baylor Little The-

Stapp as a freshman at Baylor University in 1927, the year
before tragedy changed him. (Photo courtesy of Wilford Stapp)

ater. He also spent much of his freshman year attempting to read through the Baylor library and dreaming of someday becoming a writer. He felt it was his one potentially marketable skill, though at the end of the year all he had to show for it, he said, were "reams of bad poetry." But he had something else: a girl. He'd met her at San Marcos and they'd moved on to Baylor together. Like Stapp, she'd grown up with missionary parents, and the two had become close very quickly.

It was during his sophomore year when the short arc of John Paul's life was suddenly knocked off course. Charles had always taught his boys to prepare themselves for the sudden epiphany that would illuminate the nature of their calling. In John Paul's case, it would be a pair of unrelated incidents that occurred 1,500 miles apart during Christmas vacation, 1928. He'd returned to the farm in Burnet for the holidays when, three days after Christmas, his two-year-old cousin—John Hansford Stapp, Jr., the son of John Paul's Uncle Hansford and Aunt Verna—having been left alone for a moment in front of the fireplace, tossed part of the Sunday newspaper over the grate. The paper flamed up and ignited his cotton pajamas. Someone grabbed John Jr., rushed him into the backyard, and plunged him into a tub of icy water. It was the worst kind of family tragedy. The burns were horrific, and so was the collective guilt. When the local doctor arrived at the farm, he treated the boy by applying a mixture of lime and linseed oil to the burns. John Paul remained awake for the better part of sixty hours, doing what he could for the child, rarely leaving his side—and fell into a rage when his cousin died. Stapp accused the doctor of being a "dolt" who should have immediately had the patient transported to a hospital in Austin. "Apparently his abdomen, and the viscera of the abdomen were affected by the intense blast of heat and he just didn't live," Stapp would write. "I was bitter and angry and I thought, surely something better than this could have been done for the child." He returned to Baylor after New Year's in a semi-daze. He had witnessed suffering and even death in Brazil, but never to one within the family fold.

Grief-stricken, he scheduled his return to Baylor to coincide with his girlfriend's arrival from Los Angeles, where she had been celebrating Christmas with her missionary parents, who were home from China. By the time Stapp got to Waco, however, the news was all over campus. It had happened two days earlier. She'd been thrown from her parents' car when it was speared from the side by a drunk driver who ran a red light at the intersection of Hollywood and Vine. She'd been killed, and before any of the Stapp family even learned her name. While John Paul had barely mentioned the relationship to his brothers, assuming perhaps that all would be revealed in due time, he would later refer to her as his first love. The two had even, John Paul said, talked about getting married someday and having a family.

He fell for a time into a numb rejection of everything he knew. There was no comfort for him in the faith that sustained his parents, and for the first time in his life he mocked it. He withdrew from his relatives in Burnet, his friends at Baylor. At the age of eighteen, the "tuff customer" from Bahia found himself shattered and alone in the modern world.

2

THE MOST INTERESTING PLACE

I am full of egotistical ways that are easily misunderstood, and I
tend to disregard the existence of others unforgivably, but I have
the utmost respect and appreciation for those who are broad
minded and tolerant enough to like me in spite of myself . . . I
weigh about a hundred and forty eight in my brunet complexion,
and I can slowly muscle up a hundred and eighteen pound
bar bell over my head. I eat well and I occasionally feel kindly
toward the world.

—*John Paul Stapp, unpublished memoir*

A S HE RECOVERED FROM the losses of his cousin and his sweetheart, John
Paul Stapp's focus began to change. He withdrew from his literature and
music courses, which suddenly seemed frivolous to him, and enrolled in the
Baylor pre-med program. He had casually gone out for the track team the
previous year; now he resolved to train year-round as a cross-country run-
ner, pushing himself harder and harder and—despite frequently sprained
ankles—taking on longer and longer distances. He found ways to merge a
newfound interest in physiology with his daily activities, experimenting with
diet and the basic physics of motion in an attempt to increase his speed and
stamina. "I tried," he explained, "to develop the lowest energy-cost stride possible. I figured that I could conserve my energy by swinging my arms and legs
at a minimum, by holding my back in a position that would cut down drag. I

worked out my breathing to give me a maximum expulsion of carbon dioxide." But he admitted that while he had been busy with self-research, "the other guys were just running faster."

Reorienting the focus of his studies, Stapp did his best to keep up with scientific advances elsewhere in the world, especially when they involved aviation and aeronautics. On May 28, 1931, he came across a newspaper story about the previous day's events in Augsburg, Germany, where Swiss physicist Auguste Piccard and an assistant had become the first human beings to survive a flight into the stratosphere when they rode a pressurized aluminum capsule suspended beneath a balloon to 51,200 feet. Piccard had made his flight, the story said, not in search of adventure or glory, but in the quest for knowledge. His agenda included measurements of cosmic radiation designed to provide insight into Albert Einstein's theory of relativity. It was clear that good science, Stapp noted later, need not be confined to mundane labors in a laboratory.

When John Paul received his B.A. from Baylor in June of 1931, in zoology with a minor in chemistry, he immediately enrolled in the master's program in experimental zoology. In the darkest days of the Great Depression, he scraped together just enough money from the few dollars his parents could spare and a $200 annual stipend for the children of missionaries to cover his tuition. With little left over, he slept in the basement of a condemned dormitory and later in one of the biology labs. In makeshift traps on the roof of the biology building, he caught fat pigeons suitable for grilling, and at one point lived for an entire week on handfuls of pecans gathered from campus trees. He and other down-on-their-luck students occasionally dissected biology specimens—lab rats and guinea pigs—and barbecued them with Bunsen burners or baked them in an electric oven for a midnight meal. "If it breathed it had protein," Stapp explained, "and if it had protein, I ate it." Meanwhile, he took on whatever work he could find—or dream up. He established a mosquito farm in a half-gallon jar and assembled slides of the species' life cycle, which he sold in annotated sets to biological supply houses. He collected the grasshoppers that were plentiful along the banks of Waco Creek and prepared them as study specimens for the Baylor biology department.

Stapp's energy and curiosity were his antidote for hard times. There were periods when he wanted to break away and see with his own eyes some of the amazing things Americans were doing in the face of economic catastrophe. Massive engineering projects were underway or in planning from coast to coast. Design had begun on the Golden Gate Bridge in California, and the Empire State Building opened in May 1931 after only fourteen months of construction. Perhaps the most impressive of all was the massive Hoover Dam, which had recently begun construction on the Colorado River in a canyon between Nevada and Arizona. Tens of thousands of unemployed men made their way west that year in search of work on the dam and other public projects. Stapp very nearly joined them in the summer of 1931, but elected instead to take a job as a door-to-door salesman for WearEver, an aluminum cooking utensil company, hoping to earn a little cash and conquer the shyness he felt was a legacy of his cloistered upbringing in Brazil. With his first couple of paychecks he made a down payment on a used car, an old jalopy he named Pandemonium, and hit the back roads. He would knock on the doors of up to a dozen strangers' homes a day and, despite knowing little more about cooking than what he'd learned grilling pigeons and guinea pigs, cook a meal for the woman of the house and—if all went well—sell a few pots and pans.

Back at Baylor he completed his coursework in a single calendar year, graduating in the summer of 1932, and was immediately offered an instructorship in biology at Decatur Baptist College northwest of Dallas. There, by living in the Decatur dormitory along with the students, he was able to save most of his salary for medical school. It was a pleasant place to live, as he got along well with the rest of the faculty at Decatur and found the students inspiring, even though he was made aware that his predecessor had been fired for teaching the tenets of natural selection. It was only seven years earlier that the state of Tennessee had sued a substitute high school teacher named John Scopes for the same thing in the "Monkey Trial," a high-profile bit of legal theater that had put science educators nationwide on notice: fundamentalism and modernism were incompatible, at least in the classroom. Stapp elected to steer clear of subjects such as comparative anatomy.

The more difficult problem, after two years at Decatur, was the college's anemic financial health. With more than $2,000 owed him—funds he badly needed if he was ever to satisfy his dream of medical school—he was forced to sell Pandemonium. The college reneged on its agreement and ended up paying Stapp a total of only $350 for two years' work. At the same time, younger brother Celso was now back in the States and itching to enter medical school himself. Celso had breezed through Baylor and had already been accepted into the medical program there. John Paul made the decision to delay his own plans in order to offer whatever financial assistance he could muster to help his little brother.

While at Decatur, John Paul made the acquaintance of a fellow teacher roughly his own age named Dorrel Jones. The two of them enjoyed fishing and running around in the Hill Country woods together, and they eventually hatched a business plan. John Paul had cut a deal with the South Western Biological Supply Company in Dallas to deliver prescribed quantities of snakes and lizards, and landed a second contract to provide thousands of grasshoppers, cotton fleas and scorpions. Catching reptiles and insects was relatively easy work once you knew how, and the young men grabbed every opportunity, often at night, to conduct their safaris.

The two would remain in contact for many years, though they saw each other only occasionally after Stapp left Decatur in the summer of 1934 when, without immediate prospects for employment, he enrolled in summer school in the biology program at the University of Texas. He registered for a graduate course in cell physiology supervised by Dr. E. J. Lund, who was a respected scientist of the first order. There would be no need to dance around Darwin in Austin and Stapp threw himself into his studies, spending most of his waking hours observing single-cell organisms through a microscope or absorbed in Lund's legendary five-hour lectures. Biophysics was a revelation, and John Paul worked like a dog through weekends and holidays for the duration of the initial six-week course, doing what he later estimated as four times more work than was called for. Lund was impressed.

Scientific aeronautics was in the news again that summer, and Stapp did his best to follow the progress of a celebrated project funded jointly by the Army and the National Geographic Society. A crowd of 30,000 gathered at a site in the Black Hills of South Dakota on July 28, 1934, to witness the launch of three Army officers beneath a three-million-cubic-foot balloon. The science agenda involved stratospheric radiation studies, but there was another dimension to the flight of *Explorer*. The Soviet Union had raised a balloon to a reported height of 72,178 feet that January, and while the three pilots on board had all been killed when the ropes securing their gondola to the balloon failed during the descent, the Soviets claimed a world altitude record. *Explorer* was intended to reclaim the record for the United States. Unfortunately, the rubberized cotton balloon ripped open at 60,000 feet, and the Army men survived only by jumping free and parachuting to earth. It was a demonstration of what might await scientists who took their work out of the lab and exposed themselves to the unforgiving laws of physics. A curious footnote to the near-disastrous *Explorer* flight occurred a few months later, when the chemist Jean Piccard—twin brother of Auguste, who'd made the first manned stratospheric flight three years earlier—made his own balloon flight, accompanied by his wife, Jeanette, and their pet turtle, to 57,579 feet above southern Michigan.

While the Piccards stirred his imagination and gave him a glimpse of where science might be able to take him, Stapp was, by necessity, primarily focused on more practical matters. In Austin, Dr. Lund had arranged for Stapp to receive a teaching assistantship in physiology and anatomy that paid $500. It would allow him to at least cover his tuition for as long as he kept his grades up, though it provided nothing until classes started in the fall. He again considered applying for a construction job on the Hoover Dam, where workers continued to be needed in the summer of 1934, but ended up settling for a job as a clerk-typist at the Bankhead Cotton Control Administration for $3.85 a day.

The relationship between Stapp and the demanding Dr. Lund would last for five years, despite Stapp's occasional complaints about the "tyrant" he labored under. Through it all, however, he would continue to nurture his dream of medical school and bragged to friends that he planned to attend Harvard.

By the time fall term 1934 rolled around, all of the Stapp boys were back in Texas pursuing their educations—and John Paul was helping to support them all. Robert was the budding artist, having just graduated from Baylor, Celso was on his way to becoming a doctor, and the last to arrive, Wilford Lee, had come during John Paul's second year at Decatur and had briefly lived in the dormitory there with his brother.

It was an extraordinarily unlucky time for these young men, or any young men, to find themselves without means in the United States of America. The Depression was the beginning of the end of King Cotton in the Texas Hill Country, and the collapse of cotton production created terrible hardships for the farming Burnet Stapps—one potential source of sustenance for John Paul and his brothers. Most towns had adopted stringent austerity programs, cutting services and salaries; rural communities were in even direr straits. Some federal relief money had found its way to Texas, but none of it into the pockets of the Stapps.

Of all the brothers, Wilford Lee had been by far the most reluctant to leave Brazil. He'd always been the closest to his mother and had had the toughest time adjusting to life as a teenager separated from his parents. John Paul did his best, on meager resources, to make sure Wilford was fed and clothed. Wilford accompanied John Paul on some of his midnight lizard-hunting safaris along the local bayous, and had been involved in the day-to-day drama of the Decatur science building. "I once helped Paul [the family would always call him "Paul"] move all the cadavers from the second floor to the fifth floor of the building," Wilford recalled. "The whole place smelled of formaldehyde for weeks." The youngest Stapp gradually adjusted to his new life in the shadow of a brother eight years his senior, though Wilford Lee still harbored doubts about his own ability to measure up. John Paul was an intimidating role model. "Paul embodied more different facets and aspects of a person than anybody I ever met," he would say.

Reverend Stapp and Dona Luiza kept in touch with their children through a steady stream of letters. Charles packed his typewriter with him on his by then near-constant travels throughout Brazil. By the time the last of the boys had returned to the United States, Charles had acquired a Victrola and a small library of 78s: Brahms, Mozart, Beethoven. He liked to play his records in the evenings and tap out his letters using carbon paper and onionskin sheets that he filled to the edges. He mailed copies to the whole family.

Charles was pleased by John Paul's resourcefulness and his already impressive educational achievements. "I shall be proud when I see your name in 'Who is Who in America' or when you get some of those international prizes. The kind of mind you have and the training you are giving it ought to give the world no unusual service." There was a competitive streak in Charles, who as a Baptist had fought long odds against the Catholics his entire time in Brazil, and he was heartened to see that quality reflected in different ways in each of his boys.

Though Reverend Stapp would likely have disapproved of much of it, Austin and the University of Texas were an idyll of fresh ideas and cultural opportunity—not to mention a smorgasbord of the opposite sex. John Paul wrote regularly to Dorrel Jones and Dorrel's new wife, Julia:

> It is the most interesting place I have ever been in, and I keep finding out new things about it all the time. It is the most up to date, alive place with more things going on than any other place I have been in Texas. Excuse the childish enthusiasm, but this comes nearer being the kind of a place I would like to live in than any other I know of. There is a higher percentage of good looking girls here—some of them can look at you in a most confusing manner that makes you almost forget whether it is carpals or metacarpals that you are discussing.

John Paul would never be a natural with women or with romance. He wrote to the Joneses back in Decatur: "I have to deal with three classes in which I have not less than sixty of Texas U's best coeds, and I havent given a single one of them a tumble. I have had about six dates with as many different girls since I came, and about five dates with one girl; the last time I saw her I told her not to expect me to call her any more. Not a one that I have met so far attracts me more than superficially, and that for only a short while." Though he rarely talked about it, even with Dorrel and Julia, and had resolved never to speak her name, he continued to pine for his Baylor sweetheart who'd been killed in the Los Angeles car crash, and to measure other women against his memory of her.

. . .

In the summer of 1934, the Stapps' uncle Hall Shannon, Mary Louise's surgeon brother, who lived in Dallas and was the most influential Stateside family member in the Stapp boys' lives, bought a large residence at 1709 Guadalupe near the university in Austin and offered John Paul and Robert—newly graduated from Baylor—free rent to manage it as a rooming house. John Paul remembered Uncle Hall talking about the dust storms, black clouds of biblical proportion that had swept across the dust bowl prairies in the Texas panhandle not far to the north that summer. There was talk that God had punished the people and sent this plague, Hall Shannon said. He didn't believe it, but plenty of people did. No matter how you figured it, the hard times seemed like they were here to stay. He advised the boys to look for opportunity.

Robert installed a washing machine in the basement of the rooming house and went into business as a laundry. This arrangement covered the rent and allowed John Paul to continue his financial assistance to Celso. While he was proud to be able to help his brother, his frustrations with the delay of the start of his own medical career occasionally boiled over. The Joneses heard his complaints. He said he would do almost anything to become a doctor, and it wasn't, he assured them, about money or status. This was his calling. "I was," he said later, "willing to work in a charity hospital for the rest of my life."

A shortfall in operating funds that year forced the university to cut salaries for student teachers. Stapp wrote to the Joneses in Decatur: "I am financially in close circumstances. I will have to renew a note for the first time in order to get it paid off. I borrowed another hundred dollars and sent it to my brother in Dallas. I still have face enough to fool a banker." Before his time in Austin was up, John Paul would borrow—and repay—small loans totaling more than $1,000 [more than $17,000 in 2015 dollars], mostly to ensure that Celso's tuition at Baylor was paid up on time.

In the meantime, Julia Jones had given birth to a son, and Dorrel wrote in December of 1935 to announce their decision to name the boy John Paul. In Austin, Stapp was touched. He sent a dollar for his namesake's piggy bank.

That same year, Stapp studied right through the Christmas holidays but, shortly after New Year's, he took a break to travel to St. Louis for a convention of the American Association of Science. He was unimpressed by the big city, finding it grim and unfriendly—except for the world-famous zoo, which he loved. On the return bus trip, in the early morning hours outside Chelsea, Oklahoma, he witnessed a violent car crash. He wrote the Joneses about it. "I saw a man die," he told them. Stapp had attempted to assist. "He hit his head against the left support of the windshield and cut a groove clear to the brain. When I lifted up his head I saw this hideous gash." It was Stapp's first personal encounter with a traffic fatality. Two months later John Paul received word that little brother Wilford had been the victim of an automobile accident himself, suffering two broken ribs, a short scalp wound, and a deep cut across his back.

The nation's highways and roads were becoming wicked places. Though car manufacturers were reluctant to release accident statistics, an all-time high of 35,000 Americans, a horrendous total, had died in crashes the previous year. Just that past August, 1935, *The Reader's Digest* had published an article titled "—And Sudden Death" that described, in lurid prose, the gory realities of a car crash. The magazine was besieged with requests for reprints and within weeks had distributed 35 million copies. The public was up in arms about the slaughter on its highways.

To John Paul Stapp's way of thinking there were other, more glorious, ways to travel. He had caught a glimpse of his first airplane in Rio de Janeiro as a boy, but he'd never had a chance to fly in or even get very close to one. He finally got that chance in the summer of 1936, when he joined two other graduate students for a one-dollar sightseeing tour of Austin in a Ford tri-motor flown by a traveling barnstormer. The flight only lasted about ten minutes, but the students were giddy from the experience. They admitted to being a little spooked. Civilian aviation fatalities that year numbered only in the low hundreds worldwide, but the students were familiar with the grim statistics from World War I, in which one out of every eight combat pilots was shot down or killed in a crash. Still, the chance to look down at the city from the perspective of the birds was worth the risk. John Paul talked about it for days, and told friends he intended to learn to fly himself someday.

Diversions like the airplane ride, however, were few and far between. Stapp continued to put in long hours on his toughest courses, calculus and analytic geometry, whenever he wasn't in the lab. The research that he intended to form the basis of a thesis was beginning to show results. It was a study of the electrical properties of frog tissue, and John Paul was always looking for more frogs. He wrote to Dorrel Jones for help in February of 1937: "I sure could use a crate of nice, healthy bull frogs. They would have to be alive and in good condition." Stapp offered his friend $7.50 per dozen. He simply didn't have the time anymore to roam the bayous himself. "I am weighed down with so much of so many different things to do of late that sheer exhaustion renders an increasing proportion of my work distasteful. The beautiful fishing weather might have something to do with it. I have not so much as been out in the woods for a year now. It becomes suffocating after a while to massage a second growth of callouses on your rump just from sitting in an office chair and doing lessons, grading papers, reading and writing."

During the winter term, Stapp got word that Reverend Charles and Dona Luiza were set to return to Texas in the spring of 1937 on furlough. It would be another escape from chaos and calamity. They had lived through a nasty decade in Brazil, with crashing markets, revolution and counterrevolution, and pitched street battles between Communists and forces funded by the fortunes of the coffee oligarchs. Getúlio Vargas was in the process of establishing himself as a Fascist-style dictator over what he dubbed the Estado Novo, the New State. It had been wearying. John Paul had encouraged his parents to settle in Austin during their year home. The idea of having everyone together—or at least in close proximity again—cheered him up. At the same time, the prospect of getting reacquainted with his parents worried him a little. He had changed. He'd become a man of science rather than a man of God. How would his parents react to that?

Charles and Mary Louise disembarked in May in New Orleans, where they were required to submit to two days of examinations at a tropical medicine clinic before traveling to Texas by train. Soon after, the Stapps were reunited in Austin and, by pooling their resources, were able to rent a large house in town surrounded by fig and pear trees. John Paul lived there with Robert and their folks, and they were visited regularly by Wilford and Celso. They went in together on a used piano, which livened up the house. It was a good year for the family, though everyone was required to make certain allowances: the brothers to their parents' rigid self-righteousness, the parents to their sons' tolerance of and even willful participation in the frivolities of modern America.

There wasn't, in truth, much frivolity in John Paul's life at that time. He continued to focus on little but his studies. His plans for medical school were on track and he had, by the spring of 1938, trained his sights on Tulane. Then, with the bluebonnets that his parents had always gone crazy over blooming across Texas, John Paul fell in love. The woman's name remained a mystery

even to Stapp's friends—the affair was a brief one—but he shared the painful
fallout with the Joneses back in Decatur. "I do not require a shoulder to cry
on thank you," he wrote. "For awhile I almost lost my mind, but I am alright
now. I was ditched and tossed out on my ear. I never before realized how
much of a coward and a weakling a woman could be. I ought to take up gambling. There all you lose is money. I hope my next misfortune is something
nice and pleasant like losing a leg or something."

In June, Charles and Mary Louise returned to Brazil and a morose John
Paul moved back into his Uncle Hall's rooming house with Robert. His
research was going well, but he remained otherwise somewhat depressed—
about his love life and his future and the world in general. He and Robert
had a falling out during this period when John Paul apparently offered one
too many bits of an older brother's unwanted advice. It seemed to him that
nobody he knew was happy, including himself.

"All that a person can hope to do is to cherish the moments of ecstasy
and endure the hours of hell—which is all that our damnably lopsided civilization leaves for anybody, rich or poor, to do." John Paul's private writings
were becoming more philosophical and his worldview infected by a cynicism he felt he'd earned. "Abolish rank and degree among people, and above
all, abolish the abominable vanity that drives them to seek prestige at the
expense of everything else, and maybe we could begin paving the way for
universal happiness."

The dark mood that had overtaken John Paul Stapp seemed, as 1939
began, to have descended on the world. That January, Adolf Hitler in Germany threatened the annihilation of the Jewish race and the Fascist forces
of Francisco Franco finally defeated the Republicans to seize control of the
government in Spain. Tens of thousands died in earthquakes in Chile. Optimism was in short supply. Still, by the end of summer, John Paul had completed all of the resident requirements for his Doctor of Philosophy degree in
biophysics, and had finished the first draft of a 150-page thesis. Title: *Electric
Properties of Living Cell Layers and the Application of the Iodine Coulometer to
the Measurement of Electric Energy Generated by Them.*

As if in compensation for all the hard work, good news arrived that
August in the form of a thick envelope delivered to the rooming house on
Guadalupe. John Paul Stapp had been accepted to medical school. It would

be the University of Minnesota. Wasting no time, he withdrew his life savings, $125 he'd earned coaching students prepping for final exams, and boarded a northbound train. He spent a weekend at the apartment of a friend in Rochester, Minnesota, where his friend's wife laughed at the shabby condition of his clothes, before continuing on to Minneapolis where he registered at the university medical school.

He moved into the graduate dormitory in Pioneer Hall, a four-story red-brick building with a leafy courtyard on the school's east bank campus. In addition to enrolling in freshman medical courses, he accepted a position as a research assistant and part-time lab instructor to undergrads—and set about the business of becoming a doctor. One of Stapp's enduring memories of his first year at Minnesota was of frequent passing encounters with "a tall, gaunt scientist in a long cape, his mane streaming behind his hat, his hands clasped behind him, walking in long, rolling strides, deeply preoccupied with his own thought." This was Dr. Jean Piccard, the world-renowned scientist Stapp had read about in the newspapers when he'd made his stratospheric balloon flight five years earlier. John Paul never exchanged words with the great Piccard, but he felt privileged to be practically rubbing shoulders with him.

During his early months at Minnesota, Stapp had continued to put the finishing touches on his doctoral thesis, and he returned to Austin over the Easter holiday, 1940, to sit for oral exams—with passage far from a foregone conclusion. In the demanding Dr. Lund's twenty-one years at Texas, he had approved only fourteen candidates for a doctorate in biophysics. It had been an honor studying and coauthoring research papers with Lund, whose intellectual honesty Stapp described as "uncompromising." Lund had reinforced his young scholar's faith in the scientific method, in the ultimate value of enlightened skepticism and rational analysis—even in the importance of self-criticism and the ability to take an objective view of one's own methods and results. But Lund had never made it easy. Once John Paul had completed the orals and the written final exam, he hurried back to Minnesota to resume classes, in which he'd fallen a week behind—but not before something entirely unexpected happened.

While in Austin that Easter, he met another young woman, this one named Nylah Maurine Tom. She'd graduated the previous year from the

University of Texas, where she'd been a popular sorority girl and member of the campus literary society. Though she was nearly eight years younger, she was attracted to Stapp's obvious intelligence and casual wit. While Stapp was taken aback by her flirting at first, he was flattered, and he liked Nylah enough to invite her to visit him in Minnesota. She made a brief trip to the Twin Cities a couple of weeks later, though Stapp seems to have kept their rendezvous a secret.

Meanwhile, in Brazil, in spite of her faith in the power of education, Mary Louise Stapp knew there was more to life than school. It had always been important to her that her sons find suitable wives, establish their own families, and find their own homes, *real* homes, and while both Robert and Celso had written about women they'd met, she was a little concerned about her oldest—who was now almost thirty.

In late May of 1940, John Paul wrote to Brazil and revealed his relationship with Nylah. He told his parents that he thought it might be serious, and while they were surprised by the suddenness of it, they could tell by the change in the tone of his letters that his dark cloud had lifted. There was a new confidence, a swagger. Not only did he receive word in June that, with Dr. Lund's blessing, the University of Texas had awarded him his doctorate, but he knew now for certain that he was going to be a doctor—and not just a PhD, but an MD. He believed he needed—probably that he deserved—a suitable companion.

With his parents' blessing, John Paul and Nylah announced their wedding date; it would be September 15, his father's birthday. Mary Louise scribbled her own congratulations at the bottom of one of Charles's letters. This was, she had told Charles, her greatest hope. Yet she would not live to see the day. Multiple carbons of a shocking letter dated June 15 from Campina Grande in Paraiba in the Brazilian northeast reached the States toward the end of that month:

My dear Sons;—"Tell my children to meet me in Heaven." Diga adeu a todos. A few short hours ago I received from my life companion her last message and sat by my dead in a lonely hospital room till friends came to help me do the last rites to the cold body. Your mother died as she lived fully trusting the Heavenly Father without a fear or falter. "Blessed are they who die in the Lord."

I shall try to explain right off what caused her sudden death. One of the ligaments that was sewed to the back wall of the abdomen in the suspension operation in 1915 formed a loop and a large section of the delgado intestine passed through the loop and was strangled by it. The difficulty of arriving at a complete diagnosis delayed the surgical intervention a few hours too long and she could not recover from the intoxication received from the incipient gangrene.

She was conscious till the end. I hid nothing from her and when I saw the final hour had come I told her so and we had our parting goodbye like we used to when one of us went on a trip. Her suffering was so intense that she could not talk much. She recognized you boys as the most important achievement of her life. Your mother nearly always said the last thing at night after we were in bed, "Where are my children to-night?"

Of myself only one thing I am sure of and that is that God who gave a good wife has taken her. I do not know what my life can be now without her but I am going to strive harder than ever to fulfill my calling.

By the time the family was aware of what had happened, Dona Luiza was already in the ground in Campina Grande. As with the death of his cousin in Burnet, John Paul suspected that incompetent medical advice and substandard facilities were to blame—it had taken three days on muddy jungle roads to get her to the hospital—and he suspected that Charles had likely been over-reliant on prayer and should have done more. It infuriated John Paul, but may have been hardest on Wilford, who was following in the footsteps of his brothers and was at Baylor now himself. For her part, Mary Louise had always been a stoic. Only a few weeks before she died, she'd written to John Paul: "The human suffering out here is, has & will always be terrible to see . . . You don't worry about me. I know that what can't be cured must be endured." She was sixty-one at the time of her death.

As John Paul mourned his mother, he struggled with the thought that if he'd been with her, he might have saved her somehow. At the very least, he could have supervised her medical care and made sure she got the attention she deserved. In his estimation, the world was too full of blind faith and too short of practical knowledge.

Charles soldiered on. It took him nearly eight months to construct a suitable tomb. He selected white marble slabs from Recife and had them set into a fabricated granite base. "It is not elaborate," he wrote, "but it looks very nice. It has a stability about it."

Back in Austin that September, prior to the start of fall term at Minnesota, John Paul and Nylah Tom were married in a brief ceremony attended by a small group of friends and family. While everyone who met her said she was a nice young woman, they weren't all convinced that she and John Paul were a match for the ages. Wilford, for one, sensed from the start that something was missing: "It was not a romance as such. It was more or less an 'I need you and you need me' type of thing. And her mother insisted on moving in with them right from the start. And she was mainly interested in her daughter marrying a doctor. He just didn't use good judgment."

Wilford was right. It was a miserable marriage from the start, and with John Paul's almost total dearth of funds—on several occasions he resorted to selling his blood to a local plasma bank—and his fanatical dedication to his studies, the start is all there was. After a mere thirteen weeks, at Christmastime, the couple filed for divorce. In the spirit of the self-examination that Dr. Lund had championed, John Paul came down on himself and fell into one of his periods of depression. He would not allow himself to entertain the seductive delusion that the failure was all Nylah's, or even the omnipresent mother-in-law's. A continent away, in Mary Louise's absence, Charles was left to hurt for his firstborn alone. "I pray that your domestic affairs may soon be straightened out. You deserve a nice home if ever a man did."

John Paul refused to discuss it. "It hurt him bad," Wilford believed. "It stomped his pride." Again, Stapp found himself alone, his only refuge a familiar one and the one thing he was very good at: his work.

3

COLONEL TANK AND COLONEL GAS

War is the only proper school for the surgeon.

—*Hippocrates,* **On the Surgery**

WITH WAR IN EUROPE RAGING, on October 29, 1940, the United States's first peacetime military draft lottery was held in Washington, DC. Each of the nation's draft boards had assigned a number to each of its registrants, and small slips of paper with the numbers 1 through 7,836 were inserted into small capsules, one number to a capsule. The capsules were dumped into a giant fishbowl and stirred slowly with a ceremonial wooden spoon that had been fashioned from part of a beam in Philadelphia's Independence Hall. As flashbulbs popped, Secretary of War Henry Stimson, blindfolded, reached deep into the fishbowl and withdrew the first capsule. On a podium positioned nearby, President Roosevelt leaned into a microphone and announced in his patrician lilt: "One hundred fifty-eight!" Each draft board then matched that number with the assigned name for its district, and sent that individual an order to report immediately for induction into the armed forces. And the process was repeated. Young men across the country crowded around the nearest radio set to learn their fate. In Minneapolis, John Paul Stapp, PhD, now in his second year of medical school, listened along with the rest. Stapp didn't have to listen long. His was the fifteenth number selected.

Military service had never been part of the plan, and Reverend Charles Stapp certainly had trepidations when he found out. But Charles, always well informed, understood the terrible gravity of events in Europe, which he followed through his subscription to *Time* magazine and radio reports.

"Last night I heard Roosevelt in his fireside talk," he'd written to his sons the previous winter. "He certainly gave a courageous but necessary declaration. The only hope for America is to help England down the Nazis. Let us hope that the material help will be sufficient and that the boys will not have to go."

As a student, John Paul was granted a pair of six-month deferments, and with the marriage debacle behind him, he drove himself hard. When his initial deferments were up, he was deferred again as an inactive Army reserve officer, and agreed to join the regular Army in the specialized training program for medical students. This meant that he would now be compensated by the federal government to finish school at a private's pay of $32 a month. The draft had almost come to seem like a lucky break.

It all got much more serious very quickly on December 7, 1941, following the surprise attack on the U.S. Navy fleet in Pearl Harbor. Congress declared war on the empire of Japan the following day. Military commitment was no longer an abstraction, not even for doctors. Stapp followed the progress of the war even as he tried to ignore it and bear down on his work.

And so, during the second half of the 1942–43 academic session—he'd just published a paper in the *American Journal of Physiology* titled "Responses in Size, Output and Efficiency of the Human Heart to Acute Alterations in the Composition of Inspired Air"—he volunteered to substitute for a couple of university hospital interns on vacation. While maintaining a full course load, he served as an unpaid intern-in-training for two six-week stretches, and he found out he loved hospital work more than anything he'd ever done.

At the same time he was completing his final requirements for graduation and, in October of 1943, he was summoned to take his Army physical. The examiner recorded his medical history: "Measles, mumps, chickenpox, pertussis as a child. Malaria, 1918, 1922. Tonsillectomy, 1922, adenoidectomy, 1922. No allergies. Has fallen and sprained left ankle 6–7 times since 1928. Ankles swell up for a few hours but has never given a days disability. Has participated in strenuous sports and long walks up to 35 miles without disability." His posture was "good," his figure and frame "medium," he stood five feet six inches tall and weighed 165 pounds. In spite of nearsightedness and a small umbilical hernia, the examiner proclaimed the subject to be physically qualified for limited service and approved his appointment to the Medical Corps, Army of the United States.

Four days before Christmas, 1943, John Paul Stapp received his diploma for the degree of Bachelor of Medicine from the University of Minnesota, and the very next day took the state medical board examinations. It was a proud moment for him and, by extension, for the family. All of the Stapp boys— young men now, with college educations and accomplishments of their own, thanks in no small part to their big brother's expectations and financial assistance—had enlisted in the military. Robert was with the 28th Naval Construction Regiment in Guam. Celso was already a decorated officer in Europe. Wilford Lee was a captain in the Army serving in Italy. Charles, while always a proud father, continued to fret: "Looks like my boys are departing from my ambitions for them. I never would buy them a toy gun or anything else that looked military."

Meanwhile, the dedication and work ethic of the Brazilian missionaries seemed to be paying dividends. The Baptists, in part behind Charles's lead, were becoming a force. New schools and seminaries were popping up throughout the region. There were 70,000 Baptists in Brazil by 1942 and more than 700 churches. But the most important development for his family concerned his personal life. Charles announced that he intended to marry Pearl Dunston, the daughter of a fellow missionary. "She is just as sweet and precious as I thought she was and a lot more." They were married just fifteen months after the death of Dona Luiza. In an attempt to ease the lingering sting of the Nylah Tom fiasco, Charles wrote to John Paul about his frustration with his own new mother-in-law, whom Charles described as a "grouch, just like an ingrowing toe nail." But, with both Robert and Celso having married by this time as well, John Paul must have winced at what he continued to see as his personal failure in love and domesticity.

Luckily for him, perhaps, there would always be more than enough work to get lost in. On New Year's Eve, 1943, Dr. John Paul Stapp, on suspended active duty as a second lieutenant, reported to the emergency room of St. Mary's Hospital in Duluth for his internship. During the nine-month ordeal, he was assigned 1,426 cases, performing more than 200 obstetrical deliveries, assisting in 225 major surgeries, and handling 28 cases of poliomyelitis.

In spite of the relatively meager salary—$25 a month, from which $5 was withheld pending completion of the internship—this was, for Stapp, the most meaningful work he'd ever done. Emergency medicine inspired him. Much later, remembering those draining weeks and months in Duluth, Stapp would declare them the happiest period of his life.

That period ended abruptly, however, with a call to active duty in the fall of 1944, and Lieutenant Stapp shipped out as a member of the United States Medical Corps. Having finally gained full admittance to the exclusive society of medical men, Stapp now found himself joining a new caste—one with different standards and very different expectations. The first stop: Carlisle Barracks in south-central Pennsylvania, headquarters of the Army's Medical Field Service School. The group of doctors assembled there on October 5 had no idea what the immediate future held for them. That summer had brought encouraging news from Europe as triumphant Allied forces had liberated Paris and rolled down the Champs-Élysées waving to cheering crowds. Victory on the continent was far from assured, however, and fighting in the Pacific continued at a ferocious pace. American troops on the island of Saipan in the Marianas and on carriers off the Saipan coast faced thousands of Japanese suicide troops in banzai charges that, in a single three-week period in late June and early July, had killed more than 3,400 Americans. The need for battle-ready medical personnel was still acute.

A brigadier general told the group of 525 doctors why they were at Carlisle Barracks. Their objective would be to learn and experience the practice of medicine in field combat situations. They would be quartered in barracks just as enlisted men would be. Lieutenant Stapp's barrack reminded him of a Texas cotton barn—no paint, no plaster, just bare walls and a roof. The men were instructed to take possession of their cots, spaced 30 inches apart, in alphabetical order. Dusky twilight through the dirty windows gave the scene the ambience of a graveyard.

For the next two and a half months the doctors suffered through their indoctrination into the United States Army Medical Corps. During the early days of their training, huddled together in the rain, their woolen uniforms sopping wet, Stapp and the others listened to a red-faced major explain the fine—even dainty—points of military courtesy, including how one's fingers are held in a proper salute. When an exhausted lieutenant doctor nodded off for a moment during the demonstration, the enraged major woke him with a shriek ("This is war!") and announced his punishment for a substandard salute: a $10 fine. That got everyone's attention; on their way back to their cots that evening the men practiced saluting trees and trash cans.

As their time wore on, the doctors heard lectures on the history of the medical corps and learned how to transport battlefield wounded on litters. They drilled on the parade field and sprinted from meal to meal. At night, in their barracks that stank of wet wool and lignite from the coal stoves, they struggled to sleep amid the snoring and groaning.

Toward the end of their first week of training, the doctors were introduced to Colonel Tank (it was easier to remember them by what they taught rather than by their names), a semi-cartoonish but fierce, mustachioed figure who described the organization of the Army's armored forces. Tank struck the doctors as pompous and a bit dim. Next came Colonel Gas, a slender, black-headed martinet who stomped about in a perpetual fit as he elucidated the horrors of gas warfare with a well-rehearsed series of sickening anecdotes.

The second week commenced with a morning five-mile hike with light packs, followed by a nighttime twelve-miler with heavy packs. As their training intensified, the doctors graduated to boot camp–style combat survival training. They charged up Coronary Hill wearing gas masks, crawled across open fields beneath live machine-gun fire, and marched through tear gas and the clouds from smoke grenades. Then came the stretcher-bearer's obstacle course. Two bearers would carry a 200-pound surrogate patient on a nine-pound litter up steep hills, through tunnels, muddy pits and trenches, a grove of sharpened bamboo poles, and a variety of streams and bogs. At dinner that night, the doctors could barely lift their arms to feed themselves.

The following days and weeks included orienteering exercises, field latrine construction, and combat indoctrination, much of which mirrored an infantry recruit's basic training: rifle practice, foxhole digging, and bayonet drills. The propaganda of the combat training films disgusted John Paul. In one, a terrified infantry grunt overcame his fear and made his first bayonet charge at the enemy. The film ended with the young soldier concluding that he would rather risk his own life than face a court martial for failure to perform his duties. That night, Stapp recorded his thoughts: "What a dismal state mankind has reached if the best civilization can offer is the application of advertising psychology to the abysmal business of men skewering each other with bayonets; broiling each other with flame throwers; struggling for worthless real estate; and shedding their blood for ideals refurbished for each war, with the campaign promise that it would be the last." He had a chance to reflect on another aspect of war during a weekend leave to Gettysburg. The nation had been saved on that stretch of earth by brave boys who'd been convinced to sacrifice themselves not to escape court martial but for the sake of their buddies and their leaders and their families. Stapp found it sobering.

The gravity of what the doctors were involved in gradually began to sink in. One of the men in John Paul's barracks received an emergency telegram informing him that his younger brother, a Marine and a fellow medical corpsman, had been lost in action in the Pacific theater. To make matters worse, the doctor's father had suffered a fatal heart attack upon hearing the news. It was a bleak reminder that they were all involved in something much bigger and far more consequential than their individual miseries. The near-constant complaining tapered off. The doctors, each in his own way, were taking on the mind-set and the mien of soldiers. Stapp even began making a regular donation to the National War Fund from his paycheck.

Several weeks into the course, the doctors had all lost weight; John Paul had dropped 15 pounds and felt like he was back on the cross-country trails at Baylor. They were all leaner and stronger than when they'd arrived, and they had toughened up mentally as well. John Paul had modified his own opinion of the Army, and noticed a change in most of his colleagues: "Above all, we came to the realization that we were being taught far more efficiently than in any University or College course any of us had taken, including medical school. It was all directed to integrating us into the army field combat system as effective foxhole surgeons."

At the conclusion of the training at Carlisle, Stapp joined his fellow doctors in a graduation parade around the barracks grounds, where he saw looks of almost astonished pride everywhere he turned, as if they'd surprised themselves with what they'd become. As a group, they were changed. He felt it himself. They said their goodbyes that night, and the next morning Lieutenant Stapp made the trip by train from Carlisle to Lincoln, Nebraska, to begin a three-month assignment at the resident graduate training program at the Army Air Corps regional station hospital at Lincoln Army Air Field. Whatever else he might encounter at Lincoln, he was sure of one thing: there would be airplanes.

4

SCUM JOBS

In addition to his other duties, he died.

—epitaph of the second lieutenant

HAVING SURVIVED MILITARY TRAINING, John Paul Stapp was again back in his comfort zone, where he intended to remain until his service commitment had been completed and he could get on with life as a doctor. Hospital duty in Lincoln offered Stapp a hands-on survey course in military medicine, his ward handling as many as thirty patients at a time. Stapp worked up histories based on physical examinations of his patients and studied them until he could practically recite the charts by memory. He was next assigned to the surgical ward, where he received an introduction to such topics as orthopedics, urology, and dermatology. From surgery he moved to the sanitary corps, where he learned the proper procedures for food inspection, water sanitation control, and sewage disposal.

Stapp loved the work, but found his fellow residents perplexing. All but one of the rest of the group of seventeen doctors hailed from New York City, and they formed a buttoned-up fraternity that Stapp was unable to penetrate. Of one of his more vocal colleagues, he wrote: "The champion complainer was from Columbia Medical Center. He had spent his entire life in Morningside Heights, except for an occasional safari to Manhattan to see the natives. He was like an exile from another planet, spending his off-duty hours with shades drawn, reading the *New Yorker*." Like the military, the medical caste had its own hierarchies, and John Paul was coming to understand that his place had already been predetermined by his background and his relative lack of social graces. He would always be an outsider.

But there was little time to be concerned with the social pecking order. The doctors were required to prepare and deliver lectures to air crewmen transitioning from the B-17—which was being flown in Europe—to the B-29s that were heading to the Pacific. Lieutenant Stapp did his best to instruct young pilots how to deal with in-flight injuries: how to gently handle intestines when they protruded through abdominal wounds, how to care for exposed brain tissue, and how to bandage blinded eyes. Actual war casualties were arriving at Lincoln daily, including fresh transports from Walter Reed Hospital in Washington, DC. In mid-December, 1944, German troops had launched a devastating surprise attack through the Ardennes Forest, inflicting 89,000 American casualties in the space of five weeks, the bloodiest faceoff of the entire war, which would come to be known as the Battle of the Bulge. Many of those cases came to Stapp and colleagues, along with frightening numbers of amputees from the front lines of the winter fighting in the Hürtgen Forest along the German–Belgian border. The most confusing and—for Stapp—the most fascinating cases were what the neuropsychiatric service referred to as "aerial combat fatigue." The symptoms varied: anxiety, exhaustion, depression, loss of sleep and appetite, inability to concentrate. Some patients were gripped by paranoia; others exhibited bizarre behavior that occasionally turned violent.

Observing these symptoms of psychiatric distress, Stapp believed he gained some crucial insights into human nature. Clearly, the stresses of combat affected everyone differently. When the stress was relieved, however, most people—no matter how temporarily damaged—bounced back to their "normal" selves. But it was incorrect, he believed, to classify those who broke under pressure as somehow abnormal or inadequate. It seemed clear to him that extreme reactions to the horrors of war were natural, and that only the truly abnormal personality was unscathed by exposure to the bloody hell of battle. "Indeed, it is psychopaths," Stapp wrote, "the criminally inclined and other emotionally deficient, destructive individuals, hardly an asset to a peaceful community, who are the more likely to find war to be just their dish of depravity." Stapp met a returning veteran in Lincoln who bragged about his adventures with a flamethrower: "Hit a running Jap with that flame, and he'll swell up and pop just like an onion!"

The pace of the war was frantic, and as its personnel transitioned over-seas, the Army moved its stateside resources around like chess pieces to plug the gaps. The town of Pratt, in central Kansas, about halfway between Wichita and Dodge City, would be the next stop for Lieutenant Stapp. He traveled there on a sunny Sunday afternoon in February of 1945 to report for duty. As he stepped off the train and surveyed the prairie environs, he thought it was the flattest, bleakest place he'd ever seen, and reflected that he'd come a long way from the flowering jungle paradise of Bahia, or even from the cotton farms and leafy, cottonwood-shaded bayous of the Texas Hill Country. When he arrived at Pratt Army Air Base a few miles from town, he found a haphazard clutch of unpainted huts surrounding wide blacktop runways where B-29 Superfortresses were practicing touch-and-go landings. The hospital was composed of several windowless, barnlike buildings connected by precarious-looking roof walkways.

As a junior officer, Stapp was issued a laundry list of duties: base gas casualty officer, assistant chief of the dispensary, assistant dependent clinic officer, assistant medical supply officer, and assistant medical statistics officer. "Scum jobs," in Stapp's judgment. At one point, in his role as base sanitary officer, he actually wrote a report to himself in his role as base medical inspector.

Neither Stapp's social life nor his financial situation improved much during this frantic period. He was still paying off his medical school debt and sending small amounts of money from his monthly paycheck of $116.92 to Charles and Pearl in Brazil. Still, that spring John Paul upgraded his bare room in the officers' quarters at Pratt with a long-coveted luxury item: a high-fidelity phonograph and a small collection of record albums. As it was for Charles, classical music became a calming refuge from the grind for John Paul. Hearing a familiar sonata or concerto, Stapp sometimes thought about his Baylor girlfriend, but with the Nylah disaster still fresh in his mind, he had resolved to avoid letting himself be distracted by women.

Despite the grind of his medical duties, Stapp found himself increasingly fascinated with the airplanes that surrounded him and a little envious of the men who flew them. He had never forgotten his short flight—not much more than a coffee break in the sky, really—back in his college days in Texas.

At Pratt he befriended a B-29 pilot whom he'd treated for mumps, and was offered a seat during a six-hour training mission. While the lumbering prop-driven Superfortress bomber was hardly a cutting-edge aircraft, Stapp could not have been more delighted with the experience: "No jet I ever boarded could overshadow the thrill of that moment when I went up the ladder into that instrument-cluttered aluminum dungeon." The bone-shaking growl of the four engines created its own kind of music, and Stapp felt himself in immediate harmony with it. The sight of the Kansas countryside rolling past beneath the 50 tons of metal—with other B-29s strung out alongside in formation, silhouetted against the clouds—struck him as sublime. The crisp work of the pilot and copilot, the flight engineer keeping tabs on his puzzle of dials and gauges, the radio officer cranking his radio sets and updating his logs, and the navigator with his maps, protractors, and slide rules, reminded Stapp of an experienced medical team in the operating room. They worked with a passion that could only befit men who would be heading off to war in a matter of weeks, if not days.

"I'm sold on aviation," he confided to his pilot buddy. Shortly after the B-29 experience, Stapp began to hang around the base flight surgeon, offering to assist with physical exams for aircrew members. Not much later, he submitted his application for the Army's School of Aviation Medicine.

While the airplanes were certainly part of aviation's appeal for Stapp, the pilots and crews themselves were also a major attraction. These men tended to be wild and headstrong, but they had a spirit and enthusiasm that he found exhilarating. For example, when one group of ground support personnel was ready to depart for the Pacific, the men staged a parade around the base grounds. They managed to "borrow" an elephant from a carnival that was passing through town, painted the animal red, white, and blue, and harnessed it to pull a flatbed truck carrying a small brass orchestra, followed by a weapons carrier stocked with cases of cold beer. This kind of frivolity was foreign to Stapp and contrary to his nature, but he found himself at least admiring if not exactly enjoying it.

Despite his newfound affection for planes and pilots, however, Stapp was frustrated with his assignment at Pratt. In April of 1945 he enlisted the help of a colonel he'd met at Lincoln to expedite his application to the School of Aviation Medicine at Randolph Field in San Antonio. Two weeks later he got his orders, and even through the pounding rain that greeted him as he stepped off the plane in Texas, Stapp could see that his new assignment was an upgrade from the tenement-like outposts he'd known since leaving Minneapolis nine months earlier. A few hundred yards beyond the entrance gate to Randolph Field, the base headquarters tower—known as the Taj Mahal—rose from the manicured grounds like the spire of a great cathedral. Nearby stood the brick officers' quarters nestled in an oasis of grass lawns, flowerbeds, and spreading shade trees.

The School of Aviation Medicine was the offshoot of the medical research board that had been established by the War Department during World War I. Its charge was to "investigate all conditions which affect the efficiency of military pilots, and to consider all matters pertaining to their selection and their physical and mental fitness." The school developed flight surgeons for the Army Air Corps and facilitated the latest aeromedical research. On his first full day at Randolph, John Paul was issued a syllabus of courses listing instruction in such areas as tropical medicine, war medicine and surgery, neuropsychiatry, and gas warfare. Upon completing paperwork with the school's registrar, Stapp and the 150 other enrollees were presented with an armload of technical manuals and textbooks. They were then introduced to a drill instructor who marched them sharply away from the pleasant central section of the base to a row of unshaded tarpaper shacks on the far side of the parade ground. To Stapp, the shacks looked like a line of coke ovens baking in the noonday heat. As the structures shimmered in the white glare of the sun, the men were visited by the school's assistant commandant, who explained that because a large contingent of flight nurses would be occupying the quarters that were normally reserved for student doctors, Stapp and his colleagues would be living in the ovens.

Though the conditions at the overcrowded base were less than ideal, the classes themselves—not to mention the distinguished faculty, which included Ivy League professors, celebrated specialists, and renowned medical researchers—were superb. Lectures in the morning, labs and clinical exercises in the afternoon, and, once the assistant commandant announced that the incoming class, in compensation for the degraded condition of their housing, would be afforded special admission to the base officers' club, the evenings became pleasant as well.

Late during his second week at Randolph, Stapp attended a session on respiratory physiology, with an emphasis on the special considerations applicable to high-altitude flight. The effects of a decrease in atmospheric pressure and subsequent oxygen starvation were described, and then demonstrated with simulated flights for the students in a decompression chamber. Each student doctor was subjected to the pressure differential at 30,000 feet in a mock cabin-pressure failure. The students were warned not to try to hold their breath, and they were offered the cautionary story of the medical officer who had disregarded this advice and died of ruptured lungs.

The unusual problems related to high-altitude flight again caught Stapp's attention. He knew that aircrews during winter missions in the skies of Europe suffered terribly from the cold and the bends, an insidiously dangerous form of decompression sickness. Symptoms of the bends occur when a reduction in ambient air pressure results in dissolved gases in the bloodstream—particularly nitrogen—coming out of solution, causing bubbles to form in body tissues, organs, and joints. It affects pilots and divers during ascent as they transition from a higher- to a lower-pressure environment. Nevertheless, after his first session, John Paul and nineteen of his colleagues volunteered to be sealed again into Randolph's decompression chamber configured to simulate a suffocating 40,000 feet above sea level and temperatures of 50 degrees below zero. This huge reinforced steel tank was operated by technicians who could manipulate interior pressure, oxygen level, and temperature, and who were trained to monitor their test subjects as they took them into extremely dangerous simulated environments. Electrically heated garments gave the subjects partial protection from the cold, and pressurized oxygen kept them conscious. In this condition, the doctors practiced—or attempted to practice—basic first aid. In the equivalent of what mountaineers refer to as a death zone, it was almost torture to move at all, and the experience caused Stapp to reflect on what the pilots and gunners had endured as anti-aircraft fire ripped their airplanes apart in the skies over Germany.

It was at Randolph that Stapp was first introduced to the concept of pilot ejection systems. Researchers at the school who'd been studying pilot death and injury had been calling for the development of ejection seats for military aircraft for a full year. This was still a novel idea and the only workable examples belonged to the Germans, who were by that time already using them to save pilots in distress. The doctors studied other methods of air and sea rescue as well, and sat through lectures on desert, arctic, and jungle survival. They were immersed in case studies of incidents that had occurred only weeks earlier in Europe or the Pacific.

The future flight surgeons also got some precious time in the air. Stapp received a two-hour aerobatic indoctrination flight that turned out to be pretty exciting. In the back seat of a T-6 trainer, he got to experience slow horizontal rolls and a series of stalls that resulted in gut-wrenching tail-first drops. The T-6 also completed some downward spins and spirals and loop-the-loops, and Stapp loved every minute.

When they weren't training in fieldlike conditions, the school held intelligence briefings for its students, some of whom would shortly find themselves in an actual theater of war. During their morning briefing on August 6, the students learned that an American bomber had dropped a uranium atomic weapon on the Japanese city of Hiroshima, unleashing a mushrooming fireball and high-pressure shock wave. The descriptions of the devastation caused by this new secret weapon were almost impossible for the doctors to credit at first. Neighborhoods full of human beings vaporized instantly. Temperatures on the ground of 4,000 degrees Centigrade, melting steel and concrete. At least 75,000 civilians killed or injured so severely they would die within hours, with deaths from burns and radiation sickness projected to continue for weeks. Estimates were that 65 percent of the casualties were children nine years old or younger. Then, three days later, U.S. forces dropped a second bomb on Nagasaki.

Three months earlier, on the last day of April 1945, Adolf Hitler, in his Berlin bunker, had put a gun to his temple and shot himself. By that point, the far-flung German forces had surrendered, and victory in Europe, V-E Day, had been declared on May 8. By August 15, Japan had announced its surrender and World War II was finally over. In spite of the victory celebrations at Randolph and around the country, John Paul Stapp would always recall his own ambivalent feelings during those heady days. The sickening realities of radioactive death and "nuclear leprosy" were heavy on his mind as the ten-week course at Randolph came to an end and he returned to his duties at Pratt. As a doctor, he was appalled at what had been done in his country's name. He thought of the Japanese doctors who would be working day and night with the victims, and he made a deal with himself that, for the remainder of his time in the United States military, he would refuse to participate in any program that involved weapons systems research or development.

Back at Pratt and now a bona fide flight surgeon, Stapp found the environment dull after two months of communing with the faculty at Randolph. With the war officially over, attitudes began changing almost immediately. Most of the base personnel spent their days going through the motions of their assigned duties and waiting for the day they could return home to their families. Stapp found himself waiting right along with them.

From Campina Grande in Paraiba, Reverend Charles Stapp wrote to his sons in jubilation about the end of hostilities in the Pacific. All of Brazil was rejoicing, he told them. His boys had done their parts and survived the war, and Charles felt that good times were on the horizon. He and Pearl would be making a visit back to Texas the following year, and he looked forward to reconnecting with family and getting some Stateside medical attention. "I expect to have all of my teeth out at that time and get some new ones. The ones I have are all worn out. Pearl and I keep on being good pals and make a team of our work." They often went together, Charles noted, to visit Dona Luiza's tomb.

Charles continued to be intrigued—impressed, even—by John Paul's career path, though the idea of airplanes, aviation research, and atomic bombs seemed to overwhelm him. He would never be entirely comfortable with this brave new world he read about in letters from his oldest son and followed in the pages of his news magazines. "Your higher education is quite interesting," he wrote. "I hope that you may be able to keep on foot and take care of the others at whatever level you fly. I prefer not to fly straight up so high if I am going to fly at all." Meanwhile, John Paul dreamed about rejoining the hospital staff in Duluth and eventually settling into a pediatric practice.

Instead, in early December 1945, Stapp was abruptly transferred to Davis–Monthan Army Air Base in Tucson, Arizona. It wasn't the emergency room at St. Mary's, but it was a change and at least there would be plenty of airplanes. His abiding interest in aviation was further piqued on December 9 when news of a new transcontinental speed record was announced. The previous day, Captain Glen Edwards and Lieutenant Colonel Henry Warden had

flown a prototype Douglas XB-42 Mixmaster—an experimental high-speed
bomber with fuselage-mounted engines that drove contra-rotating propel-
lers on the tail structure—from Long Beach, California, to Washington, DC,
in just five hours and seventeen minutes, recording a top speed of 434 miles
per hour. The pilots had endured frigid conditions in the unheated cockpit
and last-minute problems with the landing gear, but had bested the previous
record by three-quarters of an hour. It was the kind of thing that could only
happen now that the war was over.

Meanwhile, at Davis–Monthan, John Paul was assigned briefly to the
medical unit of a separation center where the flood of soldiers returning from
overseas was processed for release back into civilian life. The medical staff
handled as many as 600 patients a day. John Paul manned the ear, nose, and
throat station of the assembly line. He estimated that he examined approximately 9,500 eardrums and became so proficient that he could determine
where a man had served based solely on the state of his ears: New Guinea vets
could be identified by a particular chronic fungal infestation of the ear canal,
North Africa vets by the quality of sand embedded in their ear wax. When his
separation duties were done, Stapp was assigned to the base flight surgeon's
office and tasked with running a clinical group that treated prisoners of war
held at the base, most of them Germans from a captured submarine.

With the war over, the Stapp siblings were briefly reunited for Christmas that year in Burnet, and some of the frustrations that had been building
between Robert and his brothers boiled over. Robert's marriage was falling
apart, and he was on edge. He insulted both John Paul and Celso in front of
the rest of the family, ridiculing their status as reserve officers and insinu-
ating that as doctors they'd both managed to avoid the hard life of the soldier. John Paul felt bad for Robert, but left the Christmas gathering early to
return to Davis–Monthan, seething at what he regarded as his brother's willful refusal to take control of and accept responsibility for his own problems.

The clear air and winter sunshine in Arizona suited Stapp, but as fond
as he quickly grew of the desert environs, he knew not to let himself get too
comfortable, assuming that when the flood of returning troops dried up, he
was likely to be on the move again. Sure enough, on a sparkling morning just
after New Year's, he found new orders in his box at the base post office: "LT.
STAPP, JOHN P., (MC) will proceed to Kelly Army Air Depot for a seminar in

Industrial Medicine. Will report to Base Surgeon not later than 10 January 1946." Though he had no idea how long he would be there, at least he would be heading back to San Antonio (Kelly was a half hour's drive from Randolph), and that was a far sight better than Adak in the Aleutians or another stint in Kansas.

The course at Kelly, as it turned out, lasted only two weeks. The base surgeon was a full colonel who'd been director of industrial medicine for one of the big automobile manufacturers in Detroit and he ran a tight ship, augmenting his own lectures and demonstrations with those of visiting specialists. The depot at Kelly rebuilt and refurbished the Army's airplanes, and while the war was in full swing more than 45,000 civilians had worked there around the clock in three shifts. The industrial medicine curriculum covered chemical compounds, the mysterious and dangerous work in the radium dial shops, the noise hazards associated with aircraft engine tests, and more mundane nuisances such as contact dermatitis. On the last day of classes at Kelly, the colonel-surgeon produced slips of paper with all of the United States Army's principal air depots written on them, mixed them up in his cap and allowed each doctor to draw his own next assignment. Stapp drew the Midwest Air Depot—later renamed Tinker Air Force Base after an Osage Indian major general whose B-24 had gone down off Wake Island in the Pacific—a few hundred miles to the north in Oklahoma City. Later that day, Stapp received word that Tinker was going to provide a real-world crash course in industrial medicine: a maintenance hangar at the base had just erupted in flames when paint solvent vapors were ignited by drying ovens that were too near the spraying booths. Ten men had died and thirty-seven had been injured in the inferno.

It didn't take Stapp long to pack his belongings for the trip up to Tinker, but relocating wasn't as easy as it had once been. Over his months in the service, and for the first time in his life, he had begun to accumulate a few possessions: extra clothes, boxes of files, a few tools, books, an increasing collection of record albums. On the train north, he reflected on his whirlwind travels and allowed himself to imagine what a stable life as a civilian doctor might be like, perhaps in Texas or back in Duluth: his own pediatrics practice, a house. A wife. A family. He would turn forty in a few years and he was beginning to feel the pressure of time.

On the morning following his arrival in Oklahoma City, Stapp—as the new industrial medical officer for the base, in charge of environmental health support for a vast complex—was introduced to his staff: four civilian doctors, eighteen nurses, and two dozen administrative workers. Among them, they were responsible for the care of 6,000 military personnel and 13,000 civilian employees—many of them subjected daily to dangerous and deafening working conditions, deadly chemicals, and a high degree of stress. During the war, a single 3,300-foot-long plant had turned out twin-engine C-47 transport aircraft at the rate of one every hour and fifty minutes, twenty-four hours a day, seven days a week. Now the C-47 plant was busy replacing wing fuel tanks with internal tanks, a process that involved exposure to extremely toxic solvents. Chronic skin allergies were almost epidemic.

Stapp and his doctors worked hard—eighty-hour weeks weren't uncommon—touring the base shops, observing the conditions and monitoring the workers through spot physical examinations. In addition to skin problems, the base produced a steady stream of lacerations, fractures, smashed fingers and toes, sprained backs and necks, and burns. Stapp saw lots of burn cases, many of them from motor vehicle accidents. One of those cases that would haunt John Paul involved a pair of veterans from the Pacific theater who had headed out for a week of R and R riding tandem on a small motorcycle. While attempting to pass a city bus that was stopped near the Tinker main gate, the riders were hit by an oncoming bus and thrown into a ditch, their legs crushed. The motorcycle's fuel tank ruptured, engulfing the men in gasoline fire.

Stapp took charge in the emergency room, ordering morphine, plasma, and Vaseline gauze dressings. One of the men, with third-degree burns covering 80 percent of his body, died eight hours later. The surviving victim was placed in an oxygen tent. The staff labored continuously for three days to save his life. The patient received twelve units of intravenous plasma, three pints of whole blood, and six units of serum albumin concentrate to compensate for the copious fluids oozing from the burned areas of his body. Late on the third night following the accident, the patient lapsed into delirium and died. Stapp comforted the man's widow who'd sat vigil courageously—they'd been married just two months earlier—and considered the bitter irony. Two brave veterans who'd survived the horrors of Guadalcanal and the Battle of the

Philippines had returned home only to die agonizing deaths as the result of a common road accident. In spite of his staff's heroic efforts to save the two men, Stapp found himself overcome by a feeling of helplessness and pity, compounded by the memories of his cousin John Hansford, Jr., back in Burnet.

Even as he was consumed by his duties at Tinker, there were other pressing problems to deal with. Just a few days after his patient's death, Stapp received a letter from his brother Robert, who had filed for divorce and was in the midst of a custody fight over his daughter, Martha Louise. Robert had somehow become convinced that his big brother was conspiring against him with Robert's ex-wife. John Paul put the blame for the marital stalemate squarely on his brother and wrote back in frustration, accusing Robert of "psychiatric disorders." Still, he signed the letter "Your Bud," and promised to continue helping Robert financially.

Despite the challenges, both professional and personal, Stapp felt that his time at Tinker had made him a better doctor and, he hoped, a better man. On June 7, 1946, after twenty months as a general-duty medical officer, one who'd been exposed to a staggering variety of military medical problems even as he'd remained Stateside during the final months of the war, John Paul Stapp was promoted to the rank of captain, MC AUS. One week later he received invitational orders from General Charles Glenn, the command surgeon at Wright Field in Dayton, Ohio. Glenn had seen Stapp's résumé and was impressed with it, recommending that Stapp be considered for an aeromedical research post.

After completing in-depth interviews, Stapp was sent to the Biophysics Branch of the prestigious Aero Medical Laboratory at Wright. When he arrived, a branch medical service corps major gave him a tour of the facilities where Stapp met and observed scientists and engineers working on a human centrifuge that looked like a high-speed industrial merry-go-round, designed to simulate high-speed airplane turns and pullouts. It seemed like it might be fascinating work. Yet when the major asked Captain Stapp if he would like to join the Aero Medical Lab and participate in human factors research projects, there was hesitation.

It was tempting. Wright was sacred ground: Huffman Prairie, where Wilbur and Orville Wright made their experimental flights back in 1904 and 1905. Nevertheless, Stapp confessed, with only eight months remaining on his service commitment, he was reluctant to undertake an entirely new line of work. He could easily wait out the remainder of his days in the Air Corps and prepare himself for a civilian career. Still, if the Aero Medical Lab wanted him and judged that he could make a greater contribution in research, he would not object. The major was less than thrilled with the interviewee's lukewarm response and Stapp knew it, unsure whether he'd be offered the position as he boarded a transport back to Oklahoma. But three weeks later he received orders for a permanent transfer to Wright Field—followed by a glowing recommendation from the base surgeon at Tinker, Major Voris McFall. "I can truly say," McFall wrote in his letter of commendation, "that your services both to Industrial Medical Service and the Station Hospital have been excelled by no one. You are one of the most dependable and responsible Medical Officers that I have ever had the pleasure to know. Every phase of your work for me has been of a superior quality."

While Stapp was dismayed by the idea of prolonging his transition to civilian life, he packed up and headed for Dayton with the thought that he might, for the first time, be given an opportunity to demonstrate the full range of his academic training and intellectual skills, bringing together what he'd learned in the emergency room in Duluth with his work at Texas under Dr. Lund. "At last," he wrote, "I was selected for an assignment that would make use of my special training." This was still the Army, but what if this was his epiphany? His Brazil?

Shortly after arriving at Wright Field on August 10, 1946, and reporting to the Biophysics Branch, Special Projects Section, Captain Stapp was informed that he was now a project engineer assigned to the pilot escape technology program. He protested, making it clear that he was a doctor, not an engineer. The gruff branch chief, Colonel Edward J. Kendricks, explained that the unit was attached to the Engineering Division and immediately issued his new engineer a 1,200-page set of captured German technical documents, reports mostly, relating to aircraft ejection seat tests and biomedical examinations of human tolerance to the forces an explosive ejection creates. Kendricks had learned from Stapp's file that he was fluent in German.

Stapp found his office on the second floor of the Biophysics Branch building, one he would share with four fellow project engineers and a secretary, and dug in. There was a wealth of potential sources for German documents at Wright Field. Immediately following the end of operations in Europe, Major General Hugh Knerr, the commanding officer at Wright, had sought War Department permission to have several prominent German aviation designers and aeronautical scientists brought to Dayton to assist the Army Air Corps. Some thirty Germans were eventually transferred to Wright not much later as part of Operation Paperclip, an Office of Strategic Services postwar program designed to bring German scientists to America for the purpose of exploiting their knowledge and skills, and—equally urgent—to deny that expertise to the Russians. When Stapp arrived at Wright Field in 1946, five senior Luftwaffe officers were already there, laboring on a comprehensive history of the German–Soviet battles of the preceding years.

The German aeromedical community, Stapp discovered as he made his way through the reports Kendricks had given him, had been working for more than a decade on two specific aviation-related challenges: the launching of aircraft from ships and pilot ejection systems for fighter aircraft, both of which involved massive acceleration (+Gx) and deceleration (–Gx) forces. The documents described a design for pilot seats in fighter airplanes that were mounted on guide rails and equipped with a special catapult activated by gunpowder charges. In an emergency, a pilot would be able to jettison his canopy and pull a trigger that would shoot him straight up, high enough to clear the plane's tail section, where a parachute could be safely deployed. The Germans had created elaborate test facilities at Tempelhof Airport in Berlin and had constructed a 60-foot ejection test tower. As he read on, Stapp was appalled to discover that men and women—prisoners—had been used to explore the outer limits of human tolerance to g-forces and impact. The collection of documents contained reports of barbaric aeromedical experiments.

One test subject on the ejection tower who suffered a severe back injury was euthanized so that surgeons could dissect his spine and remove several sections of vertebrae for static tests to isolate fracture forces. Stapp realized that he was getting a peek into a monstrous world in which trained medical personnel had used human beings as lab animals.

5

THE CITADEL OF AEROMEDICAL RESEARCH

Humanists talk about civilization; engineers create it.

—John Paul Stapp, aphorism from **Dr. Stapp's Almanac**
and Rational Calendar, *the month "Midyear"*

THE WRIGHT AERO MED LAB had been founded a dozen years earlier by a
U.S. Army captain and doctor from South Dakota named Harry George
Armstrong. Armstrong was a hugely important figure in the development of
American aviation medical research, and he established a test environment
at Wright that rewarded creativity and teamwork. He staffed his lab with
the Army's top talent, and when the war was over he went to Europe to help
evaluate the captured aeromedical scientists and doctors there. Armstrong
was interested in one man in particular: Dr. Hubertus Strughold. Strughold
had been the chief of aeromedical research for the Luftwaffe as the wartime
head of the Luftfahrtmedizinishche Forschungsinstitut (Institute for Aviation Medicine) in Berlin, a position he continued to hold through the World
War II years. Armstrong had met Strughold in 1928 during the German's
one-year Rockefeller Fellowship in the States, had admired his work, and
even developed a personal friendship with him. Immediately after the fall of
Berlin, Armstrong went to great lengths to locate Strughold and ensure that
he not fall into the hands of the Russians. Eventually he found Strughold and
probably helped protect him from accusations of war crimes, as evidence of
atrocities committed by members of the German aeromedical community

under his direction began to surface. The defeated regime's top aeromedical man was allowed to resume a post at Heidelberg University, but he also began what Armstrong hoped would be a cooperative and fruitful relationship with the United States.

Harry Armstrong's successor at the helm of the Aero Med Lab was a man who would have a powerful influence on Stapp. Though he would be gone from Wright by the time Stapp arrived, U.S. Army Reserve Colonel Randy Lovelace had on several occasions subjected himself to punishing physical experiments of his own design. His most terrifying project, just three years prior to Stapp's arrival, had required him to make a parachute jump from a B-17 at an unprecedented 40,200 feet to simulate a high-altitude emergency. Lovelace had lost consciousness and suffered severe frostbite during the ordeal, but in the process had taught the aeromedical community a number of crucial lessons about the brutal effects of bailout at extreme heights. He had also, during his tenure at Wright Field, transformed the Aero Med Lab into a cutting-edge research facility.

In the summer of 1946, the lab was working to improve the fairly dismal performance of American aircraft during the war—not strictly performance in combat, but also the alarmingly high accident rate. Weak, poorly designed seats and inadequate restraint systems (shoulder harnesses and waist belts) were seen as major contributors to the pilot casualty rate. Conferences were convened and reports were issued calling for improved survival systems. Almost a decade earlier, Harry Armstrong had predicted that pilots could live through decelerations of up to fifty times the force of gravity "providing the belt and seat did not give way," and a Navy initiative had called for a pilot seat capable of withstanding 40 g's.

Following a review of Eighth Air Force casualties, a group of Army aeromedical doctors began lobbying for strengthened cockpit and seat standards for military aircraft. Aircraft manufacturers balked on the grounds that it made no sense to build such a cockpit if the men manning it couldn't take the g-forces involved with crash incidents.

Researchers at Wright had a tremendous challenge before them if they were going to prove Armstrong's point about the ultimate resiliency of mortal man, but Stapp found himself impressed by his new colleagues. Two of the other men on the second floor of the Biophysics Branch building were fellow medical doctors; the third was a degreed mechanical engineer and the fourth was a pilot. After lunch on Stapp's first day in the office, they all joined Colonel Kendricks and drove up a small hill to a concrete hangar-like structure. Inside was a 30-foot steel tower built the previous summer to test ejection seats, and the seat looked much like the design Stapp had seen in the German documents.

Randy Lovelace had gone to Europe shortly after V-E Day to interview some of the engineers and scientists who'd worked on the German ejection seats, and he'd returned with the specifications that had led to the design of the tower at Wright. Lovelace had also brought back an actual ejection seat which the Aero Med Lab engineers had immediately set about to reverse-engineer. These devices were clearly going to be a big part of the future of military aircraft. The Germans had begun equipping their fighter pilots with parachutes as early as 1919, but with increasing flight speeds it was becoming extremely difficult to get out of a damaged plane safely. Speed wasn't the only problem. The gyrations of disabled aircraft, the windblast, and the likelihood of hitting the tail or a propeller were all puzzles to be solved.

That first afternoon, Stapp watched as mechanics loaded what appeared to be a cannon shell into a catapult housing that bristled with electrical wires attached to recording devices. Nearby, technicians were preparing high-speed motion picture cameras and floodlights trained on the tower, at the base of which was a pilot seat mounted on vertical rails.

Stapp watched the volunteer subject strap himself into the seat before a medical officer took the man's blood pressure, pulse, and temperature. A helmet was connected to what Stapp was told was an accelerometer. Once everything was in place and the cameras were ready, the men moved away from the tower and someone called out a countdown. At zero, an engineer pulled a rope lanyard attached to a trigger and the device fired with an explosive crack. The ejection seat shot to the top of the tower where it was held tight by a set of wedge brakes.

From his vantage, Stapp could see the subject grimacing. One of the support staff scrambled up a ladder and released a cable that allowed an electric winch to gently lower the seat back to ground level. With his harness freed, the subject doubled over and clutched his stomach. A quick exam by the attending doctor revealed that all vital signs were normal, and the man gradually regained his color. Several minutes after his ordeal, he was helped out of the seat. It looked bad, but after a few minutes the man was able to stand and walk on his own power.

Stapp and his new colleagues huddled around a technician who was already extracting a strip of paper from an oscillograph. The technician announced that the test had produced 33 g's at a 500-g-per-second rate of onset. In other words, the man had been exposed to thirty-three times the force of gravity at sea level. The peak impact had lasted only a fraction of a second, but that force—if applied for the duration of an entire second—would equal the effect of 500 times the pull of gravity, a force of frightening power.

Stapp whispered into the ear of a tall captain standing nearby, asking why the subject hadn't blacked out. Stapp recalled a lecturer at Randolph declaring that a 4.5-g acceleration applied from head to foot was sufficient to cause loss of consciousness.

The captain explained that the 4.5 number applied only to exposures of more than three seconds, which would interrupt circulation long enough to deprive the brain of oxygen. In the experiment they'd just witnessed, the seat reached the top of the tower in less than a tenth of a second. The firing of the catapult caused a peak impact force of thirty-three times the subject's weight during the 40 milliseconds that the seat was propelled along the four-foot length of the catapult itself. If you multiplied thirty-three times the subject's 180-pound weight, you got 5,940 pounds—close to three tons of force—acting for two or three milliseconds at the sharpest acceleration peak.

The captain leaned over the accelerometer reading. He pointed with a pencil to a mark the technician had labeled "Start."

At that point, the captain continued, the seat went into motion. In 40 milliseconds it boosted through the catapult stroke to a speed of about 60 feet per second. On the trace's graph of acceleration versus time, Stapp could see not only the acceleration, but also the continued *increase* in acceleration. The time it took to go from start to the 33-g peak was a fraction less than seven milliseconds. With a tower tall enough to allow a continuation of the acceleration for a full second, the resulting stop would deliver 500 g's.

The charge had been a little too hot. They'd been looking for something more like 20 g's and a maximum rate of 200-g per second. Moments later, an ambulance whisked the subject off to the base hospital for further examination. Luckily, the x-rays came back negative. The subject had spit up some blood, and his back was sore, but nothing had broken. Still, Stapp could see that even a tiny error in this kind of work could have disastrous consequences.

Captain Stapp returned to his office and immediately pulled out the classified German documents. He reread them in one sitting, studying every word. One troubling image caught his eye: in the pilot seat at the base of one of the test towers was an emaciated man dressed in rags. Another image showed an x-ray of the test subject's spinal column; he had sustained a lumbar fracture at a peak acceleration of 27 g's. Their content was clearly valuable, but Stapp felt ashamed for having seen the reports. He wondered if he should be ashamed that the United States military had used them to exploit the technical knowledge they contained. He recalled the branch chief mentioning that the Army intended to destroy the documents. Stapp made a snap decision to hide them. He wasn't sure why or what he might do with them, but he had the thought that somebody should preserve the evidence of these crimes. While possession of the material could theoretically result in serious punishment, he never second-guessed his decision. The documents were now his personal burden.

Less than a week after his arrival at Wright Field, Stapp got a chance to witness another act of aeromedical courage when he was invited to join a research team gathered at base flight operations to witness what would be the Army Air Corps's first airborne ejection seat test with a live subject. A Black Widow two-seat night fighter had been modified for the experiment by having the rear canopy cut away so that the back-seater could be ejected cleanly. In spite of elaborate preparations, there were plenty of concerns preceding the test. In the wake of the slipstream—an air current created by an airplane's propeller—would the subject be able to release his seat belt harness and kick himself free of the seat? Would the shock of ejection at hundreds of miles per hour render the pilot unable to open his parachute?

On Friday, August 16, 1946, the Black Widow and two chase planes providing photographic coverage climbed up through a broken overcast to an altitude of 15,000 feet. From Stapp's vantage, the three aircraft traveling at just under 300 miles per hour were little more than gray specks in the morning sky. He wondered what it must be like for the test-subject sergeant exposed to the cold and the wind, waiting to be shot skyward like a missile. The planes were coming in from the south, lining up in perfect formation over the field. Stapp and the flight operations team gathered around a radio set where they could hear the pilots coordinating their approach. The lead pilot gave the sergeant a countdown, and moments later announced a perfect ejection.

Looking up as the formation came overhead, Stapp made out a puff of smoke, and a few seconds later the blossom of a parachute. Soon after, the test subject landed safely in a cornfield to the east of the base, and by the time Stapp and the others arrived by jeep, the pilot was already gathering up his chute. The sergeant, a combat veteran with a long string of jumps to his credit, reported that the ejection had been decidedly less violent than a ride on the 30-foot tower.

By the following Monday morning, the Biophysics Branch had a working print of high-speed motion pictures of the ejection: a profile view as well as footage shot from below by the chase planes. Stapp and his colleagues were able to study the sequence of events as they unfolded in slow motion. The seat, once fired from the Black Widow, followed an arcing trajectory as it cleared the plane's tail boom. The seat remained in an upright position, but rotated slowly to the left. After reaching its apex, it began to tumble randomly. After two and a half scrambled revolutions, the subject managed to separate himself and fall free of the seat. The entire experiment went almost precisely as planned.

On the basis of their successful test, the Wright team was granted the funds to develop an ejection-seat training program for flying personnel. A new 110-foot ejection seat tower was constructed and installed in Wright's Static Test Hangar, and it was outfitted with a recoilless cannon and a robust three-piece telescoping catapult. But even with the evidence from their tower and flight tests, the Biophysics Branch researchers would have to work hard to overcome the skepticism and distrust of Army pilots, many of whom were understandably concerned about the idea of sitting atop an explosive charge while operating their aircraft.

For his part, Stapp was already beginning to relish his new role with the Aero Medical Lab, even as he contemplated how he might make his own contributions to the improvement of aircrew protection. When he learned about a series of upcoming high-altitude studies, he filed paperwork to get himself placed on flying status at Wright.

That fall, Captain Stapp volunteered for an experimental program being run out of the Engineering and Development Branch. A team of Wright researchers and engineers had come up with a promising design for a liquid oxygen converter. The device reminded Stapp of a moonshiner's still: the liquid oxygen was contained in a spherical steel globe encircled by several coils of copper tubing which allowed the liquid to boil off at a rate that delivered pressurized oxygen gas at 70 pounds per square inch. If it could be proven reliable, it would amount to a breakthrough. World War II aircraft that operated at higher altitudes had been forced to carry bulky, heavy, high-pressure oxygen tanks. The new converter could hold about 80 pounds of oxygen in liquid form in a one-cubic-foot container that could replace 20 cubic feet of gaseous oxygen cylinders weighing twice as much. The same team had also developed a novel portable air-liquefaction system for generating liquid oxygen in the field. They had put both designs through their paces in the Wright decompression chambers to a simulated altitude of 50,000 feet and at extreme temperatures without incident. They had also proved the fire-safety of the oxygen systems by firing 50-caliber machine gun rounds through the components, also without incident.

Once his flying status was approved, Stapp got himself enrolled as a subject for the first actual flight tests with the new oxygen system. Not only was this kind of researcher participation part of the tradition at the Aero Med Lab—both Armstrong and Lovelace had put themselves on the line as subjects of their own tests—but Stapp had come to understand that if he wanted to study a phenomenon, there was nothing better than experiencing it himself. He was one of eight men who would make a series of grueling eight-hour flights. The pilot and crew would use standard-issue gas oxygen systems at 400 psi as a safety precaution, but all of the subjects would breathe oxygen provided by the new converter.

The project's maiden flight was scheduled for mid-September 1946. Captain Stapp was careful not to betray any anxiety of his own in the hours leading up to their first takeoff. He was seated, along with three other subjects, in commercial airline seats in the waist compartment of a specially outfitted B-17. Two more subjects were positioned in the bombardier's compartment in the nose, and the remaining two were placed in the rear gunner's position. All of them wore heavy winter flying gear and they all began breathing converted oxygen at an altitude of 10,000 feet.

Stapp devised his own experimental agendas for these flights and rather quickly assumed functional leadership of the tests. Working with the Nutrition Section of the Physiology Branch, he orchestrated the contents of the in-flight lunches the subjects were given in order to assess the effects of different food types in a low-pressure environment. He suspected—and was subsequently able to prove—that foods high in fat content such as beans and onions that were prone to form gas in the digestive tract caused acute distress as the gas expanded at altitude. This discovery would inform the way high-altitude aircrews—and later still, astronauts—would be fed.

The biggest problem decreasing air pressure caused for the test subjects, however, was decompression sickness in the form of the high-altitude bends. The effects could be excruciatingly painful, permanently crippling, or even fatal. Before high-altitude manned flight could become standard operating procedure, aeronautical researchers would need to learn how to prevent or at least mitigate the symptoms of the bends.

The liquid oxygen system worked flawlessly despite the annoying bite of the pressurized oxygen mask that gripped Stapp's face like a rubber vise. He had made it clear to his fellow volunteers that these would be punishing experiments, explaining that they would likely experience symptoms of the bends whenever the B-17 ascended above 30,000 feet, and he'd been right. His own knees and elbows felt as if they were being squeezed in a vise, but by concentrating on his own response to the pain rather than the pain itself— as if he was outside of his body, observing—he found that he could function quite effectively. In the course of that first flight, he was able to confirm that colder temperatures intensified the effects of altitude, as did physical exertion. He also determined that standard painkilling medications had no effect whatsoever. But his most important discovery was that the longer an airman waited to begin breathing the pressurized oxygen, the worse the painful swelling in the joints became.

The subjects of these experiments often came away with lingering symptoms that took many hours to subside. Stapp devised a procedure that had the men begin breathing 100 percent oxygen thirty minutes before leaving the ground—essentially flushing the nitrogen from the blood—and continuing to breathe oxygen for the duration of the flight. This seemed to be the key. Captain Stapp's "pre-breathing" protocol appeared to alleviate the bends. With the help of a few brave volunteers, Stapp had made an important discovery related to one of the core problems of high-altitude survivability.

By keeping the subjects on pure oxygen, feeding them low-fat foods, and outfitting them with electrically heated suits, it became possible to safely take the B-17 up to altitudes above 45,000 feet for extended periods. Even though it became a labor for the subjects to exhale, fighting against the pressure differential, loss of consciousness was no longer a danger.

Stapp's experimental work on the B-17s, and particularly his discovery of the pre-breathing procedure, won him immediate recognition throughout the Aero Med Lab, and encouraged him to seek out new opportunities in aircrew protection. By the fall of 1946, Stapp was completely absorbed in his new role at Wright, working his typical long hours and eschewing most forms of social life—though not all of his work was concerned with biomedical factors. He was finding chances to contribute to pure engineering projects as well and even managed to devise a new calibration method for the accelerometers used by the ejection seat program, allowing more accurate readings of Gx forces.

When he learned that researchers operating the Aero Medical Lab's human centrifuge were seeking volunteers for a study on pilot acceleration tolerance, Stapp was, to the surprise of no one at Wright, one of the first in line. It would be his maiden firsthand experience with the effects of rapid, massive acceleration, and he didn't expect it to be a pleasant one. Still, he saw a special glory in the self-sacrifices of great medical researchers such as J. B. S. Haldane, who suffered perforated eardrums and crushed vertebrae as a result of his oxygen saturation and deprivation experiments, and Marie Curie, who died of exposure to the radiation she studied. They had something in common with the feats of Randy Lovelace and the Piccards, and maybe even of Lindbergh. If Stapp had found his calling in the emergency room in Duluth, he was learning to apply it in thrilling, if dangerous, new ways in Dayton.

The Wright centrifuge was housed in a large circular chamber, sitting atop a massive concrete foundation. Its main structure was a 50-foot cantilevered beam anchored at its center on an axle with a one-ton flywheel on a clutch. Mounted on gimbals at each end of the beam were cabs for the test subjects. An electric motor advanced the flywheel to the desired revolutions per minute, and the radial arms of the beam began to spin, swinging the cabs outward by centrifugal force. The centrifuge was designed to duplicate the forces encountered in flight: turns, loops, dives, and pullouts. The entire system was connected to a bank of measurement and recording devices that could track both the mechanical performance of the machine and the effects of the tests on live subjects.

One of the central questions was: How much force—and what kinds of forces—could a human being endure before passing out, and could anything be done to retard or eliminate the threat to pilot effectiveness? By the time Stapp had arrived at Wright, researchers had already established that exposure to acceleration equal to four and a half times gravity for a duration longer than three seconds caused an onset of vision impairment and finally total blackout. The experience resulted in an equalization of the hydraulic pressure in the arteries leading from the heart to the brain. Blood tended to be pushed back into tissue reservoirs in the subject's abdomen—and especially the legs. The delicate tissues of the retinas of the eyes were the first to respond to the sudden oxygen deprivation. Brain functions were next to degrade, and the subject quickly became incompetent at simple manual tasks. There had been reports of this phenomenon by combat pilots in the war who had blacked out while pulling out of tight turns or during fast dives, but many aircrewmen presumed to have experienced the most extreme deceleration forces did not live to tell about it.

Passengers on a commercial airliner almost never experience accelerative/decelerative force beyond 1.5 g's—1 g being the pull of gravity on the surface of the earth. A pullout from a long drop on a very large roller coaster may approach 4 g's, the point at which color vision fades. Beyond 6 g's, lungs start to collapse and blood pools in the legs and feet. At 9 g's, a man's 10-pound head weighs 90 pounds and useful consciousness is gone.

Researchers at various Army and Navy facilities—as well as civilian scientists at the Mayo Clinic—were already working on a solution to the problem of blood flow during massive acceleration: specially shaped rubber bags were attached to the calves, thighs, and lower abdomen of the subject and inflated when activated by the g-forces induced by the centrifuge. The inflated bags pressed against the body in areas where blood tended to pool, forcing blood flow back toward the heart and the brain. This technique, which in short order led to the partial-pressure suit that would become standard issue for aircrews in high-speed and high-altitude aircraft, worked so well that subjects were able to endure extreme acceleration. The equipped pilot, instead of blacking out at 4.5 g's, was now able to remain alert and effective up to a limit of nine times the force of gravity—exceeding the structural limits of some of the day's airplanes.

Following a brief physical exam and blood test, Stapp took his seat in the centrifuge cab for the first time in October of 1946. Operators observed from a window above as the subject was strapped in and given his orientation. An electrocardiograph lead was taped to his chest and respiration-rate leads to his back. Stapp could hear—and feel—the huge motor humming beneath him as the flywheel spun up. Inside the cab was an instrument panel with a row of small lights. He was instructed to keep his eyes trained on the center light, but to press a switch whenever he noticed any of the outer lights blink off. As the centrifuge's speed was increased, his peripheral vision would begin to gray out. When he could no longer see even the center light, loss of consciousness would quickly follow.

The room went dark and all he could see was the row of five lights on the instrument panel. As the flywheel was clutched onto the great shaft, Stapp had the impression of being hurled toward the ceiling while simultaneously tumbling backward. As the centrifuge spun up, he felt the rapid onset of pressure shove him toward the floor of the cab. It was suddenly a fight just to breathe. He felt his neck muscles clench as he struggled to keep the instrument lights in sight. Then the lights on each end of the line blinked off. He pressed his switch. By this point, he could no longer raise his arms off the armrest. The pressure was immense. Finally, the light in the center of the panel went out. A moment later he saw it again, briefly, as if through a thick fog.

He realized that the centrifuge had come to a stop, though he hadn't been aware of it slowing down. He turned his head to look for help and was overcome by a wave of nausea and vertigo. As the doctor who'd examined him minutes earlier leaned over him, Stapp glanced up and managed a weak smile. He was informed that he'd blacked out at precisely four seconds at a rate of 4.7 g's. The doctor asked if he'd like to try it again. Stapp, still disoriented and drenched in sweat, swallowed hard and nodded.

The lights went out and the centrifuge began to spin. This time, Stapp concentrated on the lights and forced himself to ignore the unpleasant sensations of multiplying pressure squashing him into the seat. When the center light blinked out this time, Stapp was able to remain aware of his situation. As the lights began to return to his vision and he became aware that the centrifuge had once again stopped, he made it a point not to swivel his head but to focus straight ahead. This seemed to alleviate the nausea. He realized that by experimenting with his own movements, he could reduce the lingering disorientation and the brutal effects of the mechanical forces on his body.

In the following weeks, Stapp visited the centrifuge whenever he could spare the time from his other duties. He didn't enjoy the experience, but he was nonetheless strangely eager for it. Up until now, all of his education and research efforts had been spent objectively observing phenomena in the best scientific tradition, without bias or emotion, and reaching conclusions based on evidence and logical deduction. His experiences on the centrifuge seemed to be leading him to a new type of knowledge. "Here was a new order of things," he recalled years later in a written account of his time at Wright Field, "where I was the subject, dissociating my feelings, disciplining my insight to apply the same objectivity to evaluating things that were happening *to* me. I used the same detached logic I had employed toward data collected on frog muscles and turtle hearts. I gradually began to look at myself in the same way I looked at a patient undergoing anesthesia for surgery."

With practice, Captain Stapp became expert at controlling his own muscles and brain during the centrifuge rides, and trained himself to compile detailed mental notes on his own sensations and reactions. In a sense, he was learning how to function as his own diagnostic instrument, one that measured his own responses to the forces operating on him. He was especially fascinated by the brief blackout periods. It was like an obliteration of self, and Stapp wondered how this exotic state differed from a painless death. He understood the physiology of it: with circulation arrested, and oxygen supply to the brain blocked, the respiratory chemistry of the cells is retarded like a dying candle. If death felt like this, he wondered, what was there to be afraid of? The centrifuge had given Stapp a powerful insight into the nature of existence. "Gradually," he wrote, "I felt myself being conditioned by experience against the subjective threat of extinction, the fear of dying. At the same time, I became even more vividly aware of the link between oxygen and consciousness, between breathing and living."

He felt he was coming to a new understanding of the meaning and extremes of survival. Stapp knew from his coursework at the School of Aviation Medicine that at altitudes above 40,000 feet, without pressurized oxygen, consciousness lasted no longer than ten seconds. That's how long it takes one complete heart-full of blood to be pumped from the lungs to the brain. When aviators operated at those altitudes, they were never more than a breath or two away from obliteration. No matter how clever the oxygen delivery systems and pressurizing garments, high-altitude aircrews lived one breath, one heartbeat, at a time. Understanding survival this way led Stapp to an appreciation of what he referred to as "the glorious rebellion against dissolution achieved within the community of cells."

Thinking now of the ejection tower tests in a new way, a question began to form in his mind: precisely how much +Gx and –Gx could a human being tolerate? In other words, how much mechanical force did it take to kill a man?

. . .

As he wrestled with the conundrums of physics, in November 1946, Stapp was again summoned to Colonel Kendricks's office, where he was handed another thick stack of technical documents. About half were untranslated German. Read them, Kendricks ordered.

The first document was an exhaustive study of airplane crashes. One of its conclusions was that the casualty rate of glider pilots was significantly higher than that of light airplane pilots. The explanation, according to the German study, had to do with safety belt design. Glider pilots were restrained by belts that encircled the waist and were anchored to the backs of the seats by metal clamps. During the extreme deceleration encountered in a crash, the belts had a tendency to constrict the waist. Some of the worst crash damage inflicted on glider pilots was the result not of impact but of the abdomen surging against a tightening belt. The light aircraft pilots, in contrast, had double straps fixed to the lower outside corners of the seats and buckled across the lap. This design appeared to avoid the constriction problem. Once the gliders were reconfigured with the lap belt design, survival rates improved dramatically.

One of the American reports included in Kendricks's stack described similar crash experiments that were then underway at the U.S. Navy Aero Medical Equipment Laboratory and at the Air Corps School of Aviation Medicine. Stapp learned that as far back as 1938, General Malcolm Grow, in partnership with Harry Armstrong, had invented and personally tested the shoulder restraint harness that had since become standard in American fighter aircraft. Up until the introduction of the Grow–Armstrong system, head injuries had been the most common cause of death in airplane crashes. With straps coming over the back of the seat and fastening to the belt buckle, the head was prevented from snapping forward, lessening the chance of the skull impacting hard objects encountered during a crash. None of the American literature, Stapp was relieved to find, contained a hint of any involuntary human exposure to severe mechanical force or impact. Navy flight surgeons in 1945 had crashed eight surplus torpedo bombers, with parachute dummies aboard, at full carrier landing speeds into a sandbag barrier. Unfortunately, most of the devices installed for the purpose of recording the impact forces had been destroyed during the crashes, as had the onboard motion picture cameras. It was a damnable problem trying to gather useful information on aircraft accidents because of the difficulty of recovering any data from the airplane.

Stapp was especially fascinated by a German report that described a 900-foot narrow-gauge rail track at the Aeromedical Institute at Tempelhof Airport. The Germans had mounted an aluminum carriage on the track with room for a single test subject. Instead of wheels, the carriage was suspended by slippers surrounding the rails that allowed it to glide like a sled. Acceleration was achieved by means of a system of cables and pulleys attached to a 20-ton water tank hanging from a 60-foot tower. When the tank was released, the sled was yanked forward down the rails, achieving speeds well in excess of 100 miles per hour in a distance of just 250 feet. Braking was supplied by a steel keel beneath the carriage that was channeled between spring-loaded metal strips lined with brake-shoe metal that gradually narrowed, bringing the sliding vehicle to a stop in about 70 feet. United States bombers, however, destroyed the facility before any actual experiments could be conducted.

When he had digested the crash study documents, Stapp went back to see Kendricks. The chief informed him that the Army was planning a project that would replicate what the Germans had hoped to do at Tempelhof. In fact, the Northrop Corporation had already been awarded a contract for the design and production of a rocket-powered test vehicle on a track fitted with mechanical friction brakes. The Northrop configuration would provide more precision than the German design. Kendricks handed Stapp a pair of leather-bound volumes containing the project plans.

Leafing through Northrop's proposal, Stapp found the following paragraph:

> A 2000-foot standard-gauge track on a concrete base, a factory production replica of tracks on the coast of Holland and Belgium, used by the Germans to launch the V-1 buzz bombs, and also used to test captured buzz bombs during World War II, should be used for the sled design. This V-1 launching track is located at Muroc Army Air Base in the center of the Mojave Desert, about one hundred miles from the Northrop plant in Los Angeles.

Kendricks explained that fabrication of the brakes was already underway and that they were scheduled to be delivered to Muroc the following March. The chief leaned across his desk and told Stapp that, based on his performance at the Aero Med Lab, he was being considered as project officer for a series of tests intended to simulate a spectrum of crash forces. Years later, Stapp recalled the next question.

"If you were in charge, Captain, what sorts of tests would you propose?"

Stapp had already thought it through and replied quickly that he would start with dummies to qualify the equipment and establish protocols. Then, once convinced it could be done safely, he would move to tests with animal subjects to extrapolate safe limits for human beings.

Kendricks pulled Stapp's personnel file out of a drawer. There was a potential problem, he said. Stapp had only six months remaining in his commitment to the Air Corps. If they were going to send him to California and entrust the project to him, they'd need assurance that Stapp would be willing to extend his term of service. This was exactly the sort of dilemma Stapp had feared when he'd left Tinker for Wright Field. In the meantime, a friend from medical school had recommended him for a fellowship in internal medicine at the Mayo Clinic, one of the top hospitals and medical research facilities in the nation. It was the closest thing to a dream job he could imagine.

Stapp asked for time to think it over, to which Kendricks grudgingly agreed. Fortuitously, Stapp and one of his colleagues in the Biophysics Branch were already scheduled to travel to Rochester, Minnesota, the following week for a conference at the Mayo Clinic. Stapp promised Colonel Kendricks an answer by the time he returned.

While in Minnesota, Stapp got a chance to observe the big centrifuge at the clinic's research facility, and to speak with several scientists in the Aero Medical Department there. These were civilians involved in some of the same type of work being done at Wright Field. Stapp inquired about potential staff openings. It seemed like the perfect transition out of the Army, and a great springboard to an eventual private practice. He doubted he'd be able to turn down an offer if one was made. But upon his return to Wright, he found a letter waiting for him announcing that in fact no position would be available for him at Mayo in the immediate future. It was depressing news. Captain Stapp's dreams of the civilian life seemed to be fading.

He reported to Colonel Kendricks the next morning and agreed to extend his service with the Air Corps for one additional year and to accept the project at Muroc Army Air Base, a place he still knew nothing about other than that his brother Wilford had been stationed there briefly the year before. When he asked Kendricks what was it like out there in the desert, the chief only smiled.

Stapp was granted fifteen days of leave in December, which he spent with his brother Celso—now a respected obstetrician—and Celso's new wife, Day, at their home in El Paso. From the information he'd managed to glean from his colleagues at Wright Field, he told them, Muroc Air Base was a remote outpost populated by cowboy jet pilots and sad-sack ground personnel. His marching orders from the Aero Medical Laboratory were fairly narrowly defined: to study the design and limits of personal restraint devices for military pilots in crash or bailout situations. The official scope of Project MX-981 was also to encompass the "effects of deceleration forces of high magnitude on man." Stapp spent much of the holidays plotting his own approach and his own set of objectives. He suspected that it might be possible to turn on its head the Army's core assumption underlying aircraft cockpit design. Since

human beings, aircraft designers claimed, could not withstand severe gravitational forces, it was no use building stronger—and hence heavier, more
complex, and more expensive—airplanes. Any additional protective construction would simply be a waste.

If Stapp could disprove the Air Corps's declared limit of human tolerance, he might be able to convince manufacturers to build safer planes—and
save the lives of more pilots—and he had some novel ideas about how to do
it, ideas that he'd been careful not to describe in any detail to his superiors
in Dayton. Before leaving El Paso, John Paul confided in Celso that he felt like
everything was about to change. He'd had a similar premonition during the
dark days of his freshman year at Baylor.

In his journal, Stapp wrote simply: "I could hardly wait."

PART II

THE HOT SIBERIA OF THE AIR FORCE

6

THE BATTLE OF MUROC

I always came back to Muroc refreshed by the thought that
if people with crude wagons could challenge the unknown
and make their way without a guide or a trail, under the most
terrible hardships that heat and drought could inflict, who was I
to complain of a few minor human obstacles in the way of my
research? Whenever I read or heard the usual fatuous commentary
about the softness and weakness of modern Americans, it
occurred to me that they were descended from these pioneers
and that when the time came, they could take the exponents of
Nordic superiority on in their own backyard in the worst season
of the year and break through the best fortifications the most
demonical resistance our enemies were able to contrive or
muster. Those who failed to read our history and view the desert
mountains and plains that our people conquered in their migration
to the west paid dearly for underestimating us at Guadalcanal, the
beaches of Tarawa, and the volcanic plain of Iwo Jima.

—*John Paul Stapp, unpublished memoir*

EARLY ON A MARCH MORNING in 1947, at a bleak site on the northern edge
of Muroc Army Air Base, deep in the Mojave Desert, John Paul Stapp
introduced himself to the eight sunburned men who formed the core of
the Northrop contractor group assigned to Air Crew Deceleration Project

MX-981. Among them were chief engineer George Nichols, engineer and telemetry specialist David Hill, chief mechanic Jake Superata, and assistant mechanic Jerry Hollabaugh. Captain Stapp was in regulation uniform complete with necktie, dress shoes, and hat. The civilian crew wore ragged sports shirts and dungarees. They looked to Stapp like a band of pirates. Most of the men would later admit to finding the captain a little odd at first. To their ears he spoke with a soothing but almost stilted Texas cadence and a sermonlike fervor that seemed out of place in the backwater of North Base Muroc, as if he were trying to convert them to his vision of the future of aviation, even the future of mankind.

Stapp explained his strategy for attacking the problem of aircraft crew protection. He had a master plan, he told them. Their goal would be nothing less than the salvation of the United States Army's pilots. The project had been conceived as an effort to test and evaluate protective restraint systems for pilots forced to eject from disabled aircraft. A pilot suddenly catapulted from his cockpit, Stapp explained, had a chance to survive only if he could avoid impacting the airplane's tail structure, and then only if his neck and limbs could be kept from whipping and flailing during the massive $-G_x$ load. It was a difficult problem with many variables. The ejection tests at Wright had been successful to a little over 350 miles per hour, but a pilot leaving a jet at Mach 1 would be subjected to forces akin to slamming into a steel wall. This would be the new team's charter, Stapp announced: to learn how to protect the vulnerable human body as it encountered these horrific forces. Stapp used an unfamiliar term for what they would be studying: *biodynamics*—defined as "the relationship between force characteristics and their biological effects." They were on the cusp, he told them, of a brand new science, an applied science whose goal would be the saving of lives.

But that wasn't all. They had a chance, Stapp told them, to do something more with the rocket-sled test vehicle Northrop was building. They were going to launch it down the 2,000-foot track at tremendous speed before bringing it to a sudden stop in an attempt to simulate the effect of ejection from a speeding aircraft. They'd test a variety of belts and harnesses until they learned how most effectively to restrain and protect the parachute dummies they would employ as their test subjects. And if they could perfect the braking system they'd use to stop the speeding vehicle in its tracks, they could simulate something else: an airplane crash. Stapp truly believed, he told them, that it might be possible to keep pilots alive even in violent crashes. Nothing he'd seen in all of his medical training and observation had convinced him that pilots properly surrounded and restrained couldn't walk away.

The textbook Army line on tolerance to deceleration forces, as it was presented at the School of Aviation Medicine at the time, held that no human being could survive a sudden acceleration or deceleration force in excess of eighteen times gravity. The 18X mark was the accepted physical limit that governed aircraft design and aircrew emergency protocols of the day. Stapp told the MX-981 team he suspected the experts were wrong, and that the tolerance limit was likely higher—perhaps much higher. The only way to test his theory was to simulate an airplane crash in a controlled environment: a speeding vehicle, moving horizontally, that could be slammed to a dead stop. Stapp asked his new team to suspend its disbelief, and to think about the lives they could save if together they could make this project a success.

The Northrop guys listened politely that morning, if a little skeptically. It was hard to know what to make of Captain Stapp. They'd never encountered anyone on the base quite like him. Still, even the skeptics found his arrogant enthusiasm contagious and it was clear that whatever happened, this would not be just another research project. If Stapp was right, they were all going to make history.

Only three days before addressing his men, Captain Stapp had stepped down onto the ramp at Muroc for the first time, his ears still ringing from the thundering engines of the B-25 Mitchell bomber that had transported him and his collection of eight parachute dummies from Wright Field. Following a brief stop at the base operations office, he'd been ferried by weapons carrier to a wind-scarred, two-story wooden barrack that served as visiting officers' quarters. A sign hanging from a corner post read: DESERT RAT HOTEL. Like the operations office, the Desert Rat's dim lobby was full of sand. Stapp registered at the front desk and paid 75 cents for the night's lodging. As he climbed a creaking staircase to his 10-foot-by-9-foot cubbyhole on the second floor, he had the impression of taking up residence in a ghost town. He dropped his duffel and took a seat on a chair beneath a bare lightbulb. The only sound was the pinched banshee wail of the desert wind that bored its way right into his room, bringing with it the ubiquitous Mojave sand that would be a constant companion for the next four years of Stapp's life. At Muroc, the sand was like the wind. You could neither escape it nor tame it. It would be in your food and your bed.

The next morning, following a quick breakfast at the service club across the lane with the flight crew from Wright, Stapp requested that he and his equipment be transported to the Northrop project laboratory. The Muroc driver nodded but said he wasn't aware of any Northrop lab. In fact, he told Stapp, there wasn't much of anything out there at all.

At the Main Base gate, the driver turned toward the road to North Base, about eight miles away. Bracing himself in the back of the vehicle as it bounced along, surrounded by crates full of instrumentation and piles of dummies, Captain Stapp got his first good look at the environs of the base and beyond. It stretched out forever in all directions, an epic landscape the scale of which he'd never seen. The great Mojave Desert is 50,000 square miles of jagged, rust-colored mountain ranges, wide valleys, saltpans, and dunes. The desert floor Stapp's route traversed was pocked with sagebrush and brittle dryland grasses, yucca, creosote bush, and—every so often—a gnarled Joshua tree that looked exactly the same as it had a hundred years earlier. In just a few weeks daytime temperatures would hit 120 degrees Fahrenheit.

The Mojave is one of the lowest, driest, harshest, flattest places in the world, and it's capped by some of the clearest sky anywhere. The most dramatic features of the area were the two huge prehistoric lakebeds. Rogers Dry Lake accounted for a spectacular 47 square miles adjacent to Muroc Main Base—the largest such geological feature on planet Earth; Rosamond Lake was a little less than half that size. These hard-packed, almost perfectly level stretches of desert formed the longest natural runways in the world. The lakebeds were so flat you could play billiards on them, Stapp's driver told him. Rosamond, for example, varies in altitude only a single millimeter every 20 meters (about 66 feet). A future base commander would call them "God's gift to the U.S. Air Force."

Stapp surrendered his credentials to sentries at North Base gate. A mile and a half beyond the gate, to the east, they passed a rocket engine test stand that belonged to the California Institute of Technology. Then, almost another mile down the gravel road, he got his first glimpse of the reason he was here:

the rocket sled track. The rails were precisely four feet apart, installed atop parallel ribbons of concrete. At the far end of the track he could see a pair of unpainted shacks and a yellow-checkered water tank on a rusted trestle. Otherwise, there was nothing in view but gray desert and blue sky. Because it was Saturday and the Northrop crew had gone home for the weekend, there was not a hint of activity. The newly arrived captain strolled alone down the length of the track and considered what he'd gotten himself into.

One of the sorry-looking structures he'd noticed appeared upon closer inspection to be a small barrack. He walked it: 48 feet long and 22 feet wide, roofed with tarpaper. The building gave the appearance of having been trucked in and simply dumped in the sand. It didn't look as if anyone had even bothered to level it. The barnlike cavity of the interior was empty and unpartitioned. Stapp poked around and found that the facility had no electricity at all. He'd seen power lines at the Cal Tech site up the road, but the Army had apparently failed to extend the service to this building. Stapp ordered the weapons carrier unloaded and his cargo stacked inside. The driver was right: you definitely couldn't call it a lab.

At first, Stapp took the whole thing as a lack of respect not only for the research he intended to conduct but perhaps even for himself. His resentment built as he explored the primitive site and discovered that in addition to an absence of electrical services, there was no running water, no toilet, and no telephone. Yet by the time he stepped back outside into the wind and got another look at the stark beauty of the Mojave landscape, Stapp found himself becoming strangely energized. One thing was clear: out here he was going to be very much on his own. There was even the chance, he thought, that the Aero Med Lab back at Wright Field might just forget about him and what was referred to as his "detached service field project." Considering there was no authority or funding anywhere at Muroc to support him, he guessed he'd mostly be ignored by base personnel. By the time he directed the driver to return him to the Desert Rat, Stapp's mood had turned from suspicion and resentment to one of liberation.

On the morning he introduced himself to the Northrop team, which was now *his* team, Stapp was not only fully committed to the mission, but flushed with excitement about the prospects before them. The biggest immediate task facing the MX-981 team on its first morning was preparing the North Base track for the installation of the complex braking system for the rocket sled, and Stapp got the men working on the cast iron and steel brake fixtures that first day. Simulating pilot ejection and aircraft crashes would be a mammoth engineering challenge, as it would require a stopping mechanism of unheard-of power and precision. Only after the mechanical friction brakes were in place and properly calibrated would Stapp be able to conduct acceptance testing on the rocket sled itself, which was still in the final stages of fabrication at the Northrop plant in the town of Hawthorne, just south of Los Angeles.

That same afternoon, Stapp huddled with engineer George Nichols, a highly intelligent and sensitive man with whom Stapp forged an immediate bond, and asked about plans for bringing utilities—electrical power, most importantly—to the site so they could build a proper laboratory. They *must* have an adequate power supply, Stapp insisted. Nichols, who'd been at this sort of thing for a while, was familiar with the mind-set at Muroc and he understood the limits of the military's cooperation with the civilian defense industry. He gave Stapp the bad news. The Army would be offering them little or nothing in the way of assistance or resources. The $100,000 Northrop contract allowed them to request help only in emergencies. There would be no utility service, no telephone, and no upgrade to the facilities at the site. They would have to make do with what they already had.

As a result, March 10, 1947, marked the opening salvo of what the project officer would later refer to as the Battle of Muroc. The over card: John Paul Stapp vs. the Army of the United States of America. The first skirmish would be all about electrons.

Stapp mentioned his plans to no one, not even Nichols, but he was already scheming. After measuring out spaces for offices and a lab inside the empty barrack, Stapp got a driver to give him a tour of the rows of abandoned buildings he'd seen back at the main base. There he began rummaging around until he'd located enough wallboard to line the barrack and construct a ceiling. He found sinks, toilets, and plumbing pipe; boards for shelving and office partitions; electrical wiring and fixtures. His next stop was at the post engineer's office, where he requested equipment and assistance. The post engineer ridiculed Stapp's Eisenhower jacket.

Refusing to take offense—though he recalled the slight and the ensuing conversation decades later—Captain Stapp introduced himself as the man in charge of the Air Crew Deceleration Program. The post engineer was not impressed, and asked if Stapp had funds to pay for work and materials on a reimbursement basis. Stapp replied that while he did not have a working budget, he could if necessary request compensation from the chief of the Aero Medical Laboratory at Wright Field. Part of Stapp's strategy was to befriend as many valuable base personnel as he could, and so he kept doggedly after the post engineer until he sensed resistance beginning to flag. After being worn down by a half hour's conversation, the engineer was helping Stapp brainstorm solutions to his problems.

Stapp explained that what he needed was a temporary power line to be run from one of the base transformers at the Cal Tech project to his project's building approximately 4,400 feet away. The post engineer suggested the San Bernardino salvage depot. They might have large quantities of electrical cable, he said. The next morning Stapp requisitioned a truck and driver to take him to San Bernardino, where he was able to secure 5,000 feet of Romex #8 three-strand conductor in 100-foot rolls. He also picked up some insulators and tar tape, paid for out of his own pocket. Late that afternoon, the truck rolled into the Northrop site at Muroc with its full haul. George Nichols and the rest of the team were dumbfounded. The following day they all chipped in and helped splice and solder the rolls of cable, enough to stretch from the barrack building to the Cal Tech transformer. It took three full days of tedious handwork, but nobody complained.

When they got the homemade cable hooked up, however, they discovered that the voltage dropped from 110 volts at the Cal Tech site to less than 30 volts at the MX-981 barrack. Stapp went back into action. In further discussion with the post engineer, who was by this point becoming an ally, Stapp learned that there were numerous surplus transformers at an Army Signal Corps depot near Sacramento. After befriending some pilots at the officers' club, Stapp managed to get a B-26 bomber pilot to ferry him to McClellan Field, where he met and quickly charmed a couple of the salvage officers. A few hours later, he was headed back to Muroc with two large-capacity 440-volt line transformers suspended like ordnance in the bomb bay of the B-26. Back at North Base, Stapp had one of the transformers installed at Cal Tech and connected it to the line running to MX-981, where the second transformer was installed. When everything was in place, one of the Northrop technicians measured 108 volts between the center ground and one wire of the cable. Drawing any appreciable current in the barrack, however, caused the voltage to drop. The improvised system was better than nothing; they now had temporary power. But Stapp was not satisfied.

Stapp and George Nichols had begun spending an hour or so together after most workdays talking about the project, their families, their ideas, and one afternoon shortly after installing the transformers and electric lines, Stapp asked Nichols to show him Northrop's contract with the Army. As he studied the terms in the fine print, he learned that the United States government was obligated to provide electrical power to within 20 feet of the contractor's installation. It was there in black and white. Stapp cocked his head and smiled. The government did not appear to have lived up to its side of the bargain. He told Nichols he was issuing Northrop notification that he would require the company to meet the Aero Med Lab's specs on accuracy of instrumentation and data recording, and noted that Northrop would need a steady, dependable power supply to accomplish that.

There was nothing he or Northrop could do, Nichols protested. Stapp asked whether the lack of electric power might not be cause for a work stoppage. What would happen, he asked, if Northrop were to shut down the operation, citing the government's failure to comply with the terms of the contract? Would the company be willing to send notice through channels to Air Corps headquarters?

Nichols hesitated, unsure of sure how much trouble he wanted over this issue. He was beginning to understand that working with John Paul Stapp would be an experience unlike anything he'd known in his previous dealings with Uncle Sam. But the more he thought about Stapp's idea, the more it appealed to him. When Nichols took the concept back to Hawthorne, he found, to his surprise, support—if not exactly enthusiasm—for the work stoppage notice.

In the meantime, as they waited for the legal maneuvering to run its course, Stapp and Nichols saw to it that nobody actually stopped work: the mechanics concentrated on the brakes, while Stapp and the rest began working with borrowed tools to install the wallboard liners in the barrack. If this was going to be their working home, they were going to make it habitable.

With the men busy on what he intended to be their future laboratory, Stapp moved on to the next phase of his plan. He was aware that his frequent requests for base cars and drivers would eventually bring unwanted scrutiny on his operation, and he aimed to fly as far under the radar as he could get. He had not owned a car since Pandemonium back in Texas, and he'd been planning to buy one after he was settled in at Muroc. The problem was, his funds were thin; he was still paying back the loans he'd taken out to pay for medical school and was continuing to send money to his family when he could. So, just as construction of the project lab was going to have to rely on what he would refer to as "moonlight requisitioning and horse trading," along with his own contributions when necessary, he was going to have to be creative. His next discovery at the San Bernardino salvage yard was a lucky one: a discarded paratrooper's motor scooter—painted a sickly yellow—with a decrepit two-cylinder engine. After hauling it back to Muroc and making a few minor repairs, Stapp took it out for a test ride to determine the vehicle's top speed: 28 miles per hour. That would be perfectly adequate for traveling the few miles between the Desert Rat and North Base.

On the same trip to the salvage yard, Stapp had snagged a few field telephones and enough wire to allow him to reach and tap into the North Base phone line. The first call came in the next morning, and while Stapp could barely hear the muffled voice on the other end through the static, he could make out that the caller was the base air installations officer. The gentleman was not happy.

Stapp hopped on the scooter and took off for Main Base, stopping several times along the way to reconnect a loose spark-plug wire. When he arrived, he got an earful. The air installations officer complained that his office was being investigated by the commanding general of post engineers in Washington in response to a Northrop claim that the Air Corps was in breach of contract for failure to provide adequate power. Furthermore, Stapp was informed, Headquarters, Air Materiel Command, would be sending the air installations officer $5,000 to pay for a brand new power line to the Northrop project site to avoid a threatened work stoppage. Stapp feigned sympathy as he let the air installations officer vent, but neglected to disclose any knowledge of Northrop's motivations. He did promise to keep the office informed of any future developments. The Northrop team's growing faith in the powers of its new leader took a quantum leap the following week when electrical contractors appeared on site to install standard poles with cross-ties and real power lines.

The next day the men got busy screwing lightbulbs into the salvaged fixtures inside the barrack. Now that they had reliable power, they could finally begin laying out their facility in earnest. They might not yet have their hands on the centerpiece—their rocket sled—but MX-981 was shaping up into a bona fide medical research project.

A few days later, Stapp pulled the entire project team together and revealed some new details about his master plan. After admitting that there were a few things he had initially held back, Stapp went on to explain that while they would use the parachute dummies to perfect the configurations and components of a pilot restraint system, the dummies would only be able to tell them so much. At some point, they would need to test their solutions on living subjects. He told them what he knew about how the Nazis had done it: with involuntary subjects, with prisoners. He'd seen evidence of those barbaric tests, he said. Project MX-981 would need willing, able-bodied volunteers prepared to subject themselves to conditions no one had ever experienced and lived to tell about. It would be dangerous work, he said, and it was possible that injuries could result. Or worse.

Stapp quickly assured them that he would not be asking *them* to volunteer. What he needed were individuals practiced in making and documenting subjective medical observations during extremely stressful conditions. Furthermore, since a doctor would be the best candidate for the job and Stapp was the only doctor anywhere in sight, he announced that he intended to ride the rocket sled himself. Though none of the Northrop men were aware of it, Stapp's decision was of course squarely in the aeromedical tradition established by Harry Armstrong and Randy Lovelace before him. You didn't ask others to do anything you weren't willing to do yourself. It helped keep dangerous projects honest. One of Stapp's oft-spoken mottos would be: "If you can't do safety research safely, then it ain't safety research." The men didn't know what to make of this development. Some of them assumed that Stapp was joking.

George Nichols knew otherwise.

7

OSCAR EIGHTBALL AND THE CURBSTONE CLINIC

Applied research is short-order cooking, and it is necessary. You have the satisfaction of knowing it's not esoteric.

—*John Paul Stapp, interview for NOVA,*
"Escape! Pioneers of Survival"

TO NICHOLS, STAPP'S ANNOUNCEMENT of his plan to use himself as a guinea pig was a "revolting development." Nichols hurriedly got in touch with company founder John Knudsen "Jack" Northrop back at the plant in Hawthorne to give him the news and to seek advice.

Jack Northrop, an engineering visionary, had begun his career as a draftsman for the Lockheed brothers during World War I and in 1927 assisted in the wing design for the *Spirit of St. Louis*, which carried Lindbergh across the Atlantic. He founded the Northrop Corporation in 1932. When Douglas Aircraft acquired the company in 1939, Northrop founded a separate company using the same name. He was an outside-the-box thinker—a controversial low-drag, high-lift wing design that manifested itself in an aircraft known as a Flying Wing was his obsession—and a man not averse to risk. So when Nichols contacted him about Captain Stapp's plans to ride the rocket sled, Jack Northrop quickly signed off on the concept. If the Army was confident, he told Nichols, so was he. But neither Northrop nor the Aero Med Lab back in Dayton understood the extreme levels of biomechanical experimentation Stapp was planning. If they'd been privy to the sketches and calculations in the private notebooks Stapp kept in his room back at the Desert Rat, both Northrop and Colonel Kendricks might well have intervened before any of it got started.

In the spring of 1947, neither Kendricks nor Northrop had much appreciation for the significance of the contribution Stapp was about to make at Muroc, a contribution that wouldn't end with deceleration research. Even as he drove himself and his team to transform a barren desert outpost into a working research laboratory, the renegade doctor remained first and always, in his own mind, a man of medicine. Acquaintances in Dayton still contacted Stapp for advice before making appointments to see their own physicians. When one of the Cal Tech engineers up the road was stricken with severe chest pains, somebody thought to call down to the MX-981 site, since they'd heard the new project officer was a doctor. Stapp jumped on his scooter, his flight surgeon's kit in hand, and put-putted at full speed toward the Cal Tech project, where he found an elderly technician writhing on the floor of the workshop. The man's face was ashen. After a quick examination, suspecting angina, Stapp placed a nitroglycerine tablet under the patient's tongue. A few minutes later, the technician's color returned; he was able to sit up and, after a few more minutes, breathe relatively easily. Stapp called the base hospital and ordered an ambulance. Later, the emergency room doctor corroborated the diagnosis and admitted the man for observation.

Reports of Captain Stapp's medical prowess and diagnostic skills got around at Muroc in a hurry. The base hospital and the base flight surgeons were severely overworked, and policy dictated that, except in extreme emergencies, only active Army personnel were eligible to receive treatment. It was a cruel policy that ignored not only the civilians who worked on military projects in the desert and lived either at or near Muroc, but also the immediate families of American servicemen. The nearest medical facilities were 45 miles away in Lancaster, and those facilities had their limits. Not even the crack test pilots who were the acknowledged royalty of the base could convince the Army to provide even basic care for their wives and children. As a result, Stapp became somebody everybody wanted to know.

The role suited him well, and he came to the realization that his services might provide value not just for prospective patients, but for Project MX-981 as well. He put word out at the officers' club and wherever he went on the base that he would be pleased to provide medical care to those in need, free of charge, in the evening hours after his duties at the rocket sled track were done for the day. Thanks to his mostly reliable if underpowered scooter, if the patient couldn't come to him, Dr. Stapp was even prepared to make house calls.

Every night, then, after a long day at the track and following his regular debrief with George Nichols, he would return to the Desert Rat for a quick shower and then take off on his rounds. He would hit the base housing and residences in nearby Rosamond first. Some officers' families had doubled and even tripled up, cramming themselves into the few small ranch houses scattered about the Antelope Valley floor. Other base personnel had been moved into abandoned railroad cars, in some cases into structures that could only be described as roadside shacks. Stapp spent many an evening calling on patients at the squalid contractor housing just off the base, and at the low-rent project at Kerosene Flats, so named because of the kerosene used in the heaters that caused everything—and everybody who lived there—to reek of fuel. The project had been hurriedly built during the war for married military personnel, and the plywood walls of the duplex residences were so thin and full of cracks that they barely held off the wind and failed to keep out the sand. Stapp would typically arrive just before dinnertime, examine ailing children (most of his patients would be children), write prescriptions, dress a wound or set a broken bone, perhaps even deliver a baby. When he was done, he would gladly accept the family's invitation to join them for their evening meal—and wasn't beyond suggesting it if they failed to offer. Otherwise, if there were ever supplies or services he was lacking, he might mention them during his visits and would not be surprised when deliveries of equipment or a few hours of labor were made available to him and his team in return. He wrote in an account of his Muroc years: "The friends I made became a network of helpful sympathizers, giving me a hand when I needed it. Truly, this was a good will industry." Dr. Stapp came to call this informal barter operation his Curbstone Clinic. As his brother Wilford characterized it: "Just plain old country doctor work."

Stapp's bootleg clinic eventually brought him into contact with nearly everyone in the vicinity of Muroc, which is how he first met a rough-edged, exuberant young Air Corps captain named Charles "Chuck" Elwood Yeager, his beautiful wife Glennis, and their two sons. Yeager was both a headstrong war hero and a test pilot with a sterling reputation—despite his lack of higher education, which had led Major General Fred Ascani, a flight test officer at Wright Field, to say, partly in jest: "He barely spoke English." But not even Yeager could get his kids seen by the Army doctors at Muroc. When one of the boys or Glennis had a complaint, Yeager would invite Stapp over for a fried chicken dinner. Yeager didn't have to ask Stapp to bring his medical kit. It was understood. Once, during dessert, Stapp happened to mention some windblast experiments the Aero Med Lab might like to conduct—though the truth was that Stapp had dreamed the concept up himself. He wondered aloud whether Yeager might be willing to exert some pull to get approval for them to cut the canopy off a jet for the tests. They agreed to continue discussing the idea.

Despite his role as personal physician of the Mojave, Stapp's principal obligation was to his project, and by the first week of April, Northrop had delivered all of MX-981's essential hardware. Now the first order of business was the installation of the final pieces of the advanced braking system. Superefficient and highly adjustable brakes were essential for the proposed deceleration experiments, and work proceeded deliberately. The team took possession of forty-five independently pressurized and actuated brake units, and Stapp directed that they be laid out beside the track along a 50-foot stretch beginning 1,250 feet from the west, or starting, end of the 2,000-foot track. Each of the units comprised parallel pairs of brake shoes that could be independently activated by pneumatic-hydraulic levers. The pressurized levers pinched the shoes together at 600 psi to grab the moving keel of the rocket sled. The action was, in Stapp's analogy, like "a knife blade being pulled through a vise." Variations of braking power could be achieved by using consecutive brakes, every other brake, or any grouping the technicians chose to configure. The hydraulic pressure controlling each individual brake shoe could also be adjusted. In all, the team had a total of four million pounds of braking pressure at its disposal—enough, they hoped, to stop a speeding vehicle on a veritable dime.

Once the brakes were installed on the rails at precise intervals, the team turned its attention to the rocket sled itself, formally known as a linear decelerator. Northrop had named it the *Gee Whiz*, and had built two of them, nearly identical. G-forces were what the project would manufacture, and whiz was what they expected their decelerator would do. It was a 1,500-pound, magnesium slipper-mounted sled fabricated of chromium-molybdenum steel tubing and clad in aluminum sheeting that gave it the boxy look of a small tank. It was about 15 feet long, 6.5 feet wide, and designed to accommodate either a seat or a litter capable of carrying an instrumentation package. Propulsion for the *Gee Whiz* would be provided by previously discarded solid-fuel Monsanto jet-assisted take-off (JATO) rocket motors the Army had given them. The rockets were capable of delivering 1,000 pounds of thrust apiece and they were set to burn for five seconds. Once the Northrop team attached the telemetry instrumentation, the sled was mounted on the rails at the west end of the track. The surplus JATOs—the project had taken possession of an even hundred of them—remained in wooden crates and were stored a good 300 feet from the track out of caution.

A few mornings after installation of the brakes, Stapp and the crew learned an important lesson about working with technical equipment in the Mojave Desert. Every evening at sunset, cold air came swirling down off the surrounding mountains and displaced the hot air rising from the cooling sands of the desert floor. It made for pleasant summer nights. But every once

The Gee Whiz *decelerator, with cabin enclosure, in the brake section of the Muroc rocket sled track. Stapp never liked the boxy cabin and dispensed with it early on. (Photo courtesy of U.S. Air Force)*

in a while, morning winds would really kick up—in some cases stirring up sandstorms that filled the air with dirt, rock, and tumbleweeds. Stapp called them Muroc Paint Removers. The morning following installation of the sled, gusts reached 50 miles per hour and cut visibility down to just a few feet. By the time the crew got out to the site, the track rails were nearly buried under drifting dunes. The brakes had disappeared. The crew had to use a drag pulled behind a truck to clear the track, while going through nearly 60,000 gallons of water washing the sand out of the virgin brake system.

Stapp immediately dispatched one of his engineers to create drawings for a 50-foot-long canvas cover to protect the brakes. With the dimensions and spacing for tie-downs in hand, Stapp put in a call to Wright Field. He gave all the specifications to Colonel Kendricks, who promised to have the covering delivered within the week.

Only three days later, Colonel Kendricks himself landed at Muroc in a C-47 with a heavy tarpaulin complete with tie-downs, metal eyelets, and cotton cord. From then on, at the end of each day's work at the project site, Northrop technicians would carefully fit the tarp over the brake area, secure it and pile shot bags every few feet along the edges for good measure.

Stapp appreciated the prompt delivery of the brake tarp, but he was otherwise unhappy about the presence of the colonel. He'd gotten used to working in isolation, away from prying eyes at the Aero Med Lab. He found especially annoying Kendricks's insistence that the team pick up the pace. The chief let Stapp know he wanted to witness the rocket sled in action before he returned to Wright Field that Friday. The first run was not scheduled until the following Tuesday, but the colonel couldn't imagine why it had to take that long. Nobody on the project got the chance to so much as sit down or have a smoke as long as Colonel Kendricks was on hand exercising his particular brand of leadership, which seemed to them to mostly involve stomping around and complaining.

In the end, Stapp refused to hurry the preparations and the increasingly agitated Kendricks returned to Dayton on schedule without having witnessed anything more dramatic than a lot of sweat and a few rattlesnakes. He had, however, been able to examine the installation of the emergency backup braking system. Five hundred feet beyond the last of the friction brakes, a steel cable wound on hydraulic F-89 brake drums on either side of the track was set to snag an arresting fork on the front of the sled and pull it to a stop with a drag of up to about 10 g's—in case, somehow, the rocket sled was still going after hitting the tapering vise of the main brakes. Kendricks also got to observe the installation and calibration of some of the crucial project instrumentation, including a four-channel AM/FM telemetry system, a space-time recording device, a 16mm camera designed for mounting on the sled itself, and a battery of six Eastman high-speed (1,500 frames per second) movie cameras. The collection of visual data would be at the very heart of the project's operation; it would allow them to study the performance of the restraint systems they intended to test, and would help them understand what happened to a test subject during the brief milliseconds of maximum deceleration meant to simulate crash forces. These cameras were deployed trackside at specific intervals, spaced out to provide overlapping coverage all along the brake section where the action would be. The team also prepared a Micronex high-speed mobile x-ray unit built on a frame that could be positioned astride the track to capture the oncoming sled head-on.

Before leaving, Colonel Kendricks made a point of letting his project officer know that he had his eye on him. That he couldn't afford to relax and was expected to produce. The Air Corps would hold him personally responsible for the project's success or failure.

· · ·

On Tuesday morning, April 29, 1947, shortly after the team arrived at the site to prepare for their maiden rocket sled run, Stapp saw four late-model sedans trailing dust down the road from the entrance to North Base. Out stepped a vice president and a bevy of well-heeled Northrop executives who'd come to witness the event. It was a significant contract for the company, and its founder, Jack Northrop—though he was not present on this day—wanted eyes he could trust to evaluate the project's progress and perhaps to take the measure of Captain Stapp into the bargain.

Stapp greeted the executives, but otherwise ignored them. He called for three of the JATOs to be strapped to fitted racks on the trailing end of the sled and wired to the firing circuit. The test subject for the first rocket sled run would be one of Stapp's 185-pound parachute dummies, which was in reality little more than an oversized scarecrow. One of the crew had glued a

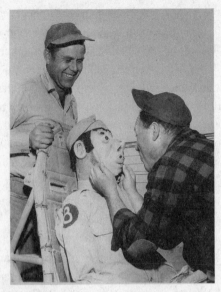

Project MX-981 chief mechanic Jake Superata has a word with Oscar Eightball prior to a rocket sled experiment. (Photo courtesy of Air Force Test Center History Office)

comical rubber mask to the dummy's head and dressed it in a surplus uniform with a number 8 insignia stenciled on its shoulders: Oscar Eightball. Stapp's offbeat sense of humor had by this time infected the team.

Oscar was strapped into the sled, in a rearward-facing position, and outfitted with several accelerometers. As the team was hustling around, running through their checklists and then rechecking what they'd already checked, Stapp moved everybody a little farther back from the rails out of caution.

A technician in a makeshift control hut about 250 feet south of the brake section of the track would activate the sled. They had rigged up loudspeakers to a two-way public address system to allow the technician to communicate with the team at the track. Once George Nichols had verified that firing circuits were connected and had completed a checklist of safety measures, the firing sequence was initiated.

Nichols closed a time relay switch on a ten-second delay, activating a 230-volt, 10-ampere circuit to the rocket igniters, and called out a countdown over the loudspeakers.

The JATOs were powerhouses designed to lift heavy aircraft off short runways. They burned potassium picrate mixed with roofing tar, and they spit flame and black smoke and propelled the sled toward the far end of the track with a fierce wallop: 3,000 total pounds of thrust. The engines were calibrated to reach burnout in five seconds, after which the sled would coast for a brief tenth of a second before hitting the brake section where, by Stapp's calculations, it should be traveling at approximately 150 miles per hour. The brakes would bite the sled's keels with a massive retarding force of 15,000 pounds, resulting in a deceleration of ten times gravity for one-tenth of a second. It was a horizontal—and much more capable—version of the ejection seat tower at Wright Field. If all went according to plan, the sled would emerge from the brakes carrying a speed of just 75 miles per hour— and simple slipper friction on the rails would bring it to a dead stop about 200 feet beyond.

Every experiment needs a theory, and that was the theory.

In reality, two of the rockets were still burning like crazy when the sled hit the brake section, at which point it was still accelerating at 210 miles per hour. The brakes were no match for the rockets. The knife slipped through the vise like it was soft butter. With only negligible loss of speed, the sled barreled through the brakes and snapped the heavy emergency cable, again without much effect on its speed. The crew and the Northrop executives looked on in horror as the *Gee Whiz* sped right off the end of the track and plowed through the Mojave gravel and sand for some 55 additional feet before coming to a stop. The bulky, bottom-heavy rocket sled, stabilized by half-inch-thick armor-plated keels which acted as runners, barely even wobbled. It stood upright and intact in a cloud of sand.

George Nichols strode out to where the emergency brake cable lay coiled like spaghetti alongside the rails. He pivoted and looked down toward the sled. Nobody had dared say a thing. Nichols turned back toward the crew and simply shook his head. It was back to the drawing board.

The MX-981 project's first big test had been an unmitigated disaster in the eyes of everyone but John Paul Stapp, who immediately proclaimed it a "smashing success." Later, he explained: "From a research and development standpoint, it was a godsend. In one test, we uncovered more weaknesses in the system and errors in procedure than could have been revealed by any five normal runs. Even the emergency system got a workout." He congratulated each member of the crew personally, smiled and shook hands with the still rattled and somewhat confused Northrop execs, and headed off to begin his postmortem.

Stapp's investigation revealed the problem right away. Two of the three JATO rocket motors performed way beyond spec. Instead of firing for five seconds and delivering 1,000 pounds of thrust apiece, they burned for seven seconds and produced 1,300 and 1,800 pounds of thrust respectively. Forty percent over-power was way too much kick for the system. Stapp's solution was radical but simple: ditch the rest of the current batch of unreliable JATOs and find rockets that were manufactured and tested to rigid specifications. He quickly phoned in a request to Wright Field before the Northrop suits had made it back to their air-conditioned offices. The project needed reliable propulsion, he explained. There could be no compromise.

While he waited, Stapp went back to work on the team's laboratory and office facilities, which meant he was again climbing around in the piles of sheet metal and plywood of the area salvage depots. He found a big wooden desk, some tables, and a few lamps. He even unearthed a medical examination table that with a coat of paint would be laboratory-ready. At this point, Stapp began making occasional trips back to Dayton to provide project updates and to confer with his colleagues in the Biophysics Branch, and he always brought back useful items: filing cabinets, office chairs, typewriters. He scoured the abandoned housing at Muroc and harvested basins and cabinets. He also collected three-quarter-inch iron pipe and, when he had enough of it, put the men to work running it beneath the foundation of the tracks to the 1,500-gallon water tank reservoir. Once he had water running to the building, Stapp was able to install an evaporative cooler he'd traded for medical services and couple it to a pump motor he got from the base supply office. This last effort managed to reduce the temperature in the new laboratory by 20 degrees, for which the men were especially grateful.

Nevertheless, every few days, Stapp would hear from his boss back in Dayton. When would the next rocket sled run happen? Colonel Kendricks wanted to know.

While the team waited for new rockets, Stapp's Curbstone Clinic continued after-hours operations, its popularity growing as word of free medical

The Project MX-981 team on the Gee Whiz, *North Base, Muroc Army Air Field. Top row: David Hill, Jerry Hollabaugh, George Nichols, unknown, Jake Superata, Ralph DeMarco. Bottom row: Ed Swiney, unknown (Schmidt), unknown, Harry Goniprouskis, John Paul Stapp. (Photo courtesy of U.S. Air Force)*

services continued to circulate at Muroc and environs. For the price of one extra seat at the dinner table, any family could avail itself of a skilled general practitioner. Stapp became a figure of near reverence at Kerosene Flats, where Celia Richards and family had settled following their move from Ohio. Ten-year-old Celia fell in love with the desert at first sight: "This entire expanse of God was magnificent. I just could not get enough of it." But day-to-day life in Kerosene Flats was another matter. "They were barracks that had four apartments per barrack and were set in this horseshoe. It was godawful miserable. The conditions were a terrible tragedy. It was lower enlisted people, mostly. They weren't paid enough money and they weren't educated. They didn't know how to treat their children. The kids all got beaten. There were a lot of pregnancies. So it was just *wonderful* for all of us when John showed up. He was so much more than a doctor; he was a world-class conversationalist and a brilliant pianist. Nobody had ever met a smarter, kinder human being. My sisters and I were in awe."

Another fan of Stapp's was Maw Green. She was an old-time desert dweller, a hulking woman known for her foul temper, salty vocabulary, and legendary hamburgers. The combination gas station-café she ran with her husband was the only one near the base. She would berate her customers for drinking too much of her coffee and was especially intolerant of ill-behaved children. But Maw baked a pie every day of the week and she always saved a piece for Stapp. If the café was down to its last piece of pie, Maw's rules were that it be set aside in case Stapp stopped in for a cup of coffee before heading out on his scooter for a night of doctoring.

Not everyone around Muroc shared in the admiration society. Stapp had always had a knack for rubbing some people the wrong way. One legendary figure who came to harbor an open dislike for him was Florence "Pancho" Barnes, perhaps the most colorful character in the history of Muroc. Pancho was a flamboyant heiress, the granddaughter of wealthy Civil War hero, inventor, and showman Thaddeus Lowe, and she grew up a tomboy in the sparkling world of Los Angeles aristocracy. She became one of the most celebrated female pilots of her day, garnering at one point in her career the designation of "fastest woman on earth."

In 1935, Pancho bought a failing 80-acre alfalfa farm on what was then the edge of the air base. She immediately had a dirt runway constructed on the property and later added a hangar, stables, and horses, a restaurant and bar, a glamorous circular swimming pool, and air-conditioned motel rooms. She became the Antelope Valley's celebrity entrepreneur and she gradually expanded her holdings until she had some 364 acres of desert dude ranch, which she rechristened the Guest Ranch Hotel and Happy Bottom Riding Club. Pancho herself was hardly a beauty—as Chuck Yeager would say, "She was either the ugliest woman you ever saw or one of the ugliest women you ever saw"—but Pancho's place was always decorated with a staff of lovely and very friendly young women known as "hostesses." There was a debate about the hostesses' true function at the ranch, and while nudity was tolerated, even encouraged—with the evening goings-on occasionally turning fairly raunchy—Pancho vigorously denied that she ran a brothel. "My girls are sugar to catch the boys," she said. When they were at work, the hostesses were all business—serving food, pouring drinks, and dancing with the customers. "But," Pancho declared, "what those girls do on their own time is their own business."

Pancho's ranch served a purpose for the base, providing high-strung pilots with a safe haven to relax and blow off steam. Pancho served great beefsteaks and occasionally took care of the tabs of some of her favorite test pilots. She knew how little money they made and she honored the skill and guts it took to fly experimental aircraft for a living. The ranch also attracted movie stars, tycoons, gamblers, and gangsters who found her desert oasis the perfect escape from the prying eyes of the press—or the law. As a result, Pancho's airport was crowded with planes some weekends. At its heyday, the Happy Bottom Riding Club was said to have had some 50,000 card-carrying members worldwide.

Captain Stapp's workhorse schedule didn't allow him much time for the partying at Pancho's. He visited once or twice, but wasn't comfortable in the anything-goes fantasyland she'd created for her regular customers. Stapp rarely drank, and when he did it was usually no more than a glass of wine with dinner. Stapp and Pancho, whom he referred to as a "desert Jezebel," mostly avoided each other.

While the details of their rare encounters are unknown, at some point Pancho made it clear that Stapp was not welcome at the ranch. When Stapp's colleagues and superiors from Wright Field would visit the base and he attempted to arrange lodging for them at the ranch, Stapp was informed that any associates of his were likewise personae non grata.

· · ·

Stapp never forgot Pancho's insult, but it was what he regarded as professional disrespect that bothered him more. Shortly after Colonel Kendricks first offered him the Muroc project, Stapp had tried to requisition one of the more than 200 crash helmets in possession of the Aero Med Lab back at Wright Field, but a cranky procurement officer with an eighth-grade education had turned him down. Helmets were reserved for pilots. No exceptions. It was the first indication of the lowly status of Project MX-981. Still, Stapp knew that if he were going to conduct meaningful human deceleration tests, he would need protective headgear, with or without the Army's support. Which brought him across the San Gabriel Mountains in the summer of 1947 to the School of Aviation Medicine at the University of Southern California (whose research was supported at that time by the United States Navy) and eventually to the offices of Dr. Charles "Red" Lombard and Herman Roth. The facilities at USC were state-of-the-art, including both an 18-foot centrifuge and a decompression chamber.

Red Lombard, an aviation physiologist and World War II veteran, was the department's director of research, and displayed on his desk was a row of crash helmets of his and Roth's own design. Lombard explained to Stapp what he knew about protecting the skull and its contents, both from fracture and concussion. The outer shell of his new ventilated Toptex helmet was made of four plies of a special fiberglass fabric impregnated with an extremely strong styrene resin. The helmet lining was a 5/8-inch thick cellulose acetate foam designed to indent when hit with a force of more than 10 pounds per square inch. Lombard wanted to distribute the force of a blow to the head over the largest possible area, hence the rock-hard outer shell, and to absorb as much of the energy as possible in the compressible plastic sandwiched between the shell and the rubber sweat lining. Yet even with all this protection, the helmet weighed only 24 ounces. A prototype had already saved the life of a Muroc test pilot when his canopy blew off at 30,000 feet and struck him in the head.

Stapp went over the objectives of his project with Lombard, and asked about acquiring one of the helmets for himself, understanding that the tolerances were so tight that Lombard's creations all had to be custom-made, fashioned from a mold of the individual subject's head. Lombard didn't hesitate. He made Stapp an appointment for a fitting the following week.

Stapp's success with Lombard and USC emboldened him to begin pressuring the Aero Med Lab for project improvements. Stapp wanted another flight surgeon to join him at Muroc and as many Army technicians as could be spared to assist the Northrop crew. But his most urgent request was for a new building to house rocket sled research equipment and facilities—he declined to be more specific about his plans than that. The response came from Colonel Kendricks almost immediately. There would absolutely be, he assured Stapp, no additional personnel assigned to the project, nor would there be a new building for a short-term field project under the direction of a reserve officer with little more than a year left to serve. The colonel ordered Stapp to withdraw the request.

While his negotiations with Kendricks continued, Stapp made sure the team stayed focused on its mission. Not long after the less-than-satisfactory maiden rocket sled run, and some repair work on the sled and the brakes, the team again strapped Oscar to the seat and conducted a follow-up test. With just two of the vintage rocket motors firing this time, the sled accelerated more slowly and the two-second overburn sent it into the brakes at only 120 miles per hour. The sled coasted out of the brake section and came to rest a good 10 feet short of the emergency brake cable. These speeds wouldn't work for the research Stapp envisioned, but what they'd managed to demonstrate was that the unreliable JATOs would never be acceptable for MX-981's purposes. With this data in hand, Stapp was now able to argue, once again, for high-quality rockets made to the project's specifications. Within a week, this time, the Army issued authority to Northrop to subcontract procurement of new rocket engines from Aerojet General.

When he wasn't busy securing new equipment or operating the Curbstone Clinic, Stapp recorded his thoughts about the project during solitary nighttime study sessions in his spartan quarters at the Desert Rat. He was pleased enough with MX-981's progress, but he was growing impatient for the day when proper equipment and preparation would allow him to climb onto the *Gee Whiz* himself and prove to the world that human beings—properly restrained—could survive events of unimaginable violence. While he was careful to maintain the image of the dispassionate scientist around the team, the whole thing was for him becoming personal. It made Stapp think of his father. He'd always wondered at the intensity of Charles's zeal for the Cause. Now he was beginning to understand it, and to realize that it had always been part of his own character as well. As he loved to say: "You might shake off the religion, but you never shake off the missionary spirit."

8

STAPP'S FIRST RIDES

The ultimate instrument for measuring the effects of
mechanical force on man is the living human volunteer.

—*John Paul Stapp,* **Impact Injury of the Head and Spine**

WHILE THE MEN WAITED for new rockets, work continued in the summer and fall of 1947 on improvements to the braking system. Using the surplus JATOs to get some baseline data, they conducted thirty-two success-ful test runs and were able to verify that the brakes could stop a 1,500-pound vehicle traveling at 400 miles per hour in just 50 feet. This was extreme decel-eration akin to that encountered in a typical airplane crash. The results heartened Stapp, but he was increasingly anxious for the day when they could strap something more lifelike than Oscar Eightball into their sled.

Back at USC for his appointment, Stapp followed Red Lombard's instructions to remove his shirt and squeeze his head into a tight bathing cap. Over the rubber cap, Lombard stretched a second cap—this one made of stocking-net material—and began slathering on layers of plaster of Paris and reinforcing them with plastic bandages. Once the plaster hardened, Lombard carefully removed the mold.

Two weeks later, Stapp held a gleaming white helmet shell in his hands and carefully guided the tight lining past his ears until the whole piece was snug on his head. The fit was perfect. Then, without warning, Lombard grabbed a big plastic mallet and whacked Stapp on the top of the head. It was a hard blow, but Stapp felt absolutely no distress or pain. The helmet's hard outer shell was neither dented nor marked. Stapp's respect for Lombard had grown with each visit.

In return for the helmet, Lombard asked only to be kept informed about its performance on the Muroc project. He told Stapp that if anything happened to him, he would want the helmet returned. He gave Stapp a look at the USC office's collection of four helmets that had been involved in accidents. They were a little beat-up, but intact. A fifth helmet was partially shattered and speckled with brown bloodstains. That one, Lombard explained, had been a fatality.

Stapp returned to the desert. It was always a bit jarring to leave behind the comforts and distractions of Los Angeles for the rough world of Muroc. The first modern-day inhabitants of the area alongside Rogers Dry Lake bed had arrived in 1910, the year of Stapp's birth. Clifford and Effie Corum, along with Clifford's brother Ralph, had built their homestead on what was no more than a lonely water stop on the Southern Pacific railroad line. The Corums worked hard to establish some basic infrastructure. By the time they had a general store up and running, and a United States post office commissioned there, the family proposed to name the little town Corum. But when a records search revealed that a Coram, California, already existed, they simply spelled their name backward and named the place Muroc.

In 1932, Army Lieutenant Colonel Henry "Hap" Arnold began the government's acquisition of land adjacent to the lakebed for a bombing range, which led to the construction of a support facility originally known as Mohave Field, and later as Muroc Field. Arnold initially declared the area "not good for anything but rattlesnakes and horned toads." Six years later, he began locating much of the Air Corps's research and development efforts there. Over the decade of the 1940s, the base was expanded to encompass 470 square miles and a 15,000-foot runway.

The nation's experimental military aviation program had taken off in a big way in the years after the war; test pilots flew hundreds of hours every month in scores of experimental planes, many with radical and controversial designs. The pace of the work was frantic and, inevitably, pilots lost their lives. As new altitude and speed records were established, the death toll rose. Not long after Stapp's arrival, the base would be losing pilots at the rate of nearly one a week. Stapp remarked that it was not uncommon to find yourself shaving alongside a man in the morning and have him reported dead that same afternoon. "The streets of Muroc," he observed, "are named for my friends."

Stapp did his best to follow the progress of the flight programs at the base, but he also kept up with work in Chaves County, New Mexico, where the Army was running Cold War research programs using high-altitude plastic balloons. One particularly interesting effort was dubbed Project Mogul, and it involved raising specially calibrated microphones into the stratosphere for the purpose of detecting sound waves generated by Soviet atomic bomb tests. When a crashed Mogul balloon and payload were discovered by a rancher named W. W. Brazel near the town of Roswell, New Mexico, in the first week of July 1947, the Army held a press conference at Roswell Army Air Field (RAAF) to announce the discovery of the debris. Stapp read the reports. Part of the description offered by the public information officer that day included mention of a "flying disk." On July 8, 1947, the *Roswell Daily Record* ran a banner headline: "RAAF CAPTURES FLYING SAUCER ON RANCH IN ROSWELL REGION." The story, in which locals claimed to have seen extraterrestrial creatures hustled away by Army officials, created a brief stir, and was then quickly forgotten.

• • •

Two months after the incident at Roswell, in September, Congress passed the National Security Act of 1947, transforming the Army Air Forces into its own branch of America's uniformed services: the United States Air Force. The attendant bureaucratic changes did nothing to mitigate the dangers for the pilots at Muroc, and the increasingly deadly nature of their work affected everyone who lived there. Stapp's team felt it. Their collective cause was, after all, to help save those lives. It lent a palpable urgency to their preparations for the rocket sled experiments.

As autumn 1947 approached, base gossip centered on one particularly challenging mission: to break the sound barrier. Specifically designed for the purpose, the top-secret six-million-dollar Bell Aircraft XS-1 rocket plane— the X stood for "experimental research," the S for "supersonic"—would be released from the belly of an airborne B-29 before shooting off into the ether. When Chalmers "Slick" Goodlin, a civilian test pilot working for Bell, demanded a $150,000 bonus to attempt a supersonic flight, the Air Force had agreed to take over the mission. General Albert Boyd, stationed at Wright Field, had reviewed some 150 Army pilot candidates that May. There were bets on which of the top test pilots would get the shot to attempt it—an honor either heroic or suicidal depending on whom you consulted. Boyd later said it had been one of the most difficult decisions of his life, but he eventually selected Chuck Yeager, naming Yeager's buddy and rival Bob Hoover as the backup pilot in case for some reason Yeager couldn't go.

Five months after being selected by Boyd and only one night before the historic flight was scheduled, Yeager and his wife had gone for a horseback ride on a couple of Pancho Barnes's horses. As they approached the stables on their return, Yeager failed to see a closed fence gate. His horse pulled up and threw him into the fence, fracturing two of his ribs.

Yeager understood that if he reported to the base flight surgeons for treatment, he'd more than likely be grounded for the Mach 1 attempt the next morning—and worse, the assignment might go to his buddy Hoover. Yeager says his ribs were examined and taped up that night by a doctor in Rosamond. Some versions claim it was a veterinarian. But according to residents of the Antelope Valley at that time, there was neither a doctor nor a veterinarian in tiny Rosamond.

There is another version of the story and it comes from Northrop's chief engineer on the rocket sled project, George Nichols. Yeager has repeatedly denied it, but others at Muroc remember hearing the same thing. In this version, when Yeager broke his ribs that night, he called the one doctor he knew and trusted, one he was sure would give him an honest assessment of his condition, and one he could count on to keep the incident quiet: Dr. John Paul Stapp of the Curbstone Clinic. Nichols, along with Dana Kilanowski, a historian for the Society of Experimental Test Pilots who knew Stapp well and who conducted multiple interviews with Yeager, believes it likely that Stapp did in fact attend to Chuck Yeager and that Stapp wrapped Yeager's broken ribs late the night before the attempt on the sound barrier.

Stapp himself refused to either confirm or deny it. He did not have the assignment to act as Yeager's flight surgeon, and his relationship with Yeager, he said, was nobody's business.

On the morning of October 14, 1947, Captain Yeager became the first person to go supersonic in level flight. At an altitude of 45,000 feet, Yeager pushed the XS-1 to Mach 1.06, approximately 760 miles per hour, and sent the first sonic boom thundering across the Mojave. Legend has it that Pancho Barnes's offer of a free steak to the first man to break the sound barrier was claimed that night by Yeager. That's not quite the way it happened. After Yeager landed, he went with fellow pilots Bob Hoover and Bob Cardenas to the house of flight engineer Dick Frost for a few martinis, and then to Pancho's. Upon arrival at the Happy Bottom Riding Club, however, they were met by an Air Force man who reminded them that the flight remained classified; they could not talk about it in mixed company. That put something of a damper on the celebration since Yeager couldn't actually lay public claim to the free steak. Nevertheless, his reputation was forever sealed, and six months later, when the news came out, Pancho presented him with a giant trophy and served him—with great fanfare attended by a bevy of her most beautiful hostesses—the steak he'd earned.

The men at the MX-981 site just a few miles away, however, had yet to achieve much worth celebrating publicly or privately. Whether it was the example of Yeager's flight or their own mission, Stapp noted in a letter to a colleague back in Dayton that the team seemed unusually focused and efficient. The weather had mellowed from the blast furnace of the Mojave summer— daytime temperatures now topped out in the 60s and the nights often dipped below freezing—and the North Base site was in high spirits.

The first of forty brand-new, high-precision rockets were delivered on the last day of November 1947. The new rockets were cased in welded steel; they measured two feet in length, six inches in diameter and weighed 125 pounds each. The solid propellant, about 65 pounds' worth in each rocket, was a mixture of potassium perchlorate and tar. The igniters—which consisted of small glass jars packed with two ounces of smokeless powder and each sealed with a stopper containing a spark plug—were packed and stored separately from the rocket motors for safety purposes.

Once the rockets had been inspected and stored, Stapp assembled his team and announced a target date for the world's first human rocket sled experiment: Wednesday, December 10. Stapp would always insist that for safety purposes human experiments be conducted either on Tuesday,

Wednesday, or Thursday. On Fridays, he reasoned, crew members were liable to have their minds on the upcoming weekend, and on Mondays they were still thinking about the weekend that had just passed. When they began human testing, the stakes would be high and Stapp wanted his team focused and on point every time they lit those rockets.

Stapp had by this time broken the news to Colonel Kendricks and the rest of the Aero Med Lab that he was preparing for a human deceleration experiment with himself as the subject. He'd gotten the expected protests and numerous warnings about the potential dangers of high g-forces on internal organs and especially on the brain. Some of the medical personnel at Wright Field believed that concussion was inevitable, and Kendricks himself gave Stapp a stern lecture about the established limit of human tolerance. In the end, though, Stapp prevailed and received approval to move ahead with his planning. He was even promised a flight surgeon from Wright Field to assist during his initial attempt.

The first human run, Stapp announced to the team, would be conducted with just a single rocket, and the sled's speed would top out at about 90 miles per hour. The hydraulic brakes would not even be needed. Once the five-second rocket burn was complete, the sled would simply coast to a stop a few hundred feet down the track using nothing more than the friction of the runners on the rails. This would be a test run to allow Stapp to get a feel for the experience and to indoctrinate the crew in the procedures they hoped to use on higher-speed runs in the future. It would also serve notice to Colonel Kendricks and everyone back at the Aero Med Lab that MX-981 was making steady progress and that the project was in capable hands. Stapp was hell-bent on moving forward according to plan on December 10.

Then, on the afternoon of the 9th, the Battle of Muroc flared up. A test pilot who served as liaison officer for Wright Field projects being conducted at Muroc was briefed on the plans for a human rocket sled run—and immediately began setting off alarms back in Dayton. He was shocked, he reported, that Stapp actually intended to strap himself into a ground vehicle atop a live rocket. It was crazy at best and, at worst, irresponsible. The pilot shot off a series of cables back to the Aero Med Lab in an effort to get the next day's sled run cancelled. He got in touch with Stapp and alerted him to his intention to shut the project down.

The next morning, Stapp and crew began prepping the sled at sunup. In the absence of any official orders to halt the experiment, he wanted to get the run completed before second-guessing could set in and before any such order had a chance to arrive. He suggested that no phones should be answered in the barrack office until the deal was done.

Stapp slipped on Lombard's helmet and climbed aboard the steel-tubing seat that had so far been occupied only by Oscar Eightball and his dummy brethren. The seat was mounted inside the aluminum superstructure, facing to the rear, toward the rocket rack. Technicians assisted in fastening a standard-issue Air Corps lap belt over Stapp's thighs and adjusted the shoulder straps so that he was fixed snugly against the seat back. Once the igniter was screwed into place on the rocket, he watched the men scatter to a safe distance. It was a lonely moment, in which Stapp had little else to do but contemplate his fate. If something went wrong, not only might the project be terminated, but he might be done as well. Yet, with no option to turn back, he willed himself to relax and trust in all the preparation and planning, and in the good work of the Northrop men whom he'd come to trust like his own brothers.

A few minutes after 9 a.m., a loudspeaker lashed to a pole at the edge of the track broadcast the countdown. Stapp watched the rocket nozzle in fascination as he listened to the voice coming from the speaker. He was aware of his accelerated breathing and quickening pulse as the count approached zero. At almost the instant that he heard the word "Fire!" he watched a jet of smoke spurt from the rocket nozzle, followed by a quick burst of flame. Even through the helmet, with its thick layer of plastic foam surrounding his ears, he could hear the sizzle of the rocket as it fired. Then came the powerful surge attempting to pry him out of his seat. It wasn't as bad as he'd feared. He judged the pull to be no worse than a short-field takeoff in a B-26. For five seconds, the rocket powered him backward down the track at approximately 90 miles per hour. Then there was an abrupt silence following the rocket burnout, a silence that was quickly filled with the rattletrap chattering of the decelerating sled grinding down the track.

Even though the speed he'd achieved was not particularly impressive, Stapp unstrapped himself at the end of the run and climbed off the sled in jubilation. He'd experienced no ill effects and the entire run had gone like clockwork. Their months of laborious preparation had paid off. Stapp immediately jumped on his yellow scooter and rode over to see the anxious liaison officer. He didn't necessarily intend to rub his success in the man's face, but he could barely keep the grin off his own. The liaison officer informed Stapp that he would continue doing everything in his power to get the project cancelled—or at least restricted to nonhuman test subjects. It was simply, in his estimation, too risky. Stapp found it an astounding statement for a man who flew experimental aircraft for a living.

Back at the track, after consulting with George Nichols, Stapp assembled the rest of the Northrop team and told them that he wanted to move forward with a follow-up test the very next morning. They had momentum, and he wanted to exploit it before Wright Field could order them to stop.

So, at 8 a.m. on December 11, Stapp arrived at the barrack office where the men were drinking coffee and sharing a box of doughnuts. Stapp refused to join them. A full stomach makes for a messy autopsy, he remarked. It was just Stapp's peculiar sense of humor, and the crew was for the most part used to it by now. But on this morning the autopsy comment unnerved them. Although Stapp would admit later that he had entertained his own doubts about the wisdom of pushing his luck this way, he spent a half hour reassuring the Northrop engineers and technicians that everything would be just fine and that there would certainly be no need for an autopsy. Still, he acquiesced to a suggestion that he have the base hospital send an ambulance over to the site, where it could sit idling alongside the track just in case.

George Nichols was one of those who'd been spooked by the autopsy remark, but Stapp refused to discuss—even with Nichols—any suggestion that they reconsider the decision to resume testing. The only change Stapp requested be made to the sled following the initial run was to add more padding to the steel seat. He'd experienced some bruising on his lower back following the previous day's test. Even a single rocket delivered a pretty formidable wallop.

The flight surgeon captain from Wright gave Stapp a routine physical exam, and Stapp climbed back up onto the seat of the *Gee Whiz*, where he was again strapped in. He squeezed the rubber handgrips, squared his feet up against the footrests, and braced the back of his helmet against the headrest in anticipation of the violent jolt. Again, he felt his pulse racing in the moments leading up to ignition. He tried to identify his own reaction. Was it fear or just anticipation? Whatever it was, he was quite sure of one thing: he was not enjoying himself and he wanted it over with. He breathed hard through clenched teeth.

The rockets fired and the sled shot down the track almost before his brain could register what was happening. He had the sense that the sled was pulling out from under him and that he might shoot out the back to be deposited on the tracks. He tightened his neck muscles and strained to keep his head upright. The five seconds of acceleration were a blur of smoke, fire, and raw power. Then the rockets shut off and all Stapp could hear was the screaming vibration of the sled hurtling toward the brakes at 150 miles per hour. Almost instantly he was slammed against the back of the seat. It felt at first as if all his bones would be crushed. The screech of the sled's keels catching the vise-grip of the brakes was akin to the sound of a high-speed car crash: rasping metal and crunching impact. The worst was over in just a fifth of second. The *Gee Whiz* slid to a screeching stop a few feet short of the emergency cable.

As he had on the first run, Stapp managed to unstrap himself and step out of the rocket sled before his team, sprinting from the barrack building, could reach him. He took inventory of himself and found no injuries beyond a slight pain in the middle of his back and some soreness in his gums. He made a mental note to work on the padding for the seat and to get some sort of protection for his mouth. But again, the sensation of triumph overwhelmed any discomfort. He'd gone from airplane speed to a complete stop in just a couple of seconds—and he'd walked away. The best part was that he knew he was nowhere near the limit of what could be endured. In that instant, he was convinced that if he could only keep the naysayers and disbelievers of the Air Force at bay, he'd be able to redefine all the assumptions about the effects of biomechanical force on the human body.

When Nichols reached him, Stapp grabbed his shoulders and looked him in the eye. He said that if this was what 10 g's felt like, there was no question they could do 35.

Stapp would later come to refer to such feelings of extreme elation following a sled run as "survival euphoria." The blast of adrenaline that flooded his bloodstream as he walked away from the *Gee Whiz* that morning, the feeling of invincibility and clarity of perception that came with making it through

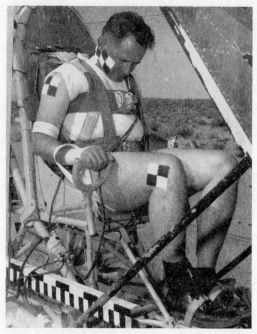

Aboard the Gee Whiz. *Stapp declared himself experimental apparatus. (Courtesy of U.S. Air Force)*

a traumatic event, was addictive. As miserable as he'd been just prior to the run, he found himself wanting more.

Stapp's sled runs would be increasingly hard on George Nichols. He'd grown close to Stapp and once human testing began he worried almost constantly about his friend's health. It seemed impossible to Nichols that you could subject a body to these sorts of stresses and not pay a price at some point.

Later in the afternoon of this latest test, on his way back to Main Base on his scooter, Stapp lost control in the gravel and was flipped into a rocky ditch. Skinned-up and a little shaken, he brushed himself off, climbed back on the scooter, and resumed his commute. The incident had been far more jarring and painful than anything he'd yet experienced on the rocket sled and that got him to thinking. His subjective impressions of the sled runs were going to be as important as any data collected from the telemetry. He needed to document and to be able to somehow analyze what a crash impact *felt* like if he was going to find the key to effectively counteracting it.

He spent that evening in the Desert Rat making careful notes in his project book:

> *Entry into the brakes resulted in a very abrupt ramming sensation against the back, with slight pain at the level of the ninth thoracic vertebrae coincident with the juncture of two cushions placed behind the back. Mental note was made to use a single flat cushion thereafter. Although no attempt at breath holding was made, the viscera plunged up against diaphragm causing a sharp exhalation. Mental note was made to hold the body as vertically as possible thereafter, so that the decelerative forces would be directed normal to the body. No sensation of jolting or concussion of the head. Vision, hearing, touch, and pain sensations remained continuous. Orientation, consciousness, and thought processes were not interrupted. There was no discontinuity or loss of memory. Immediately after the sled came to a gentle halt beyond the brakes, the sensations noted were of mild sharp pain in the ninth thoracic spine area, and diaphragmatic pain resembling a moderate blow to the epigastrium accompanied by pain bilaterally at the acromio-clavicular area which may have been referred to shoulder strap pain. The subject got out of seat and carriage without difficulty. All pain disappeared quickly after walking around for a few minutes.*

The next morning, as would become standard operating procedure for MX-981, Stapp and his team assembled to examine the data from the previous day's sled run. Velocity measurements were still as much art as science. Technicians had installed a magnet on the right front sled slipper, and the magnet passed over magnetic coils at 25-foot intervals along the track. With the circuit closed on the coils, the passing of the magnet over a coil induced a short surge that flickered an argon light that was recorded on 16mm film by a galvanometer. A second galvanometer then recorded a trace from a thousand-cycle tuning fork oscillator. By counting thousand-cycle time signals between blips on the magnetic coils, Stapp could reliably calculate the average speed from coil to coil. But decoding the results from the film records was tedious and laborious work.

The telemetry system was, in contrast, a piece of cake. At least, when it worked. Four tiny radio transmitters using four separate frequencies were mounted on the sled in a ruggedized and shock-protected metal pack. Silver-cel batteries supplied power to the transmitters, and signals from accelerometers, strain gauges, and other sensitive instruments could be transmitted back to the barrack building and recorded on photo film for examination.

A review of the data from the three-rocket run on December 11 was ultimately disappointing. Not all of the magnetic coils along the track registered the expected surges. The time-distance circuit shorted out and not a single one of the telemetry channels worked properly once the sled was in motion. Their data was a mess. As they finished sifting through what usable information they had collected, Stapp made the decision to delay all future sled runs until the project could secure reliable, state-of-the-art instrumentation. They needed to be able to prove to the Air Force—and the world—what they had accomplished. If Stapp was going to risk his life in these experiments, he wanted to at least be sure they got unassailable documentation. Stapp was never unaware of the interests, both at Muroc and back at Wright, that would like nothing more than to shut him down, and he vowed never to give his detractors that satisfaction.

The next weeks and months, then, would belong to the electronics engineers and the procurement people at Northrop. The team would keep at it until they got it right. Trial and error. The goal was to be ready to resume Stapp's human tests as early the following spring as possible. They tried out new methods of shock-mounting the sensing instruments. They tested every accelerometer and strain gauge they could get their hands on. They worked to improve the radio signals from the track to the barrack building, trying to eliminate distortion in the telemetry transmitters. The problem was that manufacturers didn't build this stuff to be mounted on top of rocket engines, or to withstand a force of even 10 g's. The technicians were improvising and Stapp was cheering them on.

Stapp himself made a series of trips to visit defense contractors and other equipment manufacturers around the country. He scheduled appointments at the National Research Council and the National Bureau of Standards in Washington, DC, and visited the Air Surgeon's Office in the Pentagon. He stopped by Eglin Field in Florida to observe aviation hardware tests. If better accelerometers and the like were out there, Stapp was going to find them.

Meanwhile, word of the tests had gotten out and *Time* magazine had run a short piece about the Muroc sled runs, and had mentioned the name of Captain John Paul Stapp as the daring medical officer who'd willingly subjected himself to punishing forces in the name of aircraft safety research. Stapp wasn't sure how the information had leaked—officially, MX-981 remained a secret program—but he reasoned that in the end it could only help his cause. Even Charles Stapp down in Brazil read about his son's exploits. "We were quite elated by finding the description of your project in Time," he wrote. "You must be going places or they would not be reporting it in Time. I pray every day that you may come safely through all these experiments. You scientists often risk too much to prove your findings. Your minds are too valuable to society to be risked." Not much later, in response to the *Time* article, a physiologist from M.I.T. paid a visit to Muroc to inspect the rocket sled site and to offer his expert insight. According to David Hill, Northrop's telemetry specialist: "He looked at our setup and said, 'If anyone gets 18 g's, they will break every bone in their body.'"

9

ANTICIPATING STRICT COMPLIANCE

> There should never be any surrender to the fallacy that an air
> transport accident is an act of God.
>
> —*John Paul Stapp, notes for a speech*
> *to the American Rocket Society*

I N LATE FEBRUARY and early March of 1948, swirling black storm clouds
descended on the Mojave, scattering thousands of acres of blowing sand
and dirt across the Antelope Valley. The blinding sandstorms buried farm-
land and uprooted trees, smothering roads, cabins, and whole towns. Stapp
flew to El Paso to see his brother Celso in mid-February, returning to the
base a few days later to find a good inch of sand covering the floor of his room
in the Desert Rat. "I debated whether to plant a date farm or sweep it out."
Winds were measured at 86 miles per hour. Several buildings on the base
lost their roofs, and Stapp's joke was that he never had to brush his teeth: just
step outside and smile, and they were sandblasted clean.

Yet despite all this, John Paul Stapp was more enthralled than ever with
the California desert. It was exactly what the little girl from Kerosene Flats
had said it was: magnificent. Especially at sunset. A magazine writer had
referred to the environs of the base as "the hot Siberia of the Air Force," and
Stapp had hated that. As spring came to Muroc and the sage and mesquite
greened up, Stapp realized that he was exactly where he wanted to be. If the
Mojave was an acquired taste, well then, he had acquired it. "To me," he said,

"the desert was a challenge. I called it freedom's last stand." Stapp's sense of liberation came not only from the vast windswept landscape but also from the invigorating isolation and the thousands of miles of separation between his project and headquarters at Wright Field.

By early April of 1948, as President Harry Truman was signing the Marshall Plan into law, freeing up billions of dollars for the rebuilding of European nations ravaged by the war, blue skies and calm weather had returned to the desert—the Paint Removers had blown themselves out—and Stapp and the Northrop men were busy preparing for a resumption of rocket sled tests. The engineers had reworked the telemetry system and were getting consistently good results on at least three of four channels. Stapp announced a series of human deceleration experiments targeted to achieve 10 g's with a rate of onset of 500 g's per second. The flight surgeon from Wright was not available, so Stapp requested that a doctor from the base hospital be in attendance, along with an ambulance.

On April 8, a Thursday, the team mounted three of the new rockets to the rack and shot Stapp down the track with an acceleration of 3.27 g's. Maximum speed hit 180 miles per hour at the instant of rocket burnout. Nothing too spectacular, but Stapp was pleased. Their new instrumentation was clearly an improvement over what they'd had to contend with in December, and Stapp had easily survived deceleration measured at 11.6 g's. It was the first well-documented, quantitative demonstration of the survivability of extreme decelerative force, and considering that the cockpits of most military aircraft of the day were built to withstand a paltry 6 g's, MX-981 was about to rock the Air Force's world.

Stapp developed one of his trademark lines at some point during these live tests. "Why," he liked to ask, "are we always underestimating Man?"

Part of Stapp's plan was to gradually increase –Gx force in order to safely approach the human tolerance limit, whatever that limit turned out to be, and the team was ready to go again on the following Tuesday with a goal of reaching 15 g's. This run went as smoothly as had the previous one. Over the next four weeks, the team completed eleven more runs in this series of experiments, incrementally ratcheting up the speed of the sled as it entered the brakes, and adjusting the brakes to deliver rates of onset as high as 1000 g's per second, culminating on May 5, when Stapp took a peak deceleration of

an astounding twenty-four times the force of gravity. This was the equivalent of a full stop from 75 miles per hour in just seven feet or, in other words, freeway speed to zero in the length of a very tall man. Stapp reported feeling the impact get stronger with each run, but he was still walking away from the sled at the end of the track. He downplayed the effects on his body in reaction to the increasing concern from his team, and especially from George Nichols. But he *was* getting beat up, and the stresses had begun to take a toll. Following the run, Stapp wrote: "Afterward I was sore and stiff in both sacroiliac joints. The lower lumbar spine and coccyx were quite sore and tender to the touch." But so far, at least, the ambulance had returned patientless to the base hospital following each of Stapp's high-speed runs.

The next series of tests, however, was brutal. On May 15, 1948, Stapp endured a 20-g stop with a 1000-g-per-second rate of onset without too much distress. On June 8 he made two 20-g runs in a single day. Nichols was convinced Stapp was overdoing it, and he argued for a more sensible schedule. By the end of the month, Stapp had hit nearly 35 g's and his back and hips were hurting. Anybody who watched him walk could see it. After the last run in the series, Stapp emerged from the sled dizzy and disoriented. He needed some steadying hands just to get back to the lab.

Stapp himself began to wonder whether the jarring stops were affecting his brain and threatening the reliability of his subjective medical observations. He had been reporting the results of each run to Colonel Kendricks in Dayton (Wright and Patterson Fields had been formally merged in January to create Wright-Patterson Air Force Base), and though he considered censoring the information about his own condition, he included notes on injury concerns.

"There was a terrific jolt," one set of notes disclosed, "and for thirty minutes after the ride I felt dull, confused and depressed. I was unable to concentrate clearly and developed a persistent, twinging headache. The state of depression lasted for about 48 hours and then gradually cleared." He suggested that another flight surgeon be assigned to the project to assist him and to be on site to conduct thorough examinations following the sled runs.

The written reply he received, dated May 26, 1948, was devastating. Kendricks wrote: "I am definitely of the opinion that a relaxation of your program is indicated ... I desire that you not act as subject on this project for a

period of 90 days." It seems a reasonable precaution given the personal punishment Stapp had been reporting, and an indication that the Aero Med Lab
leadership was reaching the limit of its own human tolerance of risk. Some of
the MX-981 team, George Nichols included, were privately relieved. "Assuring
you of my earnest appreciation and support," Kendricks closed, "and anticipating strict compliance."

Stapp was outraged. This would sideline him for the entire summer. To
make matters worse, the order came just as MX-981 was achieving its most
impressive results. He grabbed the next flight bound for Dayton and, thirty-six
hours after staggering from the rocket sled, found himself standing before
Colonel Kendricks in "one of our familiar shouting contests." Stapp made his
well-worn arguments, and pointed out that the project was racking up major
successes that had the potential to save the lives of Air Force pilots and aircrews all over the world. In his view they had come nowhere near what they
might accomplish with proper support from the Aero Med Lab. He pleaded
for a reconsideration of his status. If they would just leave him alone for a few
more months, he'd bring them results that would make it all worthwhile.

Kendricks was unwavering. The possibility of a serious injury was very
real and, in his view, likely. Such an injury would result in his having to shut
down the project permanently. Moreover, any controversy surrounding an
injury to the project officer had the potential to jeopardize the chief's anticipated promotion to brigadier general. He would not risk it—for Stapp or himself. "*You* are risking *my* neck," Stapp recalled Kendricks telling him. There
would be no further discussion.

Stapp returned to Muroc in a dark mood—dejected, but not altogether
defeated. He would, he decided, simply move the project into a new phase.
Kendricks may have grounded him, but MX-981 was still alive and there was
still work to be done. Rocket sled deceleration experiments in the Mojave
Desert would continue. Only the next time they fired the rockets, there would
be a new subject in the sled's seat. It wouldn't be John Paul Stapp, but neither
would it be Oscar Eightball.

. . .

Stapp did manage to get some help not long after his hurried trip back to
Dayton. That June, Captain Vern Mazza was dispatched to Muroc, where he

made two relatively low-speed, rear-facing runs. Larry Lambert, a master sergeant and test parachutist—in fact, the same man who'd made the first American seat ejection, which Stapp had witnessed back in August 1946—was also transferred to Stapp's laboratory the following month. Lambert, a genial character with a wild shock of hair and a quick smile, spun long, dramatic tales of his aerial exploits—not all of which were entirely believable—and along with a firecracker personality, he proved to be clever and resourceful, qualities the underfunded project could never have enough of. Like Stapp, Lambert had an intuitive genius for talking supply sergeants and military agencies out of needed hardware and services. Stapp had already located a gold mine of abandoned treasure at the recently deserted East Camp beyond Rogers Dry Lake—one of the oldest settlements in the vicinity of the base, dating back to the early 1930s when it housed the Muroc Bombing and Gunnery Range Detachment—including a disassembled airfield operations tower stacked in crates alongside an old airstrip, and a couple of small buildings that might survive a move.

Stapp pointed his new sergeant at the problem. In a matter of hours, Lambert had managed to secure a large flatbed truck and a crane, and he led them across the dry lakebed toward East Camp. Meanwhile, Stapp, with help from his Northrop team, laid foundations made of salvaged concrete blocks near the lab building. Late that afternoon, the flatbed came rolling in with a pair of buildings aboard. With everyone pitching in, the structures were eased off the truck onto planks and situated atop the foundations. The wages for the moving crew and the crane operator, negotiated by Lambert, consisted of two cartons of cigarettes and one box of cigars.

The next morning, Stapp addressed the project team and revealed the purpose of the new structures. He'd been haggling with exotic animal dealers in Hollywood, he said; the new buildings at their site were going to serve as housing for the chimpanzees they'd be acquiring in the very near future. Nobody knew quite what to make of this news, but they'd learned to trust Stapp and shortly after hearing it they went methodically to work. First, they needed to attach the two buildings, joining them to create their primate house. Then they could begin constructing platforms to support heavy-duty chimpanzee cages, which were very much needed considering that one of the chimps Stapp had arranged to buy was said to weigh 200 pounds.

Stapp had cut a deal with a contractor to provide thirty adult chimpan-zees. It was a complicated arrangement that involved authorizations from the Department of Defense as well as the French Colonial Office in Paris. At one point, as negotiations stalled, Stapp threatened to go to Africa himself to collect the chimps. In the meantime, he read everything he could get his hands on about the care and feeding of primates. The population of the North Base sled track compound was about to explode.

10

G-FORCE CHIMPS

> For both animal and man had become involved—as inevitable
> a process as history has always shown it to be . . . And then as
> always happens there was one species that turned out to be
> the outstanding subject—the most useful, and easily the most
> gifted.
>
> —*George F. Meeter,* **The Holloman Story**

ONCE THE CHIMP QUARTERS were ready, Stapp brought the men together to explain this latest development and give it some context. It was, he told them, an essential element of the project that had been part of his vision from the first. Still, it was going to be controversial and they would all, from that moment forward, be governed by a protocol of absolute secrecy. Debate over the legitimacy of using animals as subjects in research projects went back to at least the heated seventeenth-century arguments over vivisection, the practice of operating on live animals for medical experimentation. The American Society for the Prevention of Cruelty to Animals (ASPCA) was founded during the Civil War years, and by the 1940s any mistreatment of animals was frowned upon in the United States. Most states had animal cruelty laws on the books, and zoos, circuses, and pet shops were all subject to inspection. Another target of scrutiny was Hollywood. When two horses were killed during the production of a popular film, 1939's *Jesse James*, starring Tyrone Power and Henry Fonda—the animals were reportedly blinkered and run off a 75-foot cliff—the studio was publicly admonished by the American Humane Society. The following year, the society established strict guidelines for animal treatment that would be enforced on film sets by the Hays Office, generally thought of as the Hollywood censorship board. Of course, any incidents of suspected cruelty involving monkeys or apes received heightened attention given their intelligence and humanlike characteristics.

MX-981 would be using primates—and likely other animals as well—to help the project bracket the survival limits of impact acceleration. With human testing, Stapp explained, they had started their experiments at very low levels of force, gradually increasing them as they went. By using that strategy alone, however, they would only be able to approach the deceleration force a human subject could endure, and would never be able to establish an absolute limit—at least without eventually killing the subject.

For that reason, they would conduct a new series of tests—using anesthetized subhuman subjects—in which they would work backward from the lethal end of the scale, successively diminishing force exposure through levels of debilitating injury to a plateau of safety. This would establish a threshold range into which it would be considered safe to escalate the human tests. By definition, some of the chimps were going to be killed. It was, in Stapp's mind, the only way to gather the definitive data they needed. How much force was too much? Stapp didn't sugarcoat it. He let the men know that it was going to be grisly work. He wanted to prepare them for what was ahead, and tried to impress upon them both how sensitive the Air Force was about such work and how important it was to avoid publicity. The wrong story in a newspaper or magazine could shut them down for good.

Never having worked with animals before, the men were uncertain how this new development would change their day-to-day jobs, but they had no choice but to assume Stapp knew what he was doing. Once the chimpanzee housing had been secured, the team went to work on the construction of a reliable water delivery system. Maintenance of an animal compound was going to require lots of cleaning, and cleaning required lots of water—more than the makeshift water tower on the opposite side of the tracks could provide. Stapp had reconnoitered a remote corner of the base where lengths of six-inch invasion pipe were scattered about. He and Larry Lambert drove around in a truck and collected enough pipe for a 4,600-foot line from Cal Tech to MX-981. They located some valves that were just taking up space in a base warehouse and talked the Air Installations Office into sending a few men out to North Base with a ditch digger. All Stapp and team would really need to do was assemble the various lengths of pipe and wrap the joints.

The project took about a week to complete. They flushed several barrels of chlorinated water through the pipe, and then turned on the tap. It was a major morale boost for the team to see clear water flowing from their faucets.

While the water project was in progress, the Northrop men had decided to dismantle and overhaul their braking system once again. With the higher speeds they were achieving, the tolerances of the brakes became more critical. They threaded the sled onto the tracks and towed it through the brakes over and over, verifying that each component was behaving as expected. While he waited for the delivery of his chimpanzees, Stapp talked Larry Lambert into volunteering for a few rocket sled tests. Kendricks hadn't specifically prohibited anyone other than Stapp. The first ride would be Lambert's indoctrination run, and it was set to top out at 20 g's. They strapped him into the *Gee Whiz* and sent him down the tracks at 180 miles per hour. But when the sled hit the brakes, something went wrong. The *Gee Whiz* ground to a stop much more rapidly and violently than intended. The team rushed out to the track fearing the worst.

To everyone's relief, Lambert climbed out on his own power—though he was a little wobbly.

Because Lambert complained of lower back pain, Stapp had him transported immediately to the base hospital for x-rays. Luckily, they were negative, though Lambert hobbled around for the following few days. Data showed that he'd suffered close to 40 g's at peak deceleration. About a week later, Lambert was injured again while loading equipment onto a C-47 at Main Base, this time dislocating his back. It put him in the hospital for more than a month, depriving the project of his assistance at a critical time. Stapp immediately put in a request to the Aero Med Lab at Wright-Patterson for a temporary replacement. Not only was the request summarily denied, but in response to reports of Lambert's injuries—despite the fact that the more serious injury was sustained a mile and a half away from the rocket sled track—Colonel Kendricks now sent unambiguously worded orders forbidding the resumption of *any* human testing.

Kendricks's message sent Stapp into a brooding funk that was relieved only by a wedding invitation that arrived from Texas. Throughout his assignment at Muroc, Stapp had kept up with family affairs through his father's letters and occasional quick visits to see his siblings. The two middle Stapp brothers—Robert, who was now a struggling art teacher in Arkansas, and Celso—were both married by this time, Robert for the second time. They all met up in San Antonio the first week of July for the wedding of Wilford Lee Stapp to Margaret Clark. Wilford had, like the rest of the boys, graduated from Baylor. He'd served in the National Guard and the Army Air Corps with distinction. After the war, Wilford had returned to complete his master's in geology at the University of Texas. At the time of his marriage, he was working for the Sunray Oil Company as an exploration manager.

Margaret was a lovely young woman who impressed all of them with her grace and warm personality. John Paul liked her immediately. Charles Stapp, away in Brazil, had likewise formed a positive impression. He wrote to his oldest son: "I should like to have seen the lot of you standing there. From what they have written it must have been a very impressive wedding. I like Margaret's letter and every thing the others have written about her. She looks good in the honey moon kodak pictures they sent. I am glad for Wilford and feel sure that he has made a good match and is going to have a <u>home</u>."

Despite the eight years between them, the oldest and youngest of the brothers seemed to have a special affection and admiration for each other. As a result, San Antonio would quickly come to represent something of a second home for John Paul, a retreat from the mad whirl of sand, rocket smoke, and military politics.

Meanwhile, the frustrated Northrop engineers had gone back to work on the troublesome braking system, devising a method of releasing the preset brakes to avoid a repeat of the Lambert near-catastrophe. They continued to work on a remote-control solenoid that would allow the reliable and controlled dumping of pressure from the brakes. It was tedious, demanding work, and Stapp was grateful for the quality of men Northrop had provided him.

In any government research project, the relationship between a military project lead and civilian contractors can be awkward. Civilians don't formally report to the officer and as a result the officer can't issue civilians formal orders. Though the military man is the boss in most respects, he often earns less, sometimes significantly less. In 1948, George Nichols, the senior Northrop man on the project, probably made on the order of $1,000 a month. Stapp's base pay was $253 a month. That was supplemented with another $126 for flight surgeon pay, another $100 for medical pay, and $21 for "subsistence pay." After standard deductions, his take-home total was about $425 a month. But there were never any authority issues or mutinies—and only the rare frustration or disagreement—on MX-981.

Stapp was good at anticipating the men's reactions to new developments on the project, and he was quite aware of their collective trepidation as they prepared for the coming of their new test subjects. Stapp wanted to minimize the attention the chimpanzees might attract, so he arranged to have the animals delivered to an abandoned airfield in north Los Angeles County, where he met them in a C-47. He brought the entire Northrop crew along on the clandestine mission. The initial cohort of eight chimps, in individual travel cages, was arranged on either side of the plane's narrow aisle. As they were loaded in, the chimps and men eyed each other suspiciously. Stapp had warned everyone about the chimps' prodigious strength and penchant for mischief. The men were a little spooked and stayed well away from the cages on the flight back to the base. The C-47 landed on a North Base runway, where trucks were on hand to move the precious cargo to the compound building alongside the sled tracks. It was dark by the time the animals were ushered into their new home, fed, and bedded down for the night.

Another new addition to the team was a veteran animal trainer Stapp had hired as part of his contract to purchase the chimps. His name was Tony Gentry, a Tennessean, and Stapp had found him close by in Thousand Oaks, at the World Jungle Compound, an outfit that supplied exotic animals to the movie industry. In Stapp's mind, Gentry was more valuable than any veterinarian. They fixed up a cot for Gentry in the animal house, and that first night he reported having to quell some near-riots among the primates by means of two blank-firing pistols he kept with him at all times. Gentry also had an Army issue .45 loaded with live ammunition, though Stapp had forbidden him to fire it anywhere in the vicinity of the test site.

The next morning, Stapp had to visit the base supply office to try and get more blanks for Gentry's gun. The supply officer was extremely suspicious, and was already cranky with Stapp over the order for food stocks for the chimps. Stapp had pored over government regulations to determine whether he could have the chimps admitted to the base as recruits, put in uniform, and given mess hall privileges. "I reasoned," he wrote, "that they were just as much soldiers doing their duty for our country as some of the yard-birds living in the barracks." The regulations made it clear that horses were to be provided forage, but he could find nothing covering chimps. In the end, he resigned himself to using project contingency funds to feed his new subjects.

He discovered early on that the chimps were picky eaters who required as much as 12 pounds of food per day apiece, depending on the size of the individual. The typical daily ration for a chimpanzee looked like this:

- One quart of milk
- One-half loaf of whole wheat bread
- One head of lettuce
- Six bananas
- One stalk of celery
- One large onion
- Four oranges
- Four hardboiled eggs
- One large sweet potato
- One large bunch of grapes (dessert)

Once a week or so, Stapp augmented the meals with grilled liver and bacon which he cooked on a portable stove. "They loved garlic," he said. "They'd take that garlic and just weep and eat and weep and eat. They'd rub their eyes, cry, and eat some more."

One of the chimps was a particularly discriminating eater and required two or three cans of beer before meal time to jump-start his appetite. Before agreeing to sign the requisition for beer for the chimp, the supply officer insisted on witnessing the spectacle. Stapp was required to keep detailed accounting and to return the requisite number of empty cans.

Before they could conduct any rocket sled experiments with the chimpanzees, the MX-981 team needed to develop good procedures for anesthetizing them. The chimps were powerful, and anyone working with them had to be careful. On the first attempt—they selected a 65-pound female named Isabella—Gentry opened the cage door just slightly and managed to get some loops of webbing around the animal's wrists. However, before Stapp could get a syringe with sodium pentothal anywhere near Isabella's arm, she burst through the cage door, making a quick lunge at Stapp and just missing him with her outstretched fingertips. As she made her move, Gentry jumped on the webbing trailing behind her and stopped her cold. She wheeled to go for Gentry and Stapp was able to grab her by the wrists from behind. The chimp chomped at Stapp's hand, just grazing his knuckles with her big incisors, but Gentry got her by the collar and managed to hold her head back. After a brief struggle, all three of them tumbled to the floor. Finally, Stapp got one foot under her arm and the other under her neck and got one of her arms stretched out. A moment later, Gentry had the other wrist and Isabella was finally spread-eagled on the floor and in position to receive the injection.

Stapp was embarrassed by the incident, and came up with a new plan for anesthetizing chimps: a movable back cage wall that could be brought slowly forward to gently compress the occupant. It worked, but even sandwiched between the wall panel and the front of the cage, it took four men to pull one of the hairy arms out for long enough to administer the anesthesia.

Once a chimp subject was finally anesthetized in preparation for a rocket sled run, it was removed from the cage and weighed. It was then placed onto the seat of the *Gee Whiz* and securely strapped in. Every step of the procedure was carefully scripted and rehearsed. The animals were not only valuable research subjects, Stapp reminded his team, they were intelligent creatures that deserved the same respect anyone volunteering for a rocket sled ride would be afforded.

With preparations complete and animal-handling protocols defined, the next phase of Project MX-981, live deceleration studies using nonhuman subjects, began in the late spring of 1948. The first chimp runs were at modest speeds and g-force levels, but the levels were raised fairly rapidly. Stapp was astounded at the forces the chimps could withstand. Thorough examinations were conducted following each run, and the chimps were proving to be extremely resilient. On the human tests, he had been meticulous about increasing speed and decreasing braking distances in order to approach tolerance limits safely. Experiments with the chimps were suggesting that the limits might be much greater than even he had suspected. By using them to establish the upper limits, Stapp could one day approach those same limits with human subjects. He continued to believe he could find a way past what he felt were the unreasonable restrictions of Colonel Kendricks and the Aero Med Lab.

The chimps first rode the *Gee Whiz* in the rearward-facing position, as had Stapp and Lambert; this configuration was the least punishing during massive deceleration because the body is slammed back into the seat and the force of impact is distributed over as large a section of the torso as possible. Later, a special configurable litter was installed on the sled and the chimps were strapped in facing forward, as a pilot in an airplane would be. They also rode on their backs in a headfirst position, and on their sides. The variety of configurations was intended to mimic the potential orientations of aircrew members during ejection or in an actual crash. In every position the team tried, abrupt deceleration was proved harmless up to magnitudes that would easily break up structures of the day's planes. The implications for pilots and aircraft manufacture were obvious.

The Air Force's pilots, it was obvious to everyone on the base, needed all the help they could get. Working at Muroc in the late 1940s really was like laboring in the vicinity of a war zone: a sickening column of black smoke rising like a pyre in the distance, the wail of an ambulance. Stapp had gotten to know a number of the top pilots and had become friendly with several, often running into them at the Muroc officers' club. He felt a kinship with these men. Like Chuck Yeager and Bob Hoover and Bob Cardenas, Stapp risked his life on dangerous experimental vehicles. He believed he understood the pilots' motivations and learned, as he swapped stories with them, that they shared his survivor's euphoria.

He had encountered one of the base's best pilots at the officers' club on the night of June 4, 1948. It must have been a night off for the Curbstone Clinic, because Stapp was playing the piano and dropping coins into a slot machine when Glen Edwards—whom Stapp had read about in the papers three years earlier when Edwards had broken the transcontinental speed record—came through the door. Captain Edwards was the archetypal American hero, though he was Canadian by birth and maintained dual citizenship. He was a gutsy and highly skilled pilot with the intense, dark-eyed gaze of a man to be taken seriously. He'd served with great valor in North Africa during the war, winning four Distinguished Flying Crosses and six Air Medals. When Erwin Rommel's Afrika Korps assault group broke through Tunisia's Kasserine Pass in 1943, Edwards's squadron flew eleven missions in a single day, helping to slow the advancing enemy columns. At Muroc, he had been one of those in consideration for the Mach 1 flight. Edwards was the prototype of the new breed of well-educated flying man, precisely the type and quality of individual that appealed to John Paul Stapp. Stapp was fond of saying that the nation would be wise to recruit its pilots from the grad schools rather than the gridiron. Just the previous year, Edwards had graduated from Princeton with a degree in aeronautical engineering.

Stapp was happy to see his friend that night at the club. They shared a meal and talked until about nine, when Edwards left. Stapp was unaware that Captain Edwards was scheduled for a test flight early the next morning, which meant that he was at that point technically in violation of Air Force regulations, which required twelve hours of crew rest prior to a mission. Muroc pilots regularly pushed the limits; it was part of the Wild West culture at the base. Stapp returned to the Desert Rat when he'd gotten his fill of the club's slot machine.

According to Mac McKendry, Pancho Barnes's last husband, Edwards showed up at the Happy Bottom ranch sometime before midnight and started drinking. He tried to "hire" one of the hostesses. There was an argument and things got loud. Pancho cut the pilots who frequented her place a lot of slack, but she had her limits. At some point between 3 and 4 a.m., she asked Mac to escort Glen Edwards out. Mac walked Edwards to the gate, ushered him off the property, and locked the gate behind him, leaving Edwards to make the lonely two-mile hike back to Main Base for his flight as a perfect dawn broke across the mountains to the east. Edwards was slotted to fly the Northrop YB-49 Flying Wing, a radical experimental jet design of Jack Northrop's that looked like a giant boomerang. Edwards had flown it before and had judged it harshly: "It was the darndest airplane I've ever tried to do anything with. Quite uncontrollable at times."

There were indications that the reason Edwards did not end up in the pilot's seat of the Flying Wing as planned that morning was that he reported

Muroc from the test pilot's perspective. The Northrop YB-49 Flying Wing is on the ramp in foreground. (Photo courtesy of U.S. Air Force)

to the flight line in poor shape. Major Daniel Forbes, with whom Stapp had also shared conversation at the officers' club the previous night, was given the pilot's assignment; Edwards, despite his experience with that particular aircraft, was relegated to the copilot seat.

Not long after takeoff, the Flying Wing departed controlled flight—rotating and pitching rapidly up in what test pilots call a maple leaf roll—and moments later broke apart in midair northwest of the base, killing all five men on board. It was an instantly indelible black mark against the YB-49. Yet while questions arose about the ultimate viability of the Wing, with many in the test pilot community believing Northrop's design to be fundamentally flawed, the story of Edwards's final night on earth remained only a whisper among the cognoscenti at Muroc. Captain Edwards would in short order come to stand for all the brave test pilots who lost their lives in the line of a very hazardous duty.

After hearing of the accident, Charles Stapp wrote to his son from Brazil:

"When my radio told of the smash up of the flying wing, this morning, I could not help but squirm and pray that you are safe. Ever so often that desert lakebed is in the news. I pray always that you may be kept safe and be allowed to discover ways of making others safe. I realize that all worthwhile things are brought into service of man by sacrifice."

In the same letter, Charles asked whether a promotion might be in the works for his son. The answer came just a few months later. On November 2, 1948, John Paul Stapp was promoted to the rank of major. Shortly after Charles got the word, a follow-up letter from Brazil arrived on December 17: "Para bena! I wondered why they had not made you a major a long time ago. Now how long will it take you to get to be a lieutenant colonel?" Charles was gamely trying to come to terms with John Paul's high-risk career, as was brother Celso. They both tried to dissuade Stapp from undertaking any more dangerous experiments. Charles had never completely accepted the notion of any of his sons serving in the armed forces. He felt that he had raised them to answer a higher calling. "I like to think of you as Doctor rather than Major," he confessed to John Paul. "It seems to me that the vocation of Doctor is much above any thing the military can add to it."

Major Stapp was heartened by the promotion and decided the time was right to request an inspection and review of the project by Colonel Kendricks himself. The team was now confident in the performance of the braking system, and Stapp had gotten a tentative promise that if the team could demonstrate proper operation of the system—and specifically the brakes—then the chief might rescind the prohibition on human testing.

A party from Wright-Patterson, led by Kendricks, arrived at Muroc on an unseasonably steamy, windless day. Stapp had been led to believe that the brass wanted to witness a rocket sled run with a chimpanzee, and that's what the team had been preparing for. But when Stapp met the contingent from Dayton at Main Base, he was informed that the experiment they expected to witness would involve nothing more lively than a parachute dummy.

The colonel and his traveling party arrived in staff cars and stood around awkwardly in the desert heat while the team prepared the rockets for firing. Stapp could see the colonel glancing at his watch and looking as if he wanted to be anywhere else. The run itself, with Oscar Eightball aboard, which went off according to plan, seemed not to impress the visitors much.

When Stapp approached Colonel Kendricks to offer his pro forma thanks for being allowed to stage the demonstration, the colonel brushed off any talk about Stapp's future on the rocket sled. However, the colonel *was* concerned, he said, about the chimpanzees. Specifically, he was worried about bad press from animal lovers if it was ever discovered that the Air Force was subjecting the animals to dangerous experiments. He indicated that he intended to declare the chimps top secret and ordered Stapp to keep them under armed guard twenty-four hours a day. The party departed for Wright-Patterson less than an hour later, and—to Major Stapp's great frustration—without any comment on whether human experimentation would be allowed to resume.

In the days that followed, Stapp assisted the base personnel office in the assignment of round-the-clock guards for the animal compound, but he didn't want the guards inside with Tony Gentry and the chimps. He was going to need more housing. So it was back to East Camp, where Stapp found a little three-room house that he had trucked to the site and set up for both the guards and the trainer. Moving expenses were billed at the established

Stapp (in T-shirt) facing off with his nemesis, Colonel Edward Kendricks (second from right), who told him: "You are risking my neck." Colonel A. P. Gagge listens (in profile, on right). (Photo courtesy of U.S. Air Force)

rate of cigarettes and cigars. Stapp, risking the wrath of the base supply officer, threw in a half-case of chimpanzee beer as a tip.

When it was finished, the new facility was spacious enough not only for
Gentry and the guards, but for two additional animal men provided by Gentry's contractor. One of the new men was a huge American Indian carnival
wrangler and general raconteur. The guy was magic with animals and as
strong as Stapp and Gentry combined.

Another improvement to the site was necessitated by the phenomenal
quantity of sewage created by the chimps. The stuff was piling up. Drawing on lessons learned back at Carlisle Barracks, Major Stapp designed a
septic trench. With the help of the air installation officer's ditch digger,
they created a 400-foot-long, four-foot-deep trench on a slight grade. Stapp
then had the trench filled with nearly two tons of almond-sized nuggets of
coke—a solid fuel made by heating coal to eliminate volatile by-products—
and broken tile on top of the coke, and then covered the whole thing over
with Mojave sand. They fed the sewage outlet from the animal house into
the upper end of the trench. In the hot, dry climate, the animal waste
decomposed completely, and all the liquid evaporated underground with
absolutely no effluent.

Despite improvements to the site, one thing that hadn't changed much
was the rocket sled itself. Stapp, however, had never had much use for the
aluminum shell that covered the seating platform on the *Gee Whiz*. Not only
was it cumbersome, it prevented external photography of the subject at the
instant of maximum deceleration. So he asked that Northrop remove it.
Instead, a single-panel windscreen was installed on the leading edge of the
sled, presenting the same angle and coefficient of drag as the aluminum carriage. The subject would still be protected from windblast but would now be
revealed for the benefit of high-speed telephotography, spotlit by the brilliant
Mojave sunshine. A color motion picture camera was mounted on the floor
of the chassis pointing up at the subject, and stationary black-and-white,
high-speed cameras operating at 2,000 frames per second positioned along
the track would capture overlapping profile views of the subject as the sled
hit the braking section. The team erected a white-sheeted plywood backdrop
across from the cameras to provide a solid background. When they tried the
photography system on a chimp run, the results were of such extraordinary

quality that the engineers could detect the rotation of the head and even the minute stretching of the web restraints.

Stapp made one other improvement that summer: he purchased a used car, a Chrysler. No longer would he have to brave the desert nights to make a house call 30 miles away on a scooter that slowed to near walking speed whenever it encountered a good head wind. Unfortunately, though, the Chrysler turned out not to be much of an improvement after all. As Stapp described it later: "An atrocity. A real wreck of a car and I kept it about eighteen months. It had been made after a strike and it was made with lots of spite and no competence."

If his personal vehicle was less than impressive, the MX-981 braking system was undeniably state-of-the-art, almost certainly the most powerful ever constructed to that point, and it was working like a charm now. The grab-force it could exert was fantastic. Each individual brake unit could apply up to 132,000 pounds of gripping power.

From December of 1948 to the following June, the team amassed an impressive array of experimental data using chimpanzee subjects that survived stops from as much as 160 miles per hour and in a distance of only 18 feet. "It was readily determined," Stapp proclaimed, "that if there were any comparison between tolerance limits of a chimpanzee and man, very high forces indeed could be endured without serious injury." In Stapp's correspondence with his Aero Med Lab colleagues back in Dayton, he was careful never to make overt mention of the animal-subject tests. He had developed some code lingo: "We have completed this week tests in which we took successful shots at about 20 g with three Republicans head first face up on the litter." While some details of the chimp tests—including injuries and fatalities—were kept out of the project log, the team lost at least two members of the original cohort of chimps during the later months of 1948 and the early months of 1949. One female was killed in a high-speed, forward-facing run that exposed her to 59 g's. MX-981 was now working at, and sometimes just beyond, the extreme upper end of the survival bracket.

The engineers held long brainstorming sessions as they investigated improvements to their restraint devices. Proper restraint was the key to preserving lives when dealing with these kinds of speeds and deceleration forces.

Strong, pliable webbing straps were the team's preferred safety material. Three-inch-wide shoulder straps anchored to the frame behind the seat and fastened in front to the buckle of a lap belt were found to be particularly effective. The lap belt was pulled around the hipbones—never the midriff—and attached at a 45-degee angle with respect to the subject's spinal column. A pair of tie-down straps attached to the rear corners of the seat, and went around the insides of the thighs to the belt buckle. The team could watch in slow motion as chimps hit the brakes at high speed, and could see that their restraint-strap configuration arrested deflection of the body from the shoulders down while preventing the knees from diving out from under the lap belt. The chimps were experiencing brutal biodynamic forces, but even at these levels, as long as they were protected from impact with a hard or sharp surface, and as long as they were securely strapped in, they were emerging relatively unscathed.

Soon enough, the team had gotten used to working with the chimps and each man knew each chimp by name. A logbook in the barrack building kept by one of the mechanics tracked the daily activities at the North Base site throughout this period:

> *5 Jan: Made a run today. Maggie in forward-facing position. Everything OK. Cart stopped with only 19 brakes. Approx. 150 G. Animal in good condition.*

> *7 Jan: Made run with Tex. Full stop in brakes. Had Photo Phone in position. Should have very good photos.*

> *10 Jan: Storm and blizzard. Lots of snow—cold. Part of gang could not get to work from L.A. Road blocked.*

> *12 Jan: Still snowing. Whole desert covered with 5 to 6 in. of snow. This is the first deep snow in the history of Muroc. Maj. Stapp left for L.A. Water supply frozen.*

> *18 Jan: Made run today = Ruthie in parachute position. Everything OK until we started to return her to her cage. She escaped and for about 10 min. there were four very busy men. No one seriously hurt. Only scratches and bruises.*

19 Jan: *Lizzie today in parachute position. Everything went OK. Full stop in brakes. Took Liz to Hosp. for Xray. Everything OK.*

20 Jan: *Another run today. Little Maggie. Very good run.*

3 Feb: *Got ready for run, wind got so bad we had to call it off.*

Despite the ever-present challenges of working in the Mojave, the project seemed to have hit its stride. And though Stapp had graduated from Oscar Eightball to live testing, he had begun to imagine test equipment that might represent a middle ground between what essentially were scarecrows bedecked with gauges and living organisms. Stapp submitted a proposal in the spring of 1949 on behalf of the USAF Aero Medical Laboratory for a next-generation test dummy. The proposal specified a 72-inch-tall, 200-pound dummy with a center of gravity and articulation that more accurately modeled the human body in response and appearance. The Air Force awarded contracts to Sierra Engineering Company and Alderson Research Laboratories, where an extraordinarily clever scientist and inventor named Sam Alderson built a prototype, which he called "Sierra Sam," to Stapp's general specs.

Alderson had been building high-quality prosthetic limbs for ex-World War II servicemen for years and had his own unique manufacturing techniques. The new dummy had a skeleton constructed of steel, skin made from latex, and an aluminum skull—with a torso full of strain gauges, accelerometers, and gyroscopes. The Air Force put it to use immediately in ejection seat and parachute design tests at Wright-Patterson, though Stapp knew it was unlikely that production models would be made available to his project.

Then, later that summer, not long after the team had wrapped up its series of tests with the chimps and restraint systems, word made it to Muroc that Colonel Kendricks had been reassigned to the School of Aviation Medicine in Texas. Stapp thought it might be good news for his project, but he couldn't be sure. When he got in touch with his colleagues in Dayton to see if they could help him understand what the change might portend for MX-981, he found no consensus.

While he waited for some direction from his superiors, Stapp's gambling instincts—or perhaps just his zeal to prove the world wrong—led him to the unilateral decision to resume human testing, albeit on a very modest scale. He made two low-speed runs himself in order to validate the restraint harness configuration they'd worked out for the chimps, and when Larry Lambert volunteered again, Stapp authorized two additional runs, neither to be accelerated beyond 90 miles per hour nor to exert more than a 5-g force on the subject. This was almost child's play for the Northrop team by this point. The only wrinkle was that for Lambert's runs Stapp had the seat rotated so that Larry was facing forward. They had done it repeatedly with chimps, but this was a first with a man aboard and the team proceeded deliberately.

On the first run, Lambert was strapped in with the full-harness shoulder-strap setup, the lap belt, and a pair of inverted-V tie-down straps. Everything went as expected. But on the second run, the team removed the tie-downs to isolate their contribution to the system. On that run, when the sled impacted the brakes, Lambert's knees and hips slid forward and his ribcage caught momentarily on the rim of the belt. He winced as he stepped off the sled, but luckily for them all there was no persistent injury and the pain subsided that same day.

Stapp felt they'd learned a lesson about the importance of the tie-downs at a reasonable price—one he would gladly have paid himself—and he included a complete description and analysis of the incident in his report for the brass at Wright-Patterson. Shortly after it was received, he was ordered to report to the Aero Med Lab.

It was late June of 1949 when Stapp knocked on the door of Colonel A. P. Gagge, the new acting chief, and was ushered into what he later characterized as an interrogation worthy of Senator Joseph McCarthy, who had recently been hectoring actors and writers he suspected of Communist sympathies. Stapp faced a panel that ordered him to repeat and verify every detail that had appeared in his report covering the four recent human runs: the two of his own and the two with Master Sergeant Lambert. His impromptu presentation was met with a barrage of detailed follow-ups regarding the procedures used with Lambert, some of which seemed to Stapp overtly hostile in nature. He was informed, finally, that the purpose of the meeting was to prepare supporting documentation for an official reprimand. He was accused of resuming human deceleration tests against orders.

The charge, Stapp admitted, was true. Nevertheless, he told his interrogators, he was progressing with responsible caution. In his view, the minor injuries they had experienced had been caused by problems with the friction brakes. The team had since improved the system. It was now reliable and quite safe, he assured them. Using the chimps, they had demonstrated that they now had an effective restraint system design. He asked Colonel Gagge for an opportunity to plead his case in private.

For the next hour, Major Stapp recounted his efforts to date and extolled the progress the team had made toward both the determination of g-force tolerance and the reliable methods of immobilizing and thereby protecting living flesh exposed to jet-age mechanical forces. This was no time for humility or politesse. He put to work all his powers of persuasion. He laid out his grand vision for future deceleration work and made the point that no serious injuries to a human subject had occurred or were likely to occur as long as he was allowed to do his work without harassment. No insubordination had been intended, he assured Gagge; it was only an excess of enthusiasm. The project was trying to save the lives of the same Air Force pilots who were dropping like flies from the skies of Muroc. Besides, he was, he said—as he had often suggested in the past—on the verge of spectacular results that required only a few more tests, a few more weeks or months, to define the limit of human tolerance.

"From his comments," Stapp wrote later of his session with Colonel Gagge, "it was soon apparent someone who wished me no good had maliciously attempted to discredit me. The end result was that I actually acquired more respect and greater stature in the eyes of the acting chief who immediately set about getting the restraining order on human experiments revoked." He was ordered to stay under a limit of 18 g's for human tests, which puzzled Stapp, but he wasn't going to complain—at least, not now. He left Wright-Patterson with renewed hope. "By September of 1949," he wrote in his personal log, "we were once more in business."

Back at Muroc, Stapp began with a forward-facing sled run, exposing himself to 15 g's right off the bat. The experience surprised him. It was a totally different thing rushing headfirst into the windblast, and then into the brakes. As the rockets fired, he squeezed the hand rests and bit down on the bite block, cringing. There was always a mean pulse of adrenaline prior to these runs. Then the blast flattened him against the seat and a moment later wrenched him with breathtaking force into the 9,000-pound-test webbing as he flew into the brakes at 150 miles per hour.

Back at the Aero Med Lab, Colonel Gagge agreed to send two more sergeants to Muroc to assist Stapp. One was a combat-tested parachutist and the other a photographic systems specialist. Stapp and George Nichols mapped out a series of experiments for the fall. Stapp, they agreed, would always make the first run at each higher increment of speed, deceleration, and rate of onset, followed in rotation by one of the pool of three sergeants who'd by now agreed to serve as volunteers, though none of them relished his turn as a test subject. After listening to their complaints for a couple of weeks—the volunteers particularly disliked the thorough medical exams that preceded and followed each test—Stapp unceremoniously pulled the sergeants out of rotation and, in the absence of any contravening order from Gagge, announced his intent to make the remainder of the runs himself.

The team gave its preferred restraint configuration a workout. Stapp hurtled down the track time and time again over the following weeks using slight variants of the strap system. On one otherwise unremarkable run that achieved 32 g's on September 8, 1949, with the lap buckle positioned a little higher than usual, Stapp experienced acute pain in his chest as he hit the brakes. George Nichols could see the agony in Stapp's eyes as he was helped off the *Gee Whiz*. A post-run examination revealed that he had torn the rib cartilage between his ninth and tenth ribs, and it kept him on the sidelines for three frustrating weeks. But as soon as the pain subsided, he was back at it.

Stapp never stopped worrying about the brass at Wright-Patterson and the g-force limits that had been imposed on the project. To make sure that no excess of caution would once again derail MX-981, he conspired with George Nichols to introduce a "slight error" into calculations that appeared in the project reports that were conveyed back to Dayton during this period. All g-force levels reported would be divided by two. Stapp would later say that he "always followed the directives from headquarters—when they made sense." But he paid a price for his deception; there were those in the command structure who never forgot it. So for a while, his superiors had no real appreciation for the results MX-981 was actually achieving—nor for the risks Stapp was taking—and wouldn't learn the full truth of the project's scope until later.

11

BROTHER, WE ARE THE GOATS

Survival is the capacity to outlive mistakes.

—*John Paul Stapp,* **For Your Moments of Inertia**

MX-981 NEVER SEEMED to run out of technical challenges, and measuring the team's accomplishments in a way that provided defensible data was always near the top of Stapp's agenda. Determining precise +Gx and –Gx values was a case in point. Like most of their telemetry instrumentation, the available accelerometers remained barely adequate at best. Most of what MX-981 had to work with relied on war-era technologies. The team was putting new levels of stress on designs that had never been intended for such conditions. Inevitably, no matter how careful the technicians were, things went wrong. Such was the nature of *things*.

Stapp had filed a standing request with the engineers at Wright-Patterson to send him prototypes of their latest designs. He was always eager to include them in field tests in his search for more accurate results. Which is why, in the fall of 1949, an engineer from the Wright Air Development Center, Captain Edward Murphy, made a brief trip to Muroc to deliver samples of the latest strain gauges to Stapp and team. The gauges were simple electronic sensors, or transducers, and Stapp wanted to attach them to the harnesses the team used to restrain the subjects of the rocket sled tests to assist in measuring g-forces. Murphy arrived at the North Base site on November 2 with four of these unimpressive-looking gadgets.

Prior to the following day's sled run, this one using a chimp test subject, one of David Hill's assistants installed the new strain gauges on the shoulder straps. But, after the run, none of the gauges registered anything at all. All four had failed. When the team took them apart, they saw the problem immediately. The transducers had been wired backward and the electronic signals had cancelled each other out. As Hill put it: "Murphy was kind of miffed off," and Murphy loudly blamed the problem on the technicians back at Wright-Patterson who'd assembled the gauges.

Murphy's disgusted comment, as those who were present recall it, went something like this: "If there's any way those guys can do it wrong, they will." Murphy returned to Dayton the next day, taking the failed strain gauges with him. This much of the story is generally agreed upon. What happened to Murphy's casual remark in the following days, months, and years is not.

A few weeks after the incident, a reporter asked Major Stapp how such a dangerous project had managed to avoid serious injury. Stapp's reply was: "We do all of our work in consideration of Murphy's Law." He went on to state the law: "If anything can go wrong, it will." It was Stapp's way of saying that the MX-981 team wanted to anticipate failure scenarios, always assuming a worst-case outcome, and to fix mistakes before anybody got hurt. It was a simple, common-sense axiom and standard engineering operational procedure. Stapp had been coining such "laws" ever since he'd come to Muroc. One example is known alternately as Stapp's Law and Stapp's Ironical Paradox: "The universal aptitude for ineptitude makes any human accomplishment an incredible miracle."

Stapp often came across as an oddball professor, one with a theatrical flair. George Nichols recalled Stapp tossing out impromptu laws—along with puns and aphorisms of all types—around North Base almost constantly. "He had an extremely unique sense of humor," Nichols said. "He started this whole business of laws. Now you've got millions of them."

Stapp never claimed to have originated Murphy's Law, though he was clearly instrumental in popularizing it. The controversy arose over George Nichols's version of the events, in which he accused Edward Murphy of failing to properly test the faulty strain gauges before bringing them to Muroc and of then compounding the problem by ungraciously blaming the failure on anonymous technicians. Murphy reacted badly, accusing both Stapp and

Nichols of taking credit for his law, which he was dead set on branding. He even began a campaign to get a plaque erected at West Point, where he'd studied engineering, commemorating Murphy's Law and making it clear that Murphy had in fact originated it.

Nichols was offended by what he saw as Murphy's efforts at self-aggrandizement. Sure, Murphy had uttered the original phrase, but it had been a throwaway. The way Nichols saw it, the MX-981 team—or, more accurately, John Paul Stapp—was responsible for crafting the statement into a useful adage. They had rightfully named it after Murphy, and, in Nichols's estimation, that's all the credit he deserved. Murphy asked Nichols and Stapp to affix their signatures to his plaque. They both declined.

The incident remained a long-term annoyance to Stapp, who felt the whole business had become a distraction. He didn't want to discuss it, nor did he want to be associated with Murphy's Law. But he always made sure that those who worked on MX-981 never forgot it.

Though the team continued to evaluate a variety of new test equipment—restraint technology in particular—they also continued to push aggressively forward in quest of the human tolerance limit. By Christmas of 1949, in out-

Stapp hits the brakes in a high-speed, forward-facing run at Muroc. The team erected white backdrops to provide maximum contrast for the photography. (Photo courtesy of U.S. Air Force)

right defiance of Colonel Gagge's 18-g limit, Stapp had endured a full 35 g's of deceleration with a 500-g onset in a forward-facing run, and the same force with an incredible 1,000-g onset facing backward. They had *doubled* the accepted human limit of exposure to mechanical force.

When Stapp finally released the results of the tests using the actual –Gx forces he'd endured, he got the Air Force's attention. MX-981 was suddenly being referred to in the halls of the Aero Med Lab as "one of the, if not *the* most, important projects in aero medicine today." It was amusing to Stapp, seeing how the sacrosanct 18-g limit seemed to have been completely forgotten. That same December, the director of research at Wright-Patterson wrote to Stapp soliciting his opinion as to whether they ought to call a press conference to lift the veil of secrecy, to share their accumulated data with the wider scientific community, and to tell the world about what MX-981 had been doing and what they had learned. Stapp, sensing that he had suddenly acquired some unexpected leverage, fired a reply back to Dayton suggesting that they hold off. He wasn't yet ready to announce his results; the data still needed analysis. Stapp then asked for one more year to complete the tests he envisioned, one year with adequate resources and without additional restrictions.

While Stapp might have felt the timing was right to push his advantage, he could not convince the Air Force to leave him alone, nor was he given the additional time he'd requested to conduct more tests before going public. He was ordered to meet the press under the joint sponsorship of the Air Force and the Northrop Company, and was to be forthcoming about everything except the role of the chimpanzees. The animal testing was still top secret and off-limits.

In Los Angeles on January 7, 1950, a day after making a forward-facing rocket sled run, Major Stapp faced a roomful of forty reporters. He made a brief opening statement and proceeded to answer questions for another two hours. The reporters found Stapp imminently quotable. They loved the way he described the sensation of hitting the brakes on a backward-facing run: "It feels much the same as being punched in the back of the head with a boxing glove." Only one question approached the forbidden topic. A reporter, who was aware of USC's centrifuge program and the staff's occasional use of goats as test subjects, asked Stapp why MX-981 didn't simply use animals for their testing. Stapp handled it like a pro. "Brother," he said, looking the reporter in the eye, "*we* are the goats on this one."

The stories that appeared in papers around the country the next morning mentioned nothing about goats or chimps, and were overwhelmingly favorable, enough so that the Air Force ordered Stapp to make a special trip back to Wright-Patterson for the purpose of delivering a formal presentation on his studies at the Air Materiel Command General Staff meeting, and to follow up with a string of working meetings with research groups around the base. One particular major general, after witnessing Stapp's films, remarked on the bravery of a particular chimpanzee and suggested that it ought to be put up for a military decoration. Stapp replied: "General, I am sure my big chimpanzee would like to meet you. I believe the two of you have something in common." Later, he admitted to his colleagues at Muroc that he had worried for a moment the insulted general would order him shot at sunrise.

One of Stapp's meetings on that triumphant tour of Wright-Patterson research groups was with an aircraft laboratory team focused on military transport seating. Stapp had kept up a nearly two-year running argument with the officer in charge about the design and orientation of aircraft seats. Based originally on intuition that Stapp felt had, by 1950, been thoroughly validated by the deceleration tests at Muroc, he again made the case that passengers in air transports be reoriented 180 degrees. In other words, they should sit facing backward. He knew firsthand that being slammed into the back of your seat was far better than being pulled forward against the seat's restraints. In addition, the seats should be built to withstand much higher g-force levels than then-current designs provided. The research officer maintained that each 1-g increase in seat strength would require an additional 15 pounds of added weight—which translated directly into added manufacturing cost and operational expense. Considering that man's tolerance to deceleration was still undefined, the argument went, it made no sense to alter the 6-g-resistant forward-facing seats of the day.

Stapp invited the lab group to review his latest data and watch slow-motion films of his experiments. He was able to show that human subjects could absorb much more than six times the force their seating could withstand, and that the current design limitations were based on nonsense. Before leaving for Muroc, Stapp presented a formal report on the need for advanced-strength, rear-facing seats in transports. In short order, the report was endorsed by some twenty Air Force generals, and a long, sometimes frustrating process of officially altering strength standards and reorienting transport aircraft seating was begun. "Such a change in standards," Stapp remarked, "was an operation on a scale with canonizing a borderline saint for sainthood."

Following a solid year of intensive redesign work, the Wright-Patterson lab produced a new rear-facing aluminum seat. It was the first tangible result of MX-981's work on deceleration forces. Tests proved the seat capable of sustaining a full 16 g's. Equally impressive, and in spite of the research officer's predictions, it weighed only two pounds more than its predecessor. Stapp would spend the ensuing weeks and months proselytizing for the installation of the stronger seat in all forms of aircraft. Eventually, the Air Force would devote three hours in the standard flight surgeon's curriculum at the School of Aviation Medicine to a description of the research that led to the new seat.

In March of 1950, back at North Base Muroc, the rocket sled tests were interrupted when a careless garbage truck driver severed the power and telephone lines to the site. Stapp took the opportunity to have the men make some much-needed improvements to the project buildings, and to orchestrate a return trip to the abandoned East Camp. With a crane and a flatbed truck, Stapp retrieved the various parts of the 30-foot observation tower he'd seen there three months earlier. The installation of the tower a few hundred feet north of the braking section of the track gave the team a bird's-eye perspective they'd been missing and allowed them to capture valuable new camera angles of the crucial instant of deceleration impact. When the rest of Muroc found out that Stapp had an aircraft control tower out at North Base, however, questions began to circulate about how it had gotten there. The only answer anybody ever got from Stapp was that it had blown there in a dust storm.

Stapp's patchwork research laboratory in the desert had come a long way from the powerless, waterless waste of sand and creosote bush he'd first encountered three years earlier. He had even managed to secure funds for a new and improved, climate-controlled chimp house after several of the residents had come down with pneumonia. The site was now quite presentable, and visitors were beginning to show up on a regular basis. At one point in the spring of 1950, Prince Bernhard of the Netherlands was making a tour of U.S. Air Force bases in preparation for his nation's purchase of a number of American military aircraft. Stapp and the MX-981 team were asked to host a simple demonstration run for the Dutch entourage. Because the chimps were still an official secret, Stapp had his team pull Oscar Eightball out of storage. He instructed the technicians to lock down the bolt securing Oscar's right elbow and to loosen the bolts in the right shoulder and wrist. He then had Oscar dressed in a crisp Air Force uniform and a cap glued to the top of his head.

Stapp positioned the prince and his entourage in the standard visitors' observation area some 35 feet from the track and directly across from the braking section. The team used a couple of the old surplus JATOs and fired the sled at 100 miles per hour. When it hit the brakes, Oscar's right arm snapped up with the momentum, the forearm locked at a right angle to the upper arm, carrying the hand smartly up to the bill of the cap in a textbook salute as the sled passed the visitors. The prince returned the salute, delighted with the entire display.

The demo run for the prince had been mostly a lark, but it suggested to Stapp that he might stage another demonstration—this one for the benefit of personnel from the Northrop plant who were scheduled to visit the project—that would be intended not as an experiment but as a way of illustrating the sheer power of the mechanical forces facing pilots of contemporary aircraft. Stapp and his team had done a good job of showing how these forces could be counteracted, but what would it look like if gravity won the battle? Stapp selected a crash harness that had been sent to the project for testing—one in which he had little confidence—and strapped Oscar into the seat, facing forward. The team then replaced the steel and aluminum windscreen with a one-inch-thick, tongue-and-groove pine-plank panel. The brakes were adjusted so that the sled would hit about 32 g's of decelerative force maintained for a tenth of a second.

On test day, with a large group of spectators looking on, the sled roared through five seconds of explosive acceleration and smacked hard into the brakes. As Stapp had intended, the harness snapped clean and Oscar sailed forward through the pine windscreen, piercing the wooden barrier as if it were paper. Splinters rained down from a sawdust cloud as the sled came to

A demonstration run "disaster" Stapp staged for the benefit of Northrop executives. The sled hit the brakes and Oscar Eightball—restrained only by substandard belts—sailed through the pine windscreen. His head was found in the desert 240 feet beyond the end of the track. (Photo courtesy of Air Force Test Center History Office)

a grinding halt. The project notebook captured the result: "The restraining harness was broken in several places and the dummy completely dismembered. The torso traveled approx. 195 feet down the track and the head about 240 feet from the brakes." One of Oscar's legs was recovered from a pool of oil near the emergency brake cable. It was an amusing spectacle in some ways, but the project team, and the men from the Northrop plant, were chastened. What Oscar had suffered was what a human subject could expect if things ever went terribly wrong.

Stapp wrote a tongue-in-cheek article about Oscar's disastrous run for the base newspaper titled "A Tale of 'Whoa.'" The idea was to help educate pilots on the proper use of their crash harnesses, and to encourage them to strap in tight every time they climbed into a cockpit. "Have you been an Oscar Eightball lately," he asked, "or do you use your harness when you fly?"

Throughout the hectic months of rocket sled tests in the spring and early summer of 1950, the notion that his research might also be applied to ground vehicles had never been far from John Paul Stapp's mind. Airplanes had demolished the sound barrier, but the cars and trucks rolling off the assembly lines in Detroit were also moving faster and coming out of the factories at an exponentially higher rate. In the event of a crash, those cars and trucks represented a cross between a torture chamber and a guillotine: they had no restraint systems and their interiors bristled with hard, sharp surfaces, their occupants surrounded by walls of glass. Stapp had recently come across a study of car accident rates conducted by Dr. Fletcher Woodward at the University of Virginia. Woodward cited advances in pilot restraint technology as a model for the auto industry. It was time, he argued, to shift focus from the driver to the vehicle. During his nights at the Desert Rat, when he wasn't called away by some medical emergency at Kerosene Flats or elsewhere, Stapp occasionally conjured up memories of the girl from Baylor and tried to visualize the collision that had taken her life. He'd been by the very intersection at Hollywood and Vine on a couple of occasions and had studied the route her parents' car must have taken, and seen the approach route of the driver who hit them.

Now he could see how he might have saved her.

12

EVICTED

Nobody will ever get that high on experiments again. We got clear to the edge with it.

—John Paul Stapp, to an Air Force interviewer

A S STAPP WAS SCHEMING on how to expand the scope of his safety research, the world around him was again convulsing. The Soviet Union had detonated its first nuclear bomb the previous fall, and now events on the Korean Peninsula—where North Korean forces were poised to attack the South—had seemingly reached a point of no return. In April of 1950, with a commitment of support from Mao Tse-Tung in China, Josef Stalin authorized North Korean Prime Minister Kim Il-Sung to invade. Only five years removed from the end of a world war, it appeared that a new chapter of global insanity was about to be written. More immediately ominous for Stapp, however, was a new commander who had recently arrived at Muroc.

General Albert Boyd from Tennessee was a legendary test pilot who would come to be known as the father of modern flight testing. Tall, gangly, and balding, with bushy eyebrows and a chiseled jaw, he'd set a short-lived world speed record back on June 19, 1947: 624 miles per hour in an F-80R Shooting Star. He was also the man who'd selected Chuck Yeager for the first supersonic flight. Boyd would put his formidable stamp on everything that happened at Muroc, including the identity of the base itself. One of his first official acts was to have it renamed in honor of the late Captain

Glen Edwards. On January 27, 1950, Muroc Army Air Base officially became Edwards Air Force Base. Pancho Barnes raised a small stink, maintaining to anyone who would listen that Danny Forbes ought to have been given the honor in light of Edwards's behavior at the Happy Bottom Riding Club in the hours before the accident. In fact, recognition for Forbes had come seven months earlier, when Topeka Air Force Base in Kansas was renamed Forbes Air Force Base.

Before General Boyd's arrival in California, Stapp had received cautionary instructions from the Aero Med Lab to resist any efforts the new commander might make to absorb MX-981 into his organization. The caution was justified. In late March of 1950, Boyd sent word to Stapp that he intended to witness a human test on the rocket sled track. He had been informed that the tests were thought to be approaching the tolerance limits of both humans and equipment, and he wanted to see it with his own eyes. As Stapp put it in a letter to a colleague back at Wright-Patterson: "He *commanded*; he didn't even ask what we were going to do." The demonstration test Stapp planned was a challenging one, designed to produce nearly 38 g's with a 1,000-g rate of onset. When Boyd showed up for the run on April 6, Stapp—using three rockets, in a forward-facing position—was shot down the track at 180 miles per hour. General Boyd, who'd brought his wife and staff along, stood stiffly near the brake section looking a bit skeptical.

Stapp later described what happened when he hit the brakes: "I was struck by the most severe impact that I had ever encountered in any sled ride. The harness inflicted a paralyzing slam to the shoulders and hips. My hands came off the hand rests and were held only by flexible nylon loops at the wrists. Like a rubber band, the tightly stretched loop on the right wrist rebounded, slamming my wrist violently against the right-hand handhold. I felt a numbing crushing pain as though my hand had been violently struck with a crowbar."

He knew immediately the wrist was broken, but he kept that knowledge to himself as he climbed off the sled. He offered a smile and a little bow to the spectators. Fortunately, he didn't have to maintain appearances for long. Boyd and entourage left almost immediately after the test. With the general out of sight, Stapp had himself driven to the base hospital for examination. The radial artery and vein had been smashed, and the wrist had swollen dramatically, turning a deep purple. A young doctor splinted it to protect the bone until the swelling subsided enough for a proper cast to be applied.

Stapp swore the hospital staff to silence. In fact, Stapp had managed to cover up a number of personal injuries suffered on the sled to that point, including the torn rib cartilage, a bruised collarbone, a hairline fracture of the coccyx and at least two suspected concussions. Because Stapp prohibited his team from recording these injuries in the project log, the precise dates on which they occurred are mostly unknown. George Nichols, however, knew about them and he fretted almost constantly that something worse might be in store.

Nichols had Stapp's car delivered to the hospital, and Stapp drove himself to the officers' club, where he settled into a corner table sipping ginger ale and joking about the wrist. Later that night, he joined the base dental surgeon and his wife for dinner, making a point of being seen in good spirits and assuring everyone that he was perfectly fine. But he made an appointment with the dentist for the following week. By his own count, he had lost or cracked open six fillings on the latest sled run. He cut a classic Curbstone Clinic deal: Stapp agreed to provide pediatric care for the dentist's children and help him secure an electric casting furnace the Air Force had refused to authorize in return for a mouthful of brand new fillings: "enough gold," in Stapp's estimation, "to change the weight and balance of my head, not to mention increasing its intrinsic value."

Soon after, Stapp plotted his own therapeutic regimen for the throbbing wrist, and became obsessed with the problem of how to heal it properly. He decided to forgo a hard cast, suspecting that while the bones might heal, the immobilization of cartilage and tendons would cause minor atrophy that would slow down the overall healing process. Refusing painkillers on the theory that they forestalled the body's ultimate tolerance of the injury, he removed the splint and bandages every two hours to soak the wrist in hot water while gently exercising the fingers to promote circulation. He kept up most of his regular duties at the project site and had the wrist x-rayed once a week. On only the second x-ray, Stapp could see that the three small bone fragments that had come loose in the joint were already being reabsorbed into the body. The discoloration shortly disappeared and the pain began to subside.

Some three weeks after the injury, Stapp traveled back to Dayton to report on his recent results, and was able to hide the effects of the broken wrist from his superiors. His return to California was made particularly memorable by his colleagues in the Photo-Engineering Division, who secured him a coveted seat in the nose of a B-17 outfitted with the latest aerial cameras. The pilot had permission to fly at extremely low altitudes—right on the deck when conditions were right—all the way across the country. At no time did the flight exceed 1,500 feet above the ground and the B-17 crossed the Rocky Mountains practically skimming the peaks. It recalled for Stapp the excitement of his first flight back in Austin in his college days. The pilot flew so low, he wrote, "that I could see the color of the squirrels' eyes. It was a great thrill to see a mountain coming at me with all the formidable rocks, trees, rushing closer as if reaching for the undercarriage of the plane, then just beyond the crest to have the terrain fall away as fast as a great green waterfall." The experience rejuvenated him and he disembarked at Muroc with a new optimism.

· · ·

While Stapp and Project MX-981 had gained some real stature within the halls of the Aero Medical Laboratory, the Battle of Muroc—now the Battle of Edwards—raged on. In June of 1950, Stapp was summoned to a private meeting with General Boyd. Boyd, reportedly still bitter about Northrop's Flying Wing debacle and the death of Glen Edwards, informed Stapp that he, Boyd, would shortly assume responsibility for funding and operation of all work being conducted on the base, and that included all of Northrop's projects. Boyd referred to MX-981 as a "tenant organization." Stapp argued the case for extending Northrop's contract and offered glowing reports on the performance of George Nichols and team. These were extraordinary men, he said, and they were giving everything they had in the cause of saving the lives of Boyd's airmen. Boyd was implacable. He wanted Stapp to train Edwards personnel to operate the deceleration project. It would be like starting over with a novice crew, and Stapp flat-out refused. He left the meeting certain in the knowledge that the glory days of the Aero Med Lab's testing in the Mojave Desert were numbered. Subsequent negotiations between the Aero Med Lab at Wright-Patterson and the command at

Edwards were no more successful. The chief of the Aero Med Lab's Special Projects Unit reported to Colonel Gagge: "Boyd and all his people . . . were quite unprepared to compromise."

There may have been more to General Boyd's banishment of Northrop from Edwards Air Force Base. The low-drag, high-lift aircraft design represented by the Flying Wing had been Jack Northrop's singular passion, and he had lobbied hard to have his tailless design selected as the next-generation bomber platform for postwar airplanes. When, following the high-profile crash of the YB-49 that killed Glen Edwards, the federal government destroyed all the prototypes they'd ordered from his company due to lack of confidence in the aircraft's basic airworthiness, Northrop was devastated. He and others suspected that political pressures and backroom deals between Northrop's rival Convair and certain high-ranking officials in the Air Force had sabotaged evaluation of the Flying Wing. In fact, Northrop's basic concept for the Wing, based on solid design principles, would eventually find some vindication when it was incorporated into the B-2 Spirit—a stealth heavy bomber purchased by the Air Force in the 1990s. And when fossils of a giant tailless pterosaur (the largest animal ever to fly, with a wingspan of nearly 40 feet) were discovered, paleontologists named it *Quetzalcoatius northropi*.

Stapp's standoff with Boyd played out during a particularly tense period not just for the Air Force but for all the military services, as events across the Pacific heated up. On June 25, 1950, the massive North Korean army had rolled across the border into South Korea in force and captured Seoul three days later. By July 1, the first U.S. ground troops had landed on Korean soil and a new war—this one a dangerous proxy conflict of nuclear powers between the United States and the Soviet Union, with a restless China poised to intervene at any moment—was on. Anxiety was high and it affected operations everywhere on the base. Given the tensions overseas and following his meeting with an impatient Boyd, Stapp began to suspect that the series of four high-speed rocket sled tests that began in October 1950 was likely to be MX-981's swan song.

Still smarting from his shattered wrist, Stapp had asked Northrop to build him better hand rests that would allow him to resume testing even before his wrist had completely healed. The first run went off without a hitch, except for the fact that a data system failure kept the team from establishing the precise speed. The second run in the series was a bit more eventful. One of the three rockets malfunctioned, and the sled accelerated to an anemic 130 miles per hour at burnout, entering the brakes at only 89 miles per hour. The brakes, having been preset to retard a vehicle traveling at almost twice that speed, brought Stapp to a dead stop in just 18 feet. Another telemetry system breakdown kept them from recording the g-force load, but the restraints performed perfectly and Stapp walked away a bit shaken but unhurt. To ensure that the incident was never repeated, somebody had the idea to attach signal flags to the rocket nozzles. The flags would jump out at the moment of ignition and would give the team a visual warning if any of the rockets misfired—in which case they would be able to quickly release the pressure on the brakes to lessen the impact.

With the new warning system in place, Stapp moved on to the third experiment that fall, which called for him to wear a 21-pound backpack parachute to allow observation of the performance of the restraints against the extra—and redistributed—weight and bulk. Stapp pulled 30 g's, but suffered nothing more than momentary disorientation and some soreness in the back. The fourth run in the series had Stapp sitting atop a seat-pack parachute, which elevated him five inches in the seat and presented an altered profile for the restraint harness. When he hit the brakes at about 150 miles per hour, he found himself being tipped forward with a powerful force that resulted in sudden increased weight surging into his forearms and down against the hand rests. As the sled pulled up to a stop, Stapp was aware of a familiar deadening pain in his injured right wrist and a new biting pain in the left.

On the ride to the hospital, he gritted his teeth and set the broken left wrist himself, snapping the bones into place. X-rays revealed that the injury was a Colles fracture, about three-quarters of an inch closer to the hand than the break in the other wrist. Stapp made a Curbstone Clinic house call the following evening and found that even one-handed he could fill a syringe and administer an injection. A couple of nights later he managed to deliver a baby boy at one of the off-base housing projects. But when Stapp made a quick visit to Texas a week later to see Wilford and Margaret, Wilford recalled his brother being unable to lift his own luggage.

Back in the Mojave on the dark night of January 25, 1951, MX-981 experienced its first human fatality, though the incident occurred miles away from the rocket sled track. Stapp had continued to use the occasional human volunteer for his deceleration tests, and Master Sergeant William Rhea was one of the most prolific subjects, having made fourteen rocket sled runs including four in which he sustained more than 30 g's. That winter night, long after the Northrop men had gone home, a car driven by Rhea, with another sergeant in the passenger seat, was hit by a freight train at an unmarked railroad crossing not far from the Main Base gate. Neither man was wearing any sort of restraint device. Rhea was slammed violently into the steering wheel before being thrown from the car and landing on his head. The passenger survived, but Rhea died nine days later from his injuries. Years later, Stapp pointed to this incident as a watershed moment in his own thinking: "The forces he sustained in this fatal accident were no greater than he had walked away from in sled experiments. It was this accident that focused my attention on the automobile crash problem and the possibility of applying the results of Air Force airplane crash research to better protection for automobile occupants."

About this time Stapp began to think seriously about a full-scale automotive crash test program using both human volunteers and animals, but he wondered if he would ever be able to convince the Air Force to fund it. On the other hand, it was nearly impossible to imagine ever getting access to the necessary resources as a civilian researcher—not just tracks and test vehicles, but test subjects like his chimps. As time had gone on, he had developed affection for a number of the project's chimps, but for none more so than a smallish 85-pound, bronze-faced female named Lizzie, who seemed to have developed a crush on Major Stapp. "She would jump up and down in excitement and wave as I came in the room. She would reach out a hairy hand and pull my head toward her, kissing my cheek tenderly."

Even so, Stapp had interacted with the chimps long enough to understand not only their strength but also their capacity for violence. He had, for example, once witnessed Lizzie, small and docile as she appeared, pick up a pair of iron pliers and quite effortlessly bend the handles until they were at right angles to the jaws. On another occasion, Stapp saw a larger female snatch a newborn chimp away from its mother and quickly kill it. Both

Stapp and Tony Gentry tried to intervene, but it happened too fast. These were not pets.

While the Air Force continued to fear what might descend on them all if the public ever found out about the chimp experiments, those who worked with Stapp have always maintained that his respect for the well-being of any living research subject was as unwavering as was his expectation that they be treated humanely. Stapp continued to cook the chimps hot meals and to make sure they had the best medical care he could provide.

During an epidemic of bacterial pneumonia, Stapp made a trip to the site every night for three weeks to administer antibiotics. But, despite anesthetics and protocols that were in strict accordance with the American Medical Association's "Rules Regarding Animal Care," the team lost some more of its chimps on high-speed tests in late 1950 and early 1951. Some of these runs were gruesome enough that Stapp was careful to ensure that no outside personnel were in eye- or earshot of the track, while other deaths were simply attributed to mistakes.

> One subject was dead from too much pentothal about ten minutes before the run came off. Post revealed nothing gross that could have proven fatal. We had inspectors present and did not think it advisable to have any screaming at the brakes. Finally, on the last run, there were no VIP present, and our tough little Ruthie was allowed to come to just before the firing, and she survived, making outcries to indicate that the stop in the brakes did not knock her out. We took her off, got the X-rays, gave her Demerol and by night she was up and bouncing around, yipping in her usual manner.

Chimps, however, weren't the only animals the team was working with by this point. In the early months of 1951, Stapp had arranged, through his contract with Tony Gentry, to buy some pigs. In many ways, pigs made excellent human surrogates for certain kinds of experiments because their internal organs are similar in size to a human's. "We have already obtained 4 pigs," Stapp wrote in a letter to a colleague at the Aero Med Lab, "and will get 11 more soon. We are getting 40-50 lbs size to fatten up to the larger more expensive 150-200 lbs size for experiments." The team modified the litter in the *Gee Whiz* to accommodate the new test subjects. Again, the animals were used to establish lethal limits of exposure to deceleration in a variety of orientations. While the pigs were covered by the same secrecy guidelines as the chimps, the ever-practical Major Stapp added a new wrinkle. He had a charcoal pit dug alongside the rocket-sled track. Whenever one of the pigs, which were routinely anesthetized prior to a run, perished as the result of an experiment, the crew would fire up the coals and butcher the animal on the spot. Stapp called it Operation Barbecue and invited select friends from the base to join MX-981 for an impromptu feast.

One Muroc friend with whom he'd continued to discuss aeronautical problems throughout his time on the base, Chuck Yeager, collaborated with Stapp on a novel experiment in the last week of May 1951. In an F-89A fighter with the canopy removed, the two resolved to see what they could learn about the effects of windblast. Stapp took the radio-observer's back seat. Without oxygen or any special protective gear, Stapp and Yeager made several flights to altitudes well above 20,000 feet.

"I dove the airplane up to 500 knots," Yeager said, "and he was buffeting. So was I. It was noisier than hell and real breezy, everything flopping around even up front." There had been speculation in aviation circles that no one exposed to that level of extreme windblast could remain conscious. Yeager and Stapp were happy to disprove it. The shoulders of Stapp's flight jacket were sliced open by the cutting force of the wind, but he was fine.

"We both took beatings," Yeager went on, "but nothing excessive. We certainly didn't become incapacitated. Now that solved a question mark." Stapp was able to satisfy himself that the windblast ejecting pilots would have to endure—at these speeds, at least—was quite survivable. It was one more piece of a much larger puzzle.

Stapp had turned forty the previous summer, and as his concept of g-force and crash testing expanded, his ambition was growing as well. His father was about to hit his own milestone. "Perhaps I shall have time someday to get sentimental about being seventy years old," Reverend Stapp had written to his sons. "What will I do now, with no deposit of Bibles, no churches to visit, no conventions to plan, no reports to make, no long list of moneys to receive and dispatch, no long trips to fields and meetings, no Institutes? Well I don't know."

Life in Brazil was as tough as ever. Bubonic plague was on the rise. Poverty and hunger were widespread, and the political situation was increasingly unstable. With an unpopular president and underground brigades of young Communists becoming bolder, the country's future seemed uncertain. Through it all however, Charles kept tabs on his oldest son's career through radio reports and the pages of the news magazines. He understood that John Paul was not allowed to divulge any of the details of his work. "What are you going to do next? The last time I saw you, you were flying too high. The last time you wrote to me you were flying too fast. Now come clippings and Newsweek saying that you are stopping too quick. I hope you make out to die of old age. There surely are easier ways of making a living."

Meanwhile, back at Edwards Air Force Base, John Paul Stapp made his final MX-981 sled run on June 6, 1951. The Northrop team had overhauled the braking system yet again and had run some spectacular experiments with the chimps to help work the kinks out. Using four rockets and a variety of subject orientations, they accelerated the sled to 220 miles per hour and stopped it in a distance of 24 feet. Airplane speed to a standstill in less than the short side of a basketball court. Those at Edwards who heard the numbers wanted to see the sled in action. On Stapp's last run, a forward-facing test that sent him speeding into the brakes at more than 180 miles per hour and brought him to a stop in about 31 feet—exposing him briefly to a punishing 45.4 g's—the team's new helmet-restraint webbing failed and his head snapped violently forward, his chin smacking his chest. As Stapp climbed off the *Gee Whiz* for the last time, his vision was fading in and out.

He had already made several useful discoveries about the effects of massive deceleration on the eyes. On backward-facing runs resulting in stops higher than 18 g's, he'd experienced what he'd referred to as "white outs." This phenomenon occurred when, during the moment of maximum deceleration, gravity sucked the blood from the tiny capillaries in the eyeballs and pooled it up behind the irises. The forward-facing runs resulted in the opposite effect, as blood in the eyes rushed outward and smashed up against the retina. If the force was great enough, the capillaries burst and hemorrhaging occurred. Stapp called this effect a "red out."

As he walked away from the sled that final time, staggering just slightly, Stapp tried to hide it, but the men could tell something was wrong. It alarmed George Nichols enough that he summoned an ophthalmologist from the base

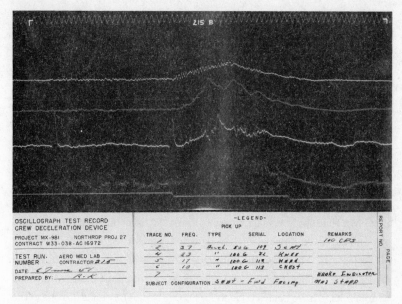

Oscillograph print from Stapp's final rocket sled ride at Edwards Air Force Base, June 6, 1951. The trace along the bottom indicates the point at which the Gee Whiz *hit the brakes; accelerometers attached to the sled and to various parts of Stapp's body provided deceleration data for the traces above. (Photo courtesy of U.S. Air Force)*

hospital to the test site. As the doctor examined Stapp, he observed intermittent spasms of the retinal veins. Stapp continued about his business, but experienced blurry vision for the rest of the day.

The men of the Northrop crew had known for a while that their time working on MX-981 was winding down. Stapp rarely made the daily entries in the project logbook himself, but the entry for June 7, in Stapp's handwriting, reads: "Preparations for turning project over to base personnel. Inventory for contractor and project equipment begun." It had been a career-highlight experience for all of them. Two of the men in particular, George Nichols and Jake Superata, had grown very close to Major Stapp and were already regretting what they assumed would be the end of their association with him. Stapp's leadership style reminded Nichols of another man he revered: Jack Northrop. "If ever two people were similar, it was Stapp and Northrop. Both of them could have gone into any field they wanted. They had tremendous mental ability, yet both were extremely humble. No one ever worked *for* them—you worked *with* them. Nothing ever sounded like an order, even if you knew that's what it was. It always came out as a request. And when you did it, they said, 'Thank you.'"

Stapp's colleagues had frequently come to see him as an inspirational figure. Nobody believed in the worth of his mission like John Paul Stapp and, though he might on occasion appear distracted or lost in thought, and he was apt to lose his temper from time to time, when he showed up for work he came in with a palpable enthusiasm that fired everybody up. He made it fun with his sideways sense of humor and combative "us vs. them" mind-set. "You had to be aware of his humor in order to communicate with him," a project engineer explained. "But those who worked with him really admired him." Not that Stapp was ever easy to work for. He demanded professionalism and expected anyone, civilian or military, reporting for duty on one of his projects to be on time and ready to go. A couple of the sergeants he'd used briefly as test subjects had chafed under Stapp's command, but the Northrop team seems to have been a good fit for Stapp's style. Only once did one of the MX-981 contractors appear at the test site obviously suffering from a hangover. Stapp's punishment was effective: he assigned the individual to dig up the carcass of one of the chimpanzees who'd died as the result of a rocket sled experiment and ordered the man to spend the morning boiling the chimp in a large tub. Stapp had decided he wanted to analyze the bones for stress fractures and needed to separate the skeleton from the tissues. "There was a lot of gastric activity while he was doing it," a chuckling Stapp said of the hung-over contractor, "and he never did it again."

The whole crew gathered for a year-end celebration the night after the final run at a restaurant in Palmdale, a small desert town about 30 miles from Main Base Edwards. They ordered steaks with all the trimmings and toasted the project's successes. Stapp had talked his dentist friend into using his new mold furnace to create silver castings of a miniature rocket sled Stapp had carved himself and rectangular medallions for every member of the original ten-man crew, each medallion embossed with a raised image of the *Gee Whiz*. One additional casting of the miniature sled and medallion had been made of gold, and Stapp presented those mementos to George Nichols, in Stapp's estimation the most indispensable member of the team. It killed Stapp to see this group disbanded, but he was powerless. At least the

Northrop men had good jobs to return to. As for Tony Gentry, the one non-Northrop project employee to whom he felt a real debt, Stapp helped arrange a position for him with the USAF Primate Center in Austin. As the Palmdale celebration gained steam and impromptu speeches droned on into the night, Stapp found himself stricken with a severe headache. He left the party early, driving himself back to the Desert Rat, where he tried to sleep off the pain.

The next morning, however, he woke to find that a charcoal-colored blind spot had developed in his right eye. He tried not to assume the worst, but as the day wore on his vision seemed to be deteriorating. When Stapp went for a follow-up exam, the ophthalmologist identified a hemorrhagic blot on the surface of his right retina, and several lesions in the margin of the left retina. While Stapp returned to work, keeping any worries to himself, by early evening half the visual field of his right eye was covered by the growing dark spot, and the headache from the night before had returned. As Stapp drifted off to sleep that summer night, he counseled himself to be prepared. If things kept getting worse, the doctor had advised, he was in real danger of losing his right eye.

The following day the team got the not unexpected official word that the jurisdictional dispute between Edwards Air Force Base and the Aero Medical Laboratory at Wright-Patterson had finally been resolved. The Battle of Muroc/Edwards was over. General Boyd had prevailed, and Project MX-981 was defunct. Stapp was immediately reassigned back to Dayton and the Northrop men returned to the plant in Hawthorne. While General Boyd would eventually lose his command at Edwards, he had managed to close the book on Aero Med Lab testing and return the Antelope Valley to the exclusive provenance of the test pilots. As legend would have it, near the end of his tenure, he was apprehended by base security police in a Cadillac convertible on Rogers Dry Lake bed. There, in the early hours of the morning and with one of the beautiful hostesses from the Happy Bottom Riding Club, Boyd was found stark naked and happily drunk.

Though they'd been vanquished, the MX-981 team left Edwards with heads held very high. During their time at the base, Stapp and his men had amassed an impressive body of research culled from ninety-three high-speed rocket sled runs with either Oscar Eightball or one of the other dummies aboard, and eighty-eight with chimpanzees and pigs. More impressively, they'd completed seventy-three runs using twelve different human beings as subjects: nineteen in the rear-facing configuration, and fifty-four in the more punishing front-facing configuration. John Paul Stapp had made twenty-six of the manned runs himself, all in the space of fifty months. He'd exposed himself to more than 45 g's and had shown that the primary factor influencing human tolerance to deceleration forces was the rate of onset, or how rapidly the g-force loads were applied. He'd also demonstrated that the eyes might be leading indicators that the tests had at least approached the limits of what the human body could endure. In the days following the final run, his vision appeared to stabilize. He still had blotches in his visual field, but they didn't seem to be getting any worse. He knew he'd been lucky.

Stapp had rewritten much of the School of Aviation Medicine's textbook on human tolerance and had thrown a gauntlet at the feet of the nation's aircraft industry. What Stapp and his Northrop team had accomplished defined how researchers would attempt to answer a number of important questions relating to crash protection in the years to follow. All subsequent impact acceleration studies would owe a profound debt to the once obscure Project MX-981. A future chief of the Aero Med Lab clarified the team's achievements: "There is now no question of human limitation in the consideration of higher strength seats and cockpits for aircraft." As one of the top medical scientists in the generation of researchers that would follow and take up Stapp's cause would express it: "It all came from Stapp. Everything."

By the last week of June 1951, Stapp was still suffering from recurring headaches, but he was now increasingly optimistic that he would retain the use of both eyes. Despite intermittently blurry vision, he drove a new DeSoto from California to Ohio in four days. He had not wanted to return to Wright-Patterson in the inferior Chrysler and so had invested what little savings he'd accumulated in more substantial wheels. It was a journey that included a brief stopoff to visit his brother Celso in El Paso and another side trip to Holloman Air Force Base outside Alamogordo, New Mexico. Holloman had, he knew, a gleaming 3,500-foot rocket sled track—almost twice as long as the track at Edwards—that had been built a couple of years earlier to test cruise missiles.

When he arrived at the base late on the afternoon of July 3, he steered through a grid of dirt streets, weatherbeaten hangars, and concrete block-houses. He eventually found his way to a rutted lane that led him four and a half desolate miles out to the Holloman test track site, which held a thing of sheer beauty. It was an 84-inch-gauge track—seven feet rail to rail—mounted on parallel white concrete piers and looking from his perspective as if it ran to the horizon. It wasn't hard to imagine what he could do with such a facility if the Air Force could be convinced to give him the chance. He had not yet, he knew, determined the absolute limits of human tolerance to deceleration forces, and while he'd told his superiors in Dayton that he'd be content to walk away once the work in California ended, he found himself struggling with the idea of leaving so much undone.

As Stapp studied the track, the sun was setting on the steel ribbons. It was a moment he wished he could share with George Nichols. The place was as stark and raw as the Mojave, only with a white gypsum sand that gleamed pink and orange in the sun's dying rays. It was lonesome-looking, forlorn, and utterly empty. Which is to say, it reminded Stapp of Muroc, back when he'd first arrived in 1947. Of freedom.

From Holloman, Stapp headed for Albuquerque and then points east along Route 66, driving through the night, Dayton-bound. The rest of the cross-country trip following the exile from Edwards was uneventful, which was somewhat surprising considering John Paul Stapp's skills as a driver. "It wasn't that he drove too fast," George Nichols explained. "He had a problem with getting distracted. He was a horrible guy in traffic, in and out of his lane. He was always thinkin' about something."

13

SOLO ON LOVERS LANE

I would like to make it clear that I place Air Force business and
my research above and beyond any personal life or obligations
to people.

—*John Paul Stapp, letter to pen pal Nancy Whitley*

MERE WEEKS AFTER General Douglas MacArthur, who had defiantly
demanded decision authority over the use of tactical nuclear weapons
in Korea, had been relieved of his command by President Truman on April
11, 1951, the United States conducted a successful test of a thermonuclear-
assisted atomic bomb in the Marshall Islands. This latest demonstration
proved that the concept of a fusion weapon—hundreds of times more devastating than the Hiroshima bomb—was workable. It was with these developments on his mind that John Paul Stapp did his best to re-acclimate himself
to the traditions and routines of military life at Wright-Patterson Air Force
Base that July. "Wright Field was its usual self," Stapp wrote of the Air Force's
greatest installation. "A tremendous conglomerate of laboratories, wind tunnels, shops, and warehouses surrounding a level expanse of runways with all
parking areas crowded by a bewildering array of aircraft, of all vintages . . . A
grim world of weapons of the future was cut, formed, assembled, and tested
in this gigantic armory."

Shortly after his arrival in Dayton, the Aero Medical Wives Wing mis-
takenly sent an invitation to "Mrs. John P. Stapp." Stapp responded with a

letter saying: "I regret that it is impossible to accept this invitation due to the fact that there is no Mrs. Stapp. If there is ever a Mrs. Stapp, she would be honored to belong to the Wing." He enclosed a check for $1.00 to cover the contingent membership. Stapp found desirable bachelor lodging, a one-room, second-floor apartment in the nearby town of Yellow Springs, where Antioch College had made some of its off-campus housing available to Air Force officers. It was a pleasant upgrade from the sand and solitude of the Desert Rat, and Stapp availed himself of the college's library, the lectures, and the concerts. Antioch brought him back to his Austin days, but the workaday world of the Aero Med Lab offices took some getting used to: "The process of adapting to the strictness and punctuality of a large headquarters organization after so many years of complete freedom in the desert, I had numerous painful encounters, but I soon learned the folkways and mores, the taboos and oracles of this great technical community." One anodyne for Wright culture was Stapp's newfound and very welcome status within the aeromedical community there. His work at Edwards had earned him an acknowledged authority on matters of aircrew protection. Not long after his return to Dayton, the National Air Council selected him as its Air Force Scientist of the Year, an award he accepted at a Pentagon ceremony that October. The newly formed Air Research and Development Command (ARDC) nominated Stapp for the Legion of Merit, which he would be awarded four years later.

Stapp was also presented with the John Jeffries Award for Outstanding Contributions to Aeronautical Science by the Institute of Aeronautical Sciences the following January, 1952, where he presented a paper that called for aft-facing air transport passenger seating. Even though Stapp's recommendations had already been implemented on an experimental basis in a couple of Air Force aircraft models, and the Navy would shortly mandate rear-facing configurations for all its air transports, the paper still caused controversy among a contingent that publicly disputed his results, charging broadly that MX-981's data required more scrutiny. To Stapp, it was simple prejudice and laziness, a knee-jerk reaction to new ideas and ways of thinking. He referred to his critics now as the SOS Club: "Stamp Out Stapp!" But despite the low-level hum of criticism, Stapp was promoted to lieutenant colonel that fall and given command of the Special Projects Section of the Biophysics Branch of the Aero Medical Laboratory.

Within days of his promotion, Lieutenant Colonel Stapp announced that the mission for his new organization would be the study of "the escape of living and uninjured personnel from airplanes, whether those airplanes are disabled in flight or forced to crash-land." He organized his team of some two dozen scientists, engineers, technicians, and medical doctors into three groups. The first would focus on abrupt deceleration, including special categories of crash injury such as concussion. The second group would study escape from aircraft in flight, examining the physics of free fall, in particular the tendency of the human body to tumble and spin randomly as it falls, as well as the equipment needed to protect airmen in distress, with emphasis on helmets and ejection seats. The third group was an auxiliary medical unit made up of pathologists and neurological consultants whose services were made available to the other two.

In addition to the ambitious research work for which he was responsible, Stapp also spent hours on end holed up in his windowless office, immersed in detailed review and documentation of the paper mountain of data the MX-981 team had produced during its years at Edwards. He went through stacks of boxes full of printed oscillograph recordings trace by trace, painstakingly calibrating and tabulating the results. Without the electronic analyzers and computers that were still on the technological horizon, his tools consisted of a three-inch section of a ruler with 100th-of-an-inch scale markings, a handheld lens, and a pair of calipers. It was slow, tedious work uninterrupted by the distraction of the sled rides, and he relished any opportunity to break away from the drudgery.

While in New York City to collect the Jeffries Award, Stapp decided to put some of his $200 honorarium toward the pursuit of high culture, something he'd missed during his Mojave years. He purchased a seat in the number one box at the Metropolitan Opera for a performance of *La Gioconda*. His panoramic perch overlooking the orchestra and stage reminded him of a seat in the nose of a B-17, and he found it every bit as riveting as his cross-country trip at treetop level a couple of years earlier. The production's attention to the smallest detail, from the perfection of the Dance of the Hours ballet to the subtle command of the orchestra's colors, gave him the exhilarating sense, for the first time in his life, of being in the presence of the very highest artistic ideal. "All the Saturday afternoons that I had devoutly listened to the Metropolitan over the radio were pale facsimiles compared to the rich experience of meeting the Metropolitan in the flesh. I returned to Dayton still enchanted, feeling that I had gotten a high moment in return for all the lonely days and nights in the desert."

Back in Dayton in the spring of 1953, just as his work on the MX-981 data was wrapping up, Lieutenant Colonel Stapp received very welcome news: he was being reassigned to Holloman Air Force Base. The high-speed test track he'd seen on his drive two years earlier had never been far from his mind since. He would take command of the small aeromedical field laboratory previously attached to the Aero Med Lab in Dayton but now reconstituted as an independent research facility. There were also reports that the Air Force intended to extend the 3,500-foot Holloman rocket-sled track to a length of 15,000 feet, nearly three miles.

Stapp arrived in New Mexico in mid-April, happy to leave behind his desk at Wright-Patterson and eager to resume his biodynamics field research. He was, as he told his colleagues in Dayton, ready for sand, sweat, and blue skies. Holloman, sandwiched between the dusty little town of Alamogordo and the White Sands Proving Ground, had been established in 1942. By the time Stapp arrived, the area was the acknowledged ground zero for space biology research in the United States.

Project High Dive was already in operation when Stapp got there. This was a balloon program that dropped Sierra Sam dummies—Sam Alderson's creations based on ideas of Stapp's from the Muroc days—from high-altitude balloons to study the problems of extreme free fall. Some forty-three of these flights were made and sixty-seven Sierra Sams were released into New Mexico airspace. A few were never recovered. Individuals in the desert environs near Holloman claimed to have witnessed alien beings and described them to reporters as hairless and clad in shiny one-piece suits that were silver-gray in color. A curious but common detail in these descriptions was that the aliens had only four fingers to a hand. All of this perfectly describes the Sierra Sam model dropped from the High Dive balloons.

The Air Force base, of course, had access to more sophisticated aircraft than balloons. Captured German V-2 rockets had been shipped to White Sands for testing immediately after the war, and the Air Force headquartered its guided missile program there not much later. When Captain David

Goodman Simons first came to New Mexico from Wright Field to conduct radiation studies, he'd articulated the gist of the Aero Medical Laboratory's larger space biology challenge: "But what are the problems of space flight in a rocket? By theorizing, the various possible dangers and limiting factors can be appraised and appropriate means of protection against each surmised. However, only by actually performing the experiment can one prove or disprove the validity of the hypothesis, learn better ways of protecting against known hazards and realize for the first time, the existence of unsuspected dangers. Only the recovery of a live animal showing no demonstrable ill effects will permit the claim that no major difficulty has been overlooked."

Initially, the V-2s were used for a variety of high-altitude experiments. Simons, working under the supervision of Lieutenant Colonel James Henry, an early supporter of space research, started with fungus spores, and graduated to fruit flies and mice. Together, Simons and Henry worked toward the goal of sending a primate into near-space and returning it to earth alive. They designed a small capsule capable of providing stable pressure, breathable oxygen, and tolerable temperatures. On June 18, 1948, a nine-pound rhesus monkey named Albert had been sealed into the capsule, loaded into the nose cone of a V-2, and launched into the extreme upper stratosphere. In spite of the fact that he'd been anesthetized, the experiment did not go well for Albert. In fact, data recovered following the flight indicated that Albert had suffocated in the capsule before the V-2 had even left the ground. Albert II was launched the following summer. With an improved capsule, this monkey survived his extraterrestrial journey. Unfortunately, the parachute recovery system malfunctioned and Albert II was killed on impact. Follow-up flights using mice were equally disappointing. Eventually, Simons would begin using plastic balloons to raise his subjects into space-equivalent conditions inside pressurized aluminum gondolas, and with much better results.

Another important figure at Holloman involved with the biology flights recalled his first meeting with John Paul Stapp. Legendary meteorologist Bernard "Duke" Gildenberg was, like Stapp, a man of great intellect and catholic interests. When he first ran into the new commander at the base cafeteria, Gildenberg had a Shakespeare paperback sticking out of his back pocket. Stapp pulled him aside and began describing how he intended to rewrite *Hamlet*, and that he meant to start on it as soon as he could find the time. "I didn't know if he was crazy, or if this guy really had something on the ball," Gildenberg said of Stapp. "But he did a brilliant psycho-analysis of all the characters, and I decided this is somebody I wanted for a friend." One of Stapp's traits that impressed Gildenberg was his extraordinary memory. "Occasionally," Gildenberg said, "he would tell me verbatim about something he'd learned on his first day of medical school." Stapp even claimed to remember feeding at his mother's breast. "I don't necessarily buy that one, but it's definitely possible with his memory."

Meanwhile, as the Korean War came to a close—an armistice agreement between the United Nations Command, North Korea, and China was finally signed on July 27, 1953—and American troops began coming home, organizational confusion shrouded the early days of Stapp's tenure at Holloman. At first, it was not entirely clear what the relationship was—or should be—between the Biophysics of Abrupt Deceleration project officially based at Wright-Patterson and the deceleration work the field laboratory in New Mexico would undertake. Stapp received his orders and direction from the Aero Med Lab in Dayton, but, as Air Force historian David Bushnell wrote in his history of the Holloman programs, "in practice Colonel Stapp was largely on his own from the moment he reached Holloman." Within weeks of his arrival, Stapp was able to have the Wright-Patterson deceleration project broadened and incorporated into a Holloman program that would now be completely under his control: Project 7850, Biodynamics of Human Factors in Aviation. Stapp identified five research problems with which his program would concern itself: 1) tolerance to impact forces, 2) tolerance to total pressure change, 3) tolerance to abrupt windblast, 4) tolerance to aircraft crash forces, and 5) automotive crash forces. The last of those would raise more than a few eyebrows among Stapp's superiors, who questioned the logic of applying the resources of the United States Air Force to car crashes.

. . .

Early in the summer of 1953, Lieutenant Colonel Stapp settled into a three-bedroom ranch-style house on a corner lot in Alamogordo just off White Plains Boulevard across from the New Mexico School for the Blind. It was the first piece of property he had ever owned. He bought some furniture

The house in Alamogordo was the first piece of real estate Stapp had ever owned, and he tried to transform his backyard into an oasis. (Photo courtesy of Wilford Stapp)

that the Holloman officers' club was auctioning off and built custom cabinets for the hi-fi system that he had managed to upgrade to state-of-the-art during his time in Dayton. He took up gardening and in fairly short order transformed his yard into a little sun-drenched oasis: a rustling bamboo hedge on the west side, a grove of a dozen fig trees on the east side, along with tamarind and apricot trees and beds of gardenias. He found that the climate and soil were ideal for roses, and so he began planting a variety of species with which he would continue to experiment.

Almost immediately upon occupying his new house, Stapp began hosting regular dinner parties. They were his social laboratories. He loved to play piano, show off his audio system, and cook for his guests to the strains of a string quartet or a rousing world-famous orchestra. A standard entrée was chicken stuffed with apples and raisins and seasoned with his own secret mixture of spices and herbs. Another was a dish Stapp called Siberian Tiger

Stapp took his first piano lessons from his mother in Bahia, battling the Baptist college's perpetually out-of-tune piano, and continued playing until his last days. (Photo courtesy of Wilford Stapp)

Stapp's audiophile obsession was enduring. Here he works on the Lovers Lane installation of the ever-evolving hi-fi system. (Photo courtesy of Wilford Stapp)

Steak, which was really just a broiled beefsteak rubbed with his own preparation of ginger, mustard, garlic, and thyme. He developed a reputation as
a master conversationalist, and on occasion would get so wrapped up in a
discussion that he would neglect his kitchen duties. General Malcolm Grow
recalled an evening at Stapp's house during which the host became so
engrossed that he simply forgot to serve the meal altogether.

Stapp's obsession with his work was seemingly unaffected by environment or by the months and years rushing by. As he had grown into middle
age, he had become a servant of his cause to no lesser degree of dedication
than had his father of his. It was a single-mindedness that now formed the
very core of his personality. Red Lombard described the Stapp fervor: "His
determination was almost a hate in his life. He hated the fact that not enough
was known about acceleration to save the lives of jet pilots in bail-outs, so he
went to great extremes to learn more, then educate the world. I think this
hate is one component of the true Stapp." The intensity of the "hate" that
Lombard describes may be the crucial difference in attitude and approach
between the affable Reverend Charles and his oldest son. Its ferocity could at
times be offputting and may even suggest why the son had—so far, at least—
struggled to find a compatible wife.

During the years in California and Ohio, Stapp had managed to form a
few casual friendships with women. He'd been taken with a beautiful blonde
whom he met occasionally at the officers' club at Edwards; she was one of
Pancho Barnes's hostesses. He'd corresponded with women who'd seen him
in the newspapers or in the newsreels at their local movie houses, some of
whom had written him seductive letters. "Doc," one of them wrote, "you are
a gentleman and a scholar. I'm going to invest in an elephant gun and go big
game hunting after you ... Wow! What a man." He'd also maintained a correspondence with Celia Richards, the redheaded girl from Kerosene Flats
to whose family the Curbstone Clinic had provided medical care. Celia was
now a high school senior in Lancaster, California, and considered Stapp the
greatest man she'd ever met. She sent him pictures of herself. Stapp was
always willing to offer advice through the mail to friends and strangers alike,
and he encouraged Miss Richards to spread her wings and see the world. She
wrote back that she wanted to go to Harvard and that he had inspired her to
become either a doctor or a lawyer, perhaps both.

Yet despite his long-suffering desire, Stapp seemed no closer to what his parents and his brothers—all married now, with Robert and Celso having begun their own families—had long wished for him: a companion and a stable lifestyle. The letters of Reverend Stapp, in his final days as a missionary, offered both motherly and fatherly advice to his oldest son. "I do sincerely hope that you will be able to find the only woman ever created for you and establish a happy home. Though it is better to wait and be sure than to be unhappy afterwards."

Stapp occasionally went on blind dates with friends. Wilford and Margaret Stapp recalled an incident in which they arranged for John Paul a date with a young beauty queen who would go on to become Miss Texas. In Wilford's car, on their way to her house, Stapp rode in the front seat with Wilford and Margaret. John Paul walked up to the door of the house and politely escorted the woman to the car. He opened the rear door, allowed his date to slide in . . . and then shut the door and got back in front with his brother and sister-in-law. "He just had no clue whatsoever!" Margaret laughed. "The poor girl sat in back by herself. And Paul had no idea that there was anything odd about that." Wilford's explanation was that his brother was simply too consumed by his work to have time for anything else. Yet there was something in Stapp's makeup that made it hard for him to connect with certain people—particularly women—on a personal level. Years later, the wife of a close friend would observe that Stapp was sometimes "naïve about people."

Through his professional successes, John Paul had finally managed to acquire the home he'd never had. His Alamogordo address was certainly promising: 300 Lovers Lane. But until he found a woman with whom to share it, he would never be satisfied. He fell periodically into depressions that recalled the time of his failed marriage. Only a few months after settling in, Stapp received a letter from his ex-wife Nylah, who'd seen one of the many news reports about the rocket sled experiments. It was the first contact since their divorce fourteen years earlier. Nylah had been married and divorced again in the meantime. She suggested a rendezvous. Stapp tore the letter up. He had no interest in reopening that painful chapter in his life, even wondering if Nylah's mother might be behind it all, scheming to reunite her daughter with the now-celebrated doctor.

While there is no way to know much of what went on during the brief marriage of Stapp and Nylah, they surely at least discussed having children. Wilford Stapp recalls conversations with his brother during their Texas years together in which John Paul talked about his desire to raise kids someday. At some point in the intervening years, however, Stapp had discovered that he was infertile. He blamed it on overdoses of the malaria treatments he'd been given as a boy in Brazil. Published studies have since documented quinine's effects on sperm count in mice, and clinical trials have shown a link between quinine and lower sperm motility in humans. In Wilford's estimation, learning that he would never be a father himself had been a major blow to his brother. Wilford and Margaret Stapp, also unable to have children of their own, would adopt four siblings.

Despite the trials of his personal life, Stapp always had his work to occupy him, and one of the most gratifying benefits of his new role at Holloman was being able to team up again with his old Northrop crew. Northrop held the contract to provide the next-generation rocket sled and associated services for the latest series of projects, and in short order both George Nichols and Jake Superata were on site in New Mexico preparing for the next phase of Stapp's research. The Holloman high-speed tests would attempt to determine the useful limits of the latest ejection seats in terms of deceleration, windblast, and pilot seating configuration.

In the late summer of 1953, they took delivery of a new rocket sled and began a series of low-speed proof runs. The new sled—actually two separate sleds—was a generational advance over the boxy hulk of the *Gee Whiz*. The Holloman decelerator would employ what they called a pusher sled capable of accommodating a dozen solid-fuel rockets. The pusher would propel the 2,000-pound test sled, including instrumentation and the test subject tucked in behind a windscreen. At rocket burnout, the pusher would decouple and the test sled would race on into the brakes. Both sleds were mounted on slippers with bearings made of stellite—a stainless cobalt–chromium alloy—surrounding the railhead. In contrast to the *Gee Whiz*, the dual-vehicle package was low-slung and aerodynamic. It looked fast and, in theory, was capable of delivering much higher g-forces. Northrop had also provided a completely redesigned telemetry system accurate to one millionth of a second. But the most striking change in the project's approach was evident in the ingenious new braking system that dispensed with the notion of friction brakes altogether. At the speeds Stapp had in mind, temperatures generated by the friction of traditional clamp-style brakes would melt steel. Stapp's metaphorical knife sliding through the vise would liquefy and the system would seize up.

The answer they came up with was a concrete ditch between the rails, 60 inches wide and 18 inches deep. The ditch, along any section of the track, could be filled with variable quantities of water. The system used movable and easily penetrated Masonite dams at 10-foot intervals to control the precise location and depth of the water. Both sections of the decelerator were equipped with fixed scoops mounted under the chassis that forced any water encountered to be ejected to the side or back. It was described as a "momentum exchange" system. When a scoop hit the water at a 90-degree angle, a retarding force of one pound for each pound of water ejected would be applied to the sled. The system offered almost limitlessly reconfigurable stopping capacity, which, along with the powerful sleds themselves, would allow the team to achieve unheard-of speeds and still bring the vehicle safely to a stop in not much more than the blink of an eye. Preliminary calculations showed that using only half of the propulsion sled's twelve-rocket capacity could produce velocities in the range of 430 miles per hour and, with a water-brake force of 300,000 pounds, bring it to a halt in a distance of about 30 feet.

The question that was not very openly discussed was whether a human subject could really expect to emerge unscathed from the sheer biomechanical outrage of such an event. Were they still approaching the limit of tolerance, or had they already passed it? Later, in an interview with an Air Force historian, Stapp explained his methodology, one that was unchanged since his California days: "With anaesthetized animals, we started at the highest levels of mechanical force where we knew they would be lethal and then successively diminished the force of the runs so that with the fewest possible experiments, we could come down to the lethal point and then go below that through injury levels to a level where there would be no injury. But with humans, we did it the other way. We started low and built upward toward injury levels."

Stapp reached back into sacred history to find a proper designation for his work. The Saint Simeon Stylites School of Research, he called it. Saint Simeon Stylites the Elder was a Christian ascetic who is said to have lived for thirty-nine years atop a 40-foot column in the Syrian desert. "He had his food and everything else sent up to him on ropes," Stapp wrote, "and he never bathed and he baked himself day and night, apparently doing a sort of long-term religious marathon. It seemed to be pretty much of an example of what we were doing."

In October 1953, Stapp announced that the target date for the first high-speed test run of the new deceleration system would be the following month. They would use a dummy strapped in with their latest harness configuration. It would give the team a chance to evaluate the performance of the rockets they'd acquired, to exercise the new telemetry setup, and to observe the dual sled in high-speed action. Northrop had christened its shiny new red-and-white hot rod *Sonic Wind*. Most everyone working on the project in the Holloman outback was optimistic and anxious for the first live-subject runs that they knew were coming.

Everyone that is, except George Nichols. Nichols had never gotten over the eye injuries and broken bones of MX-981's final months at Edwards. He believed that if Stapp continued to court these severe g-forces he was going to blind himself, and Nichols didn't want to be around for what he feared could be an even worse tragedy. "It doesn't make sense to me," he later recalled telling Stapp. "It plain doesn't make sense." But Nichols, like Jake Superata, knew he *had* to be there, both for Stapp and for Northrop. Neither was willing to entrust the job of chief engineer of this project to anyone else.

The shakeout run on November 23 turned out to be a bust. Only three of the six rockets fired and what was being touted as the world's fastest ground vehicle entered the water brakes at a relatively lumbering 200 miles per hour. They tried again a couple of weeks later. On that second test run, the sled's scoop hit the water brakes at just under 400 miles per hour, treating the handful of spectators to a dramatic side effect of the new braking system: a powerful twin fountain erupting from the sides of the sled's slippers and forming a spectacular arcing rainbow spray against the intense blue of the high desert sky. The dummy pulled a solid 20 g's. The team repeated the test with similar good results in the first week of January 1954.

Next up, just two weeks later, would be an anesthetized chimpanzee. With Tony Gentry's help, Stapp had gotten his colony of chimps transported from Edwards to a new upgraded facility at Holloman. Stapp was happy to be able to reunite with his favorite chimp, Lizzie. She still went nuts whenever he walked through the door of the primate house, and was notably spared duty on the more dangerous experiments.

For the chimp run, Stapp positioned himself near the camera station about 250 feet from the track, adjacent to the water brakes, where he could observe the critical moment of deceleration. As *Sonic Wind* came to a full stop, he waited for the ordnance specialists to fire a flare indicating it was safe to approach. When he saw the signal and made it to the sled, not only did the chimp seem in fine shape, but the team had gotten astonishing color photographs at 128 frames per second. They had some failures in the time-distance recordings, but there was no doubt that the subject had endured—apparently without damage—higher deceleration levels than had ever been achieved at Edwards.

Stapp was now confident enough to schedule what he termed an "all-out proof test" for February. For this run, the chimp subject was restrained by double-thickness 6,000-pound-test nylon webbing arranged in dual three-inch-wide shoulder straps and a lap belt. It was the identical configuration they would use for human tests. Once the test subject was secured, *Sonic Wind* was fired and, seconds later, hit an 18-inch-deep wall of water at a speed of 315 miles per hour—and was slowed to just 60 miles per hour in a distance of 30 feet. As many of these runs as Stapp had witnessed, this one almost took his breath away. "It is almost impossible to describe the visual shock," he wrote. "There was a brief instant of quivering telescoping retardation in the midst of showers and columns of spray thrown up higher than the 60 foot power poles in the background, followed by an optical illusion of the sled appearing to lunge unrestrained out of the brake section." Their data showed that the sled—and its passenger—had survived a force of more than 100 g's for a brief instant with no resulting damage.

Within hours of the February run, Lieutenant Colonel Stapp was already making plans for a trip to ARDC headquarters in Baltimore. Due to the new post-Korea reorganizations, approval for a resumption of human rocket sled tests would now have to come from the Human Factors Office, and Stapp—fresh from his team's successes—was eager to make his case.

To Stapp's totally unexpected delight, the chief of Human Factors, Brigadier General Donald Davis Flickinger, a fellow doctor and war hero, exhibited a sympathetic comprehension of the Aero Medical Field Lab's objectives and methods. "Above all," Stapp recalled, "his complete confidence in my judgment was the most heartening experience of my entire military career up to this point. Even my own brother had hinted more than once that he suspected me of suicidal trends, much to my annoyance. Here was complete and enthusiastic endorsement that gave me security and confidence to go ahead."

Stapp was equally delighted when, during the final week of February 1954, he was contacted by executives at 20th Century Fox who had gotten word of the dramatic research going on at Holloman. Fox wanted to discuss the possibility of a feature film based on Stapp's programs. Stapp invited producer William Bloom to New Mexico to witness the next *Sonic Wind* run and provided Bloom with specifications and data relating to the deceleration work. He also agreed to provide, confidentially and without Air Force permission, project film shot by George Nichols ostensibly for Northrop's own use. Stapp wasn't sure how the Air Force was going to react to his cooperation with Hollywood, but he was sufficiently concerned that he decided to keep the details of his arrangements with Bloom to himself, at least for now.

14

WAITING FOR A TIGER TO SPRING

[Man] even entertains such irrational concepts as that what his
earth needs is more tigers and fewer people, because tigers, no
longer a threat, turn into underdogs of nostalgic preciousness to
those who have never been roared at or bitten.

—John Paul Stapp, "Man Between Environments"

NOW, WITH THE UNQUALIFIED support of his superiors—however fragile or
fleeting it might prove to be—Stapp wasted no time planning for his
own ride on *Sonic Wind*. Receiving formal approvals to conduct high-speed
human tests from ARDC headquarters on March 17, 1954, he gathered his
team and declared his intention to make a run just two days later—a highly
unusual decision for Stapp given that the 19th would be a Friday. It was a
measure not only of an almost paranoid eagerness to go before General
Flickinger had time for second thoughts, but also of his trust in the Northrop
men. He set Holloman abuzz when he announced that his target would be
400 miles per hour. The world land speed record in any type of vehicle stood
at 402 miles per hour, and base personnel liked the idea that Stapp just might
put Holloman in the record books. While the Air Force mandated lock-
down security and imposed a news blackout on anything having to do with
the rocket sled tests, a lone journalist named Jim Haggerty—who was well
regarded by Stapp and others at the base as the author of numerous magazine articles on military aviation and naval topics—was invited to be present

and was given exclusive access to Stapp and the crew. The Air Force didn't want to completely embargo the story, but it did want control over how it would be presented to the public.

Early on Friday morning, Stapp donned a set of pilot's blue winter coveralls and drove himself out to the test track blockhouse for the standard preliminary physical examination. In spite of his enthusiasm for the project, he confessed to Haggerty to feeling mildly depressed. "As usual," Stapp wrote later, "I studiously cultivated a fatalistic attitude toward the run, compelling myself to consider only the experimental results and the intent introspection required to record and remember all sensations and impressions during this run." When Haggerty asked Stapp how he felt about things that morning, Stapp replied that he was most definitely *not* looking forward to the experience. "You know it's going to hurt," he said, "but you don't know how much."

The dental laboratory at Holloman had fashioned a new rubber bite block from castings of Stapp's own teeth and jaws. It was decided for this run that Stapp would not wear a helmet, but that his head would instead be cradled and strapped into a thick, padded headrest built into the seat. His team secured him with nylon straps that had for the first time been coated with anti-abrasion wax, which decreased their pliability but would help alleviate the friction they would be subjected to at impact.

As he sat in the sled that morning, contemplating the ordeal before him, Stapp conferred with George Nichols during a final inspection of the system. Stapp admitted that he had real reservations about what was to come. It had been almost three years since his last ride. He was about to experience the wallop of a full 6 g's of acceleration, to be followed by twice the duration of abrupt deceleration he'd ever experienced. The increased duration of exposure was one of the key differentiators between the Edwards experiments and what Stapp intended for his sled work at Holloman. The chimp had come through it, but that didn't prove a man could take it.

Nichols slipped the bite block into Stapp's mouth. Though Stapp's gloved hands were positioned beneath his knees and strapped firmly to the seat, his fingers pinched a lanyard he would use to trigger a high-speed Fastex camera pointed directly at his face. He would trigger the camera when the countdown reached three.

At two minutes till firing, a technician switched the telemeter transmitter from an external power supply to the batteries aboard the sled. Following final checks, most of the crew scrambled back toward the safety of the blockhouse. Jake Superata swatted Stapp's knee for luck and screwed the fuse into the electric circuit that would ignite the seven rockets. Then Jake was gone.

At X minus thirty seconds, a red flare arced across the morning sky. As if he was trying to sync himself to the low thrum of the electric circuits, Stapp's breathing came faster and faster. He bore down, chomping on the bite block.

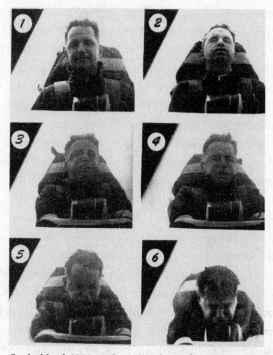

On the March 1954 run, Sonic Wind went from zero to 421 miles per hour and back to zero—all in less than eight seconds. The mechanical force was akin to an airplane crash. This series of photos shows Stapp 1) at rest, 2) at rocket ignition, 3) accelerating, 4) at peak velocity, 5) hitting the brakes, and 6) at the dead-stop point. (Photos courtesy of U.S. Air Force)

"It was," he said later as he tried to describe the experience, "like crouching behind a bush waiting for a tiger to spring."

At X minus three seconds, Stapp jerked the lanyard to trigger the camera and clenched for the blow. Then: the explosion. It was nothing like MX-981. This was a new dimension of physics: launched—fired, actually—at what would turn out to be an astonishing 421 miles per hour.

Five seconds later, *Sonic Wind* hit the water. When it did, the sled appeared to observers to be transformed instantly from a ballistic blur into the fountains of foam and then, weirdly, to jump from the spray carrying no speed or momentum at all, and quickly slide to a full stop. Mind-numbing velocity meeting an invisible wall.

Lieutenant Colonel Stapp, who'd just survived man's fastest trip across the surface of the earth, struggled to relax the muscles in his neck and allow his head to sink into the padded headrest. In the intensity of his relief—it was over and he'd managed to survive once again—he could feel the raw burn of the windblast's lash on his cheeks and forehead. He was only vaguely aware of the sound of voices approaching. His field of vision was occluded, and that was the worrisome thing. He tried to rub the clouds out of his eyes.

The thought came to him: he'd just lived through the force equivalent of an airplane crash. He could only guess at what the telemetry would say, but he didn't have to examine the data to know he'd once again proved something about what human beings could endure. All you had to do was protect them and restrain them effectively, and they could take almost anything. There could be no denying it now and Stapp felt the euphoria returning, juicing his veins, a sensation he hadn't experienced since Edwards.

As he was unstrapped and helped from the sled, his legs were rubbery. His mind, however, was clear. He knew they were still well short of the survivable limit. There was more work to be done. Before they'd even gotten him to the ambulance, Stapp was already thinking ahead. The men saw him smiling as the attendants got him to lie down and then slid him inside.

Back in his office the next day, Stapp was able to write his report to General Flickinger claiming triumph. Data showed that he had actually been exposed to 12 g's during the five-second acceleration to peak speed as he was slammed into the back of his seat. At the moment the sled's scoops burst the first dam and hit the trench full of water, and his weight surged forward into the straps and harness, he was exposed to an immediate force of 22 g's at an onset rate of 600 g's per second. The event seemed even more astonishing as they studied it than it had been as seen from the blockhouse or alongside the track. In the final analysis, the restraint system had performed beautifully and Stapp had, albeit with a little help, walked away. He'd suffered a set of deep bruises where the shoulder straps had chewed into his flesh, deeper strap bruises on the insides of his thighs, and a few small lacerations where grains of sand had nicked his face. Though his eyes still felt sore, his vision had mostly cleared up. Stapp had set a new land speed record, but more impressive was how fast he'd come to a stop. Stapp had gone from zero to 421 miles per hour and back to zero, all in less than eight seconds.

Writer Jim Haggerty had witnessed it, but he wondered how he was going to describe it. How could it sound like anything other than science fiction? The Air Force greenlighted the story on June 8 and the newspapers had it the next day. The *Austin Statesman*: "ALAMOGORDO AF COLONEL TRAVELS 421 MILES PER HOUR IN SPEED SLED." The papers had the facts, but Haggerty wanted to capture something of Stapp himself. His version appeared in the June 25 issue of *Collier's* under the title "Fastest Man on Earth." It introduced Stapp to a new audience of American readers: "He's a mild-mannered, self-effacing gentleman with the air of a college professor." A cover drawing for the issue showed Stapp grimacing as *Sonic Wind* hit the water brakes. "Every research scientist likes a finite problem," Haggerty quoted him, "one that has a definite end in sight. This is such a problem. When I complete the programs, I'll have a perfect graph of the survivable force limits of the human body."

With the appearance of Haggerty's story, Stapp had become the very face of the intrepid man of modern science, the lab-coated genius with slide rule willing to expose himself to unheard-of danger in the quest for otherwise unattainable knowledge. In Brazil, the papers claimed the "Homem Mais Veloz do Mundo" as one of their own. Reverend Charles Stapp, who'd gotten a copy of the *Collier's* article, wrote: "Congratulations. Para Bena! We saw your picture and write-up. A good many people read that magazine. Walker saw it in Liberia. People still talk a lot about you. I am now the father of 'The fastest man on earth.' I tell them I am glad you broke your record without breaking your neck. You should hear the comments that are made by men women and children." In the weeks following the *Collier's* story, Stapp was invited to appear on the television shows *I've Got a Secret* and *Arthur Godfrey and His Friends*.

By early summer, producer William Bloom at Fox already had a writer working on a film treatment and was lobbying on behalf of the Air Force drama he envisioned with his boss, Fox president Darryl Zanuck. Meanwhile, Stapp remained extremely circumspect with his own superiors. He was concerned, particularly, about General Flickinger. Don Flickinger had by this time been appointed the director of research for the ARDC, and Stapp was learning that the general's endorsement of the sled research did not extend to the publicity barrage that Stapp sensed was coming, and which he was in many ways courting. Flickinger, who understood that his own control was likely to diminish to the degree that Stapp's lionization in the press unfolded and built upon itself, knew he would have to apply his own leverage.

Stapp arranged for a meeting between Bloom and Flickinger, but counseled Bloom not to reveal too much about the degree of cooperation and access Stapp had promised Fox. At the same time, Stapp had opened up another confidential dialogue with an editor at *Life* magazine about an exclusive story that he thought might provide him with some of the material compensation he would never get from the Air Force. He wrote to Frank Cockrell, the scriptwriter Bloom had assigned to work on the film: "When I cross the speed of sound on the sled, I mean to make Life Magazine pay as dearly as possible for a signed article. For science, my all; from Life a paying basis."

A mostly pleasant side effect of his growing notoriety was a flood of correspondence from far-flung family, unknown admirers, and, on occasion, from old friends with whom he'd lost touch. In July 1954, Stapp received a letter from Dorrel and Julia Jones, his friends from back in the Texas Hill Country. They had gone half-crazy when they'd seen Stapp on television. Julia had jumped up and tried to alert the world: "I began screaming for Dorrel. He was way out back and I was standing on the front porch yelling as loud as I could, and if you will remember there is nothing wrong with my lungs, so I roused the neighbors, but good." Stapp was especially glad to get news of the Joneses' first child, his namesake. "John Paul is 6' tall," Julia wrote. "He is interested in anything concerning Science. Do you suppose just naming him for you could have caused that? If you and Dorrel hadn't hardened me to it many years ago, I would have gone nuts last summer when he kept a big collection of snakes out in the Garage."

Despite the hoopla, Stapp's oversight of the work going on at Holloman kept him as busy as ever. David Simons's space biology research, now using high-altitude balloons, was continuing at its own steady pace. Simons, a physician from Pennsylvania who'd joined the Army Air Corps just as World War II ended, had been assigned to Wright-Patterson in 1947 just about the time Stapp left for Muroc. Tall, lanky, with thinning hair and pale blue eyes, Simons was an intense man who gave off a whiff of arrogance. Beginning back in the summer of 1951, Major Simons had conducted a total of thirty-nine balloon-borne experiments, some of them reaching heights above 100,000 feet. Twenty-six of the flights, using a spherical capsule design originally developed by the Navy, had carried live payloads: fruit flies, mice, hamsters, guinea pigs—and finally, cats, dogs, and monkeys.

Working with Minnesota contractors General Mills and Winzen Research, Simons was, by the summer of 1954, launching the first of a series of cosmic radiation research flights from Sault Sainte Marie, Michigan. The location's northern latitude—where extraterrestrial radiation is found at lower altitudes—provided an opportunity to expose the test subjects to primary cosmic rays. High-altitude pilots traveling beyond Earth atmosphere would be exposed to radiation emanating from deep space, and no one knew how dangerous this might be. How much damage would pilots be expected to subject themselves to? Black mice were used in some of these flights because when a radiation particle struck a hair follicle, the pigmentation was destroyed and the hair turned white. It was a perfect marker that allowed follow-up examiners to target a very specific location in the search for radiation effects on tissue and organs. On four of these flights, Java monkeys occupied the capsule, allowing Simons and team to study the effect of super-stratospheric altitude on the general behavior and performance of primates. If space travel were ever to become reality, the questions would all require answers.

The Balloon Branch at Holloman had been involved in top secret work such as Project Mogul well before Stapp's arrival. During Stapp's tenure, at the height of the Cold War, Project 119L, also known as Project Moby Dick, sent stratospheric balloon-borne cameras drifting over the USSR to photo-graph clandestine military sites. The balloons held the promise not only of high-quality surveillance, but also of valuable upper atmospheric research. Still, the General Mills-built plastic balloons were tricky to manufacture and launch, and even trickier to control. Remains of a Moby Dick crash on Russian soil discovered in 1956—the crash would have occurred at least two years earlier—generated a formal complaint about U.S. spying from the Soviet Union.

Stapp stayed in close touch with Simons the summer and fall of 1954. The two men shared a fascination with the effects of extreme conditions on living creatures. What Stapp had not yet shared with Simons was the thought that with a bigger capsule and adequate protection systems, he might support sending a human subject up above the atmosphere for an extended period. But they clearly weren't ready for anything like that yet. Two of the Java monkeys Simons sent up in balloons that year died in flight due to overheating.

Meanwhile, Stapp scheduled the next rocket sled run for August 20, 1954. This time it was to be an extreme windblast experiment intended to dupli-cate the effects of pilot ejection at full jet aircraft speed. As a result, *Sonic Wind* was equipped with a set of double doors in place of the windscreen. At peak speed, the doors would spring open, tripped by a track-mounted cam, to expose Stapp to, as he put it, "a slap of wind on my belly and chest." This time he would wear one of Charles Lombard's state-of-the-art helmets. Stapp traveled to Lombard's Impact Research Institute in Los Angeles and had new latex molds of his head and face created: "And there was every hair and wrinkle of my ugly countenance in pitiless, snow white relief." Lombard personally delivered the finished helmet, this model including a heavy-duty faceplate, to Holloman just in time for the run.

On the morning of the 20th, Nichols and Superata strapped Stapp into the sled with a shoulder harness and a chest belt, a lap belt, and leg straps. They helped him with the helmet and stuck the bite block between his teeth. They cinched the helmet to a set of restraints built into the headrest on the seat. His knees and ankles were tied together, and his feet were bound to the metal footrest. His hands and sleeves were secured and reinforced with loops of pressure-sensitive tape. A special nylon neck curtain had been fabricated to protect Stapp's throat from the worst of the windblast. It was a warm morning and Stapp, perspiring heavily, felt like a lobster on the boil: "I squirmed in the purgatory of sweat and re-breathed carbon dioxide."

When the countdown reached zero, eleven rockets—the most they'd ever used—fired simultaneously, slamming Stapp back into the seat with a lung-sucking fury. His eyes were fixed on the two gray doors directly in front of him, and—four and a half seconds after rocket firing—they sprung open. The windblast pounded Stapp just as *Sonic Wind* hit its peak speed of an incredible 509 miles per hour. Half a second following rocket burnout, the sled hit the water. Stapp surged against the restraints and the padding inside the front of the helmet. But everything held tight. The sled came to rest 88 feet from the end of the track.

Superata, who was the first to reach *Sonic Wind* after the run, got the impression that Stapp was suffocating inside the helmet, and sliced open his own hand in his struggle to free the helmet straps and get the faceplate off. Stapp was conscious and alert: "I will remember that particular breath of fresh air as long as I live." Stapp smiled broadly for the photographers who were already clicking away. With the exception of the withering heat, Stapp felt as well as he'd ever felt following a sled run and said it was the "easiest" of them all. He'd managed to avoid the dullness in his brain that he'd experienced on the March run. "I burst out laughing," he recalled. "I felt wonderful but woefully thirsty."

Since the purpose of the experiment had been to study the effects of massive windblast, the braking configuration had not been as radical as on the March run. Stapp sustained an average of only 10 g's—but for a prolonged duration of 2.7 seconds—as *Sonic Wind* came through the water. In addition, telemetry showed that Stapp had been exposed to 750 pounds per square foot of windblast pressure—a fearsome force in its own right. The post-run physical examination at the laboratory found no serious medical problems. Stapp had the familiar bruising from the shoulder straps, but the only other injuries appeared to be blood blisters, "half the size of a dime," where individual grains of sand had penetrated the flight suit, most of them on the arms and shoulders.

Flush in the glow of his survivor's euphoria, Stapp began talking almost immediately of a new sled that might permit him to clock 900 miles per hour. There was always more to learn, he told his team. The new design would be "light, simple and beautiful," and would come equipped with an aerodynamic Plexiglas canopy as on a fighter jet. In fact, he confided, he already had engineers at work on specifications so that he could get a proposal out to Northrop that fall.

George Nichols, relieved that Stapp had cheated serious injury once again, bit his tongue. He wasn't sure how much more of this he could stand.

15

SUBJECT HAD CONSIDERABLE APPREHENSION

> I considered myself to be expendable, just as any combat
> soldier is expendable in the service of the country . . . My life
> was no more precious to me than anyone else's was to them.
>
> —*John Paul Stapp,* **The San Antonio Light**

STAPP HAD CONTINUED to correspond with Celia Richards, the adoring young woman from Kerosene Flats who'd sailed off to Europe following her high school graduation. So when she wrote him that after more than a year of studying ballet in Paris and touring the Continent, she was returning to the States, Stapp—who was scheduled to be in New York at that time for a television interview—agreed to meet her ship and drive her across the country, back home to California. The trip would not turn out to be what either of them had imagined.

"John met me on the dock," she recalled. She was eighteen, and carefree. Stapp had just turned forty-four. "I had no luggage—just sacks. I had a couple of Gouda cheese balls and wooden shoes from Holland. He didn't recognize me because I had lost so much weight and matured so much. Changed. It was a moment of . . . he was smitten, right then and there. It was scary, because I'd never had that effect on anybody in my life."

Later, Stapp described his astonishment in a letter to a friend. "When she left for Europe she was 5 ft 1 inch in height, weighed 140 lbs and was a demure child. After ballet dancing, dieting and the excitement of glamorous

185

experiences, she returned weighing 109 lbs, a radiant, marvelous woman. Her shoes had holes and she was a little threadbare from budgeting, but she would look good in a paper bag."

Stapp announced that he would get them a hotel room for the night. Richards hesitated, wanting to make sure there wasn't any misunderstanding. As she would put it: "I was afraid. I had not had any ... experience." Stapp quickly assured her that they would have separate rooms. So why, she wondered, was he pulling out all the stops? Lobster in the hotel dining room, with Champagne. He certainly hadn't planned it this way, but it was as if all his years of deprivation and longing had come crashing down on him. He wasn't even sure he knew what he was doing. He had always cared about this young woman and now he thought, suddenly, that he might be in love with her—as if she were the reincarnation of his long-gone Baylor girlfriend.

"He was just very excitable. Somebody that has that much creativity inside him ..." Richards had to rebuff Stapp's advances. "It was all very stressful for me," she said. "He had known me as a child and seen me grow up. It was very confusing for us both."

It seemed to him as if most of the bad decisions he'd made in his life had involved women, and in some ways he was reluctant now to come down off Saint Simeon's 40-foot column. But he couldn't deny what he was feeling for Richards, despite the difference in their ages and despite the fact that he was supposed to be a father figure.

"He apologized the next morning," Richards recalled. "He apologized sincerely. He didn't have to but he did."

Stapp and Richards drove from New York to Ohio to visit Richards's aunt and uncle. They continued on to Alamogordo, and then to El Paso. Stapp and his brother Celso were close, but his relationship with Celso's wife, Day, was one he seemed to mostly tolerate.

"We knocked on the door," Richards said, "and she opened the door and said, 'Oh, John. And this must be your whore.'"

Richards was mortified. Stapp put his arm around Richards's shoulders. He leaned over and whispered, "Dismiss it."

Richards could only guess at what caused the "whore" comment. "A lot of women," she said, "looked at me and hated me. They looked at me in my turquoise skirt—which I was wearing that day; it *wasn't* provocative—and they hated me. I was confused constantly by life."

Stapp found himself increasingly resentful of the reception he and Richards had received, and annoyed by little things such as his brother's new television set. "They were huddled in the darkness like Neanderthals," he observed. "Dinner was in little tray-stands, grabbed and grunted over in the darkness. Conversation? Why even bother to teach the children how to talk? Came Liberace, nickering and wheedling like an old maid in heat, lasciviously titillating the piano with his phony virtuosity. My sister-in-law smirked like a proper plip, and gushed 'Isn't he wonderful?' I restrained a gush of projectile vomiting and bolted out into the clean, cool night. What a bilge box of prostituted electronics! New opiate of the people!"

The short stay in El Paso would turn out to be a turning point in the relationship between the two accomplished Stapp brothers. Notwithstanding their loyalty to each other, they had always been competitive—and now an undercurrent of resentment began to surface. Stapp opened up to William Bloom, the Hollywood producer: "Sure he [Celso] is a good doctor, but why doesn't he give my work as much understanding and appreciation as I give his? I'll admit the rivalry was intense between us four boys all through childhood, but I gave up being a child a long time ago." John Paul would spend years seething over Day Stapp's treatment of Celia Richards. "My sister-in-law," he wrote, "is a homely Mississippi farm girl with an inadequate personality." In a particularly churlish judgment—Celso would shortly become the chief of obstetrics and gynecology at Providence Memorial Hospital—he referred to his brother as a "somewhat fat headed he-midwife with medical delusions of omniscience."

Happy to put El Paso in the rearview mirror, Stapp and Richards continued across the desert and on to California. At some point during the drive, Stapp asked Celia Richards to marry him. He had never met anyone, he told her, who affected him the way she did. He wouldn't pressure her, he promised, but he wanted to know if there was a chance. She told him she didn't think so. She admired him—revered him. Still, she couldn't imagine them as man and wife. "That just wasn't how I thought of John." She didn't completely slam the door—at least in Stapp's mind, she didn't—but she was trying in her way and with all her might to push it firmly shut. Reluctantly, Stapp left her at her parents' house in Lancaster and returned to Holloman. He did not, however, abandon hope.

Over the next few weeks, Stapp lost 16 pounds on a diet of coffee, buttermilk, and carefully rationed orange juice. He lost two inches from his waistline. "I am making a studied effort," he wrote to William Bloom in Los Angeles, who had become his confidant during this period, "to rid myself of microscope slouch, to sit and stand in good posture. This morning a married lady I know asked me what I had done to make myself suddenly appear so much taller. I told her I was in my second adolescence. Honestly, I have never felt better. As for looks, it is a comfort to be less repulsive."

In the meantime, Celia Richards began to blossom. She landed a job teaching dance—"ballet and ballroom and jazz and everything!"—at a well-regarded studio in Sacramento. She quickly became a minor sensation. She auditioned for Gower Champion and was accepted into Bob Fosse's Music Circus theater group. One day, however, she returned to her apartment from rehearsal and found that her parents had arrived and donated all of her possessions to the Goodwill while she'd been at work. Richards's mother was a manic-depressive who'd had a crush on Stapp back in the Muroc days, and her father was a bitter, abusive man; both of them disapproved of the entertainment business. They told her they were taking her home to Lancaster, loaded her into their station wagon, and drove away that same day.

"My whole world," she lamented, "was crushed." She couldn't help but think about what a life with John Paul Stapp might be like. He could be, she believed, her ticket to another world, one of brilliance and importance and excitement, far from Lancaster and the smothering attention of her parents.

As always, Stapp had his work to distract him. Rocket sleds and balloons weren't the only projects he was pursuing in 1954. Project 7850 had been established to attack a broad scope of issues related to human space travel, and one of those was gravity—or the lack of it. German aeromedical researchers such as Hubertus Strughold and Heinz von Diringshofen had done some theoretical work prior to the end of World War II, but the most important subgravity experiments had been conducted by the U.S. Air Force with animal subjects in V-2 rockets launched from White Sands in 1948. Those flights had exposed rhesus monkeys to as much as three minutes of weightlessness, and found "no serious ill effects" that could be ascribed to the condition.

The next slate of Air Force subgravity experiments in the early 1950s employed high-altitude Aerobee research rockets using both monkeys and mice. Russian researchers were also conducting subgravity flights with their preferred test subjects, dogs. Interestingly, while the USAF animal experimentation—including all of John Paul Stapp's research—used anesthetized subjects, the Russian dogs were apparently awake and aware throughout their flights.

A breakthrough in subgravity exposure test design had occurred in May of 1950, when two German scientists, Fritz and Heinz Haber—working by that time at the Air Force School of Aviation Medicine—published a paper detailing their development of an airplane flight profile they termed a "Keplerian trajectory." The Habers' trajectory was a great parabolic arc in the sky in which "centrifugal force would exactly offset the downward pull of gravity and engine thrust would counterbalance air friction." The technique took advantage of the phenomenon an elevator passenger experiences when an ascending elevator comes to a stop: that momentary sensation of floating as the force of gravity briefly dissipates. The problem was, these subgravity profiles were devilishly tricky to fly. Skilled pilots such as Chuck Yeager and Scott Crossfield had both flown the trajectories at Edwards in 1951, but neither had managed to achieve more than a few seconds of zero gravity.

The Habers weren't the only Germans working for the U.S. on problems relating to space travel. Another was Wernher von Braun, a leading figure in rocket technology who'd been brought to the White Sands Proving Grounds as part of Operation Paperclip. And then there was an Austrian-born German researcher named Harald von Beckh, who had been conducting airborne subgravity orientation experiments with turtles in Argentina after the war, attempting to discover whether body position with respect to the Earth's surface might affect a space traveler's ability to perform tasks in a weightless environment. Von Beckh found that his turtles were able to improve basic physical performance through repeated exposure. He would shortly be invited to Holloman to assume direction of the Air Force's Aero Medical Field Laboratory's subgravity program, and would establish himself as the most flamboyantly bizarre of the group of German researchers at Holloman. Von Beckh liked to raise the Nazi flag above his house in Alamogordo, blast German martial music from his hi-fi, and shoot fireworks at the night sky.

Despite the value of some of his research, he seems to have been regarded more as a clown than as any sort of threat to peace and good will in the town. According to one American researcher who knew him in Alamogordo, Von Beckh was "a screaming screwball."

Stapp had no problem working with the German aeromedical men at Holloman, but he seems to have resented them. He continued for years to refer to some of them as Nazis. Guests at Stapp's dinner parties recall him bristling about the policies that had brought what he believed was, for the most part, second-rate talent to the U.S. Some of the Operation Paperclip Germans, in his estimation, probably deserved prison sentences.

One case in particular was Hubertus Strughold. While Stapp had been studying bullfrogs at the University of Texas, Strughold was already the top aeromedical man for the Luftwaffe. At about the time Stapp was taking up residence in the Desert Rat at Muroc, Strughold arrived in the United States and accepted a position at Randolph Field. Two years later—despite Stapp's judgment: he'd met Strughold and thought him a fool—he was named the first Professor of Space Medicine at the Air Force's School of Aviation Medicine. Strughold and von Braun, only a few years removed from the German war effort, were becoming the celebrity faces of American space science and engineering. Stapp was determined not to let them own it.

Stapp had vented to dinner guests about his frustration with the funding and recognition the captured Germans got for results he judged to be overrated. He believed he could do better. By 1954, subgravity research was seen as some of the most important biomedical work at Holloman, and so Stapp authorized David Simons to ramp up his own zero gravity studies that summer. At this same time, while the difficulties of flying the technically challenging parabolas had yet to be mastered, a talented Holloman test pilot named Joe Kittinger had come to Stapp's attention.

Floridian Joseph William Kittinger flew anything with wings. He'd been at Holloman less than a year, but he already had a reputation as one of the hardest-driving and most competitive pilots on the base. Stapp called Captain Kittinger into his office and explained how the Habers' flight pattern worked. Great discipline was required. It was critical that the mission be executed precisely. At about 12,000 feet, the pilot would put the plane into a 10-degree power dive, achieving a predetermined airspeed—at which point he would pull up into a 35-degree climb while reducing power. This is where the parabola shape of the profile begins: the aircraft easing over the top of a huge arc, eventually curving back down into a 35-degree dive. At the crown section of the parabola, the occupants would—if the profile was flown correctly—experience anywhere from fifteen to sixty seconds of something very like the weightlessness of space flight.

The twenty-five-year-old Kittinger had trained as a fighter pilot, but—to his intense frustration—had just missed combat in Korea. He'd heard the warnings that space research was a dead end for a pilot's career. The conservative Air Force brass remained convinced that outer space was the stuff of Buck Rogers, comic books, and sci-fi movies, not a serious problem of practical engineering. Yet after listening to Stapp describe the peculiar challenges of subgravity research and the lab's need for a dedicated pilot, Kittinger was intrigued. "Hell, I didn't care," he said of the advice not to get mixed up in the world of space research. "It sounded like it would be a hell of a lot of fun out there with Stapp." Kittinger had never met anyone quite like Lieutenant Colonel Stapp, and it took the young pilot only a few minutes to make up his mind. He volunteered for the assignment on the spot, and within a week Stapp managed to get him reassigned.

In September of 1954, Kittinger began practicing the Keplerian profiles in a T-33 and quickly mastered the technique. The goal was to provide the maximum zero gravity interval at the top of the arc, and Kittinger's flights began consistently producing zero-g durations of greater than thirty seconds. The first piece of test instrumentation for the subgravity flights was a golf ball suspended by a string from the roof of the cockpit. When Kittinger saw the ball "floating," he knew he'd hit it right. One of the early researchers in the back seat of the T-33 took a cat along to observe its reaction to the loss of gravitational pull, and found that the cat adjusted calmly to the weightless environment. During one of the parabolas, Kittinger recalled, the cat floated freely up into the front cockpit, and a gentle push reversed its momentum and sent it sailing rearward again. Unfortunately, not all of the human subjects reacted so well to the sensation of weightlessness; disorientation and nausea were common complaints. Even more unfortunately, the program's project officer, Major David Simons, was one of those who had a tendency to get sick—making him the target of some ribbing from Kittinger.

The subgravity work at Holloman under Stapp was essential to qualifying human beings for space travel, and the following year Stapp would transfer the space biodynamics project to the control of Captain Grover Schock,
who would go on to become the first American scientist to receive a PhD in
space physiology.

. . .

In October of 1954, seven months after Stapp's inaugural rocket sled run at
Holloman, a Northrop pilot in California became the first man to eject from
an aircraft at supersonic speed. Unfortunately, he did not survive. Stapp had
been waiting for an opportunity to investigate just such an accident, and he
interrupted his work at Holloman, flying west to perform the autopsy. What
Stapp wanted to know was whether the pilot had suffered major harm from
windblast, tumbling, massive deceleration, or some combination of the three.
His post-autopsy conclusion was that the pilot had been killed instantly
when he'd struck the tail surface of the aircraft. The investigation revealed

*Stapp (on right) experiencing simulated zero gravity at the apex of a Keplerian
trajectory. These weightlessness studies were an important prerequisite to
manned space flight. (Photo courtesy of U.S. Air Force)*

nothing to indicate that the biomechanical forces involved had contributed to the fatality.

Meanwhile, even as Stapp's attention was drawn temporarily westward, Hollywood's interest in the scientific goings-on in southern New Mexico had intensified. Stapp had completed his seduction of the contingent from 20th Century Fox that summer, allowing producer William Bloom and screen-writer Frank Cockrell to fully immerse themselves in the world of Holloman and the work of Stapp and his team. They read the technical reports and inspected the equipment, they interviewed the various crews and shot rolls of test film, and they sat spellbound for hours as Stapp regaled them with stories of his Edwards days and teased them with the coming attraction of what they all began to refer to as the Big Run. Stapp promised Bloom it would top 600 miles per hour, and possibly go as high as 800: "I am betting that in spite of considerable gut juggling and brain bouncing . . . I will get away with nothing worse than bruises, perhaps 2 or 3 days of that sickening concus-sion headache, and not more than two weeks of feeling under the weather. My feeling about it? I have none." As Cockrell worked on his screenplay for the movie, he wrote Stapp every few days with questions about how things worked. Stapp's replies often ran to several pages. The Fox men gave him something he valued greatly in return for his cooperation: a willing audience. "How refreshing," he wrote, "to get letters from people who have something besides orderly rows of nuts and bolts lining their souls."

Stapp's superiors were, of course, aware of the film project and had agreed to offer the Air Force's cooperation, but General Flickinger in Bal-timore was never aware of the extent to which Stapp had been sharing his thoughts (and more) with Hollywood. Not only had Stapp been clandestinely forwarding copies of Northrop's films of the deceleration tests to William Bloom (and even allowed Bloom to specify the type of film to be used), on occasion he sent Cockrell sketches of designs for a new rocket sled. "My own ideas are not classified," Stapp explained, "until I start putting down perfor-mance figures and dimensions. But for the sake of all that is holy, don't say one word of this to my bosses. They are going to be told one step at a time of all this. Please don't sell me down the river."

Darryl Zanuck, head of 20th Century Fox, had imposed a deadline of December 1, 1954, for Frank Cockrell's final screenplay for the movie they had originally titled "Windblast" but were now calling "Space Medicine." This meant that William Bloom would be free to travel to New Mexico as Lieutenant Colonel Stapp's guest to witness the Big Run, which had been scheduled for the 2nd, very likely as a courtesy to Bloom. As November waned, however, Stapp was forced to postpone the run a few days. He finally settled on December 10, seven years to the day after his first rocket sled ride in the Mojave Desert.

George Nichols dreaded the approach of the Big Run, fearing for Stapp's eyes: "It was an extremely tense time for me. I had nothing but sleepless nights for a couple of weeks leading up to it. I fully expected total blindness coming out of that run. I didn't think the eyes could possibly withstand it." Stapp prepared himself for the worst by practicing his household rituals in the dark: showering, dressing himself, moving around his house, teaching himself to prepare simple meals.

The Holloman-based team, meanwhile, worked around the clock to make sure nothing would be left to chance. They had been running chimps at speeds beyond 600 miles per hour to ensure they had the restraint configuration exactly right. While they wanted to collect more windblast data on the Big Run, their jettisonable windscreen had proven unreliable and the swinging-door system they'd used for the August run simply added too much weight. The team, with Stapp's approval, made the decision to go without any windscreen or barrier at all. Stapp would wear his new helmet and would make sure his limbs were tightly secured, but he would otherwise face near-Mach 1 velocity exposed to the elements. George Nichols continued to argue for some sort of canopy to enclose Stapp, but found himself overruled.

Stapp didn't sleep well the evening of December 9, filled with both anticipation and the familiar pre-run depression. He'd stood alone in the backyard of the house on Lovers Lane and watched the sunset, in case it turned out to be his last. The temperature dropped down to near freezing overnight, but the forecast was for a pleasant day in the mid-sixties.

As usual on the morning of a rocket sled run, Stapp ate nothing for breakfast. Waking at 4:30, he spent an hour answering correspondence and thumbing through medical journals. At six, he strapped himself into his Cadillac, which he'd had his technical team outfit with the latest seatbelt technology, and headed for the base. On the drive, it occurred to him that he ought to have spent some time learning Braille. The School for the Blind was just across the street from his house and he saw the students practicing with their white canes all the time. He could easily have taken a course.

As he pulled through the Holloman main gate, he could feel the familiar foreboding. It was, in Stapp's words, a "rational, fatalistic depression with nearly total divorce from affect." In spite of his obsession with these experiments and his insistence on appointing himself as their test subject—this would be his twenty-ninth rocket sled ride—he truly hated these pre-run hours. Even his trusted colleagues were reluctant to approach him prior to a run. "I relax inner discipline afterwards," he had written to Bill Bloom that summer, "but it is grim and purpose without feeling beforehand."

It was only barely light by the time Stapp got to the track, fresh from his pre-run physical. He noticed a dozen or so jeeps lined up at the far end. The Air Force had invited the press in for this one, and the photographers were setting up tripods and trying to figure out the morning light. The Aero Med Lab had its own crews in place. There were two Wright-Patterson photographic systems experts operating three rare high-speed 35mm color cameras with special Wollensak lenses. Another team was operating a high-speed, ribbon-frame camera that would produce film fit for engineering analysts to study. A Holloman film crew was in position with multiple cameras, and Northrop had its own engineer onsite with the most advanced of the company's photo systems.

The experiment, however, was on hold due to a thin band of cloud overhead. They needed bright sun for the filming. In the meantime, there were a thousand other details that needed attending to. One of the peculiar problems with the Holloman braking system was that coyotes and desert birds were drawn to the pools of water between the rails. Two men with binoculars were assigned to monitor the water and alert the team to any intruders. Nobody wanted to find out what might happen if Stapp hit a Mexican chickadee at several hundred miles per hour.

Even after years of study, none of them—Stapp included—had any idea where the human limit of tolerance to g-forces really was. There was a belief among a contingent at the Aero Med Lab that they were all preparing for a slaughter, but there had been doubters from the start. As Stapp had explained it to the magazine and newspaper people in the days leading up to the Big Run, their mission was to probe, very deliberately, what he referred to as the "onset of injury"—that point at which the human vessel begins to break down. This was the only way they could establish the point of onset definitively, which Stapp was honest about: there was always the very real possibility that they could overshoot that limit. The math and the chimpanzees could only result in an educated guess. Short of the Nazi approach to human experimentation, the only answer was a willing volunteer.

The clouds hung around through the morning hours, and it wasn't until a few minutes after noon that the hold was finally lifted. Stapp had been seated on *Sonic Wind*, positioned just eight inches from the southern end of the track, for almost an hour and a half at that point.

Stapp had assigned Major David Simons as flight surgeon for this run, and Simons assisted Jake Superata with the restraints, deliberately securing each of the 6,000-pound-test nylon straps. Stapp sat still as he was slowly immobilized, bound tightly around his ankles, his knees, thighs, elbows, and wrists. They attached and tightened the four 2,000-pound-test nylon straps that would hold Red Lombard's helmet firmly to the headrest. Lombard had equipped this particular model with a Plexiglas face shield and an improved nylon skirt that encircled Stapp's neck to protect against sand and debris. They left the chest straps for last because once they cinched those down, Stapp would have to struggle just to fill his lungs with air.

Though the sun had yet to break through the overcast, Stapp was already heating up in the wool flight suit and wishing they could get on with it. This was what it must be like, he thought, preparing to face a firing squad. Several weeks earlier, Stapp had summoned his new favorite pilot, Captain Joe Kittinger, to his office and explained that he wanted professional aerial documentation of the Big Run. He wanted Kittinger to get a jet lined up and to come screaming overhead a couple hundred feet off the desert floor just as the rocket sled was launched down the track. It would be an extremely tricky bit of timing, and Kittinger had been practicing for days in a T-33. He would need to be over Stapp's starting position at an airspeed of 350 miles per hour and accelerating to allow him to track with *Sonic Wind* as it rifled toward the water brakes.

As Stapp sat trussed up like a mummy, his knees and hips aching, waiting impatiently for the sun, he knew that Kittinger was already airborne and circling somewhere to the south.

At the precise moment the sun finally broke through and bathed them all in a welcome glow, Stapp saw a flare shoot skyward signaling the start of the countdown. It was now 12:40 p.m.

"Two minutes!" The voice from the loudspeakers was clear even through the fiberglass helmet. Superata positioned the gum-rubber bite block between Stapp's teeth, then he lowered and secured the helmet's face shield. Stapp winced as a pair of strong hands gave the double-thickness chest straps a final tightening tug. The straps compressed his lungs so tightly that Stapp labored to breathe. George Nichols, who'd been hovering nearby, gave all the restraints one final check. When he was satisfied, he slapped Stapp on the knee. Rather than sprint the hundred feet to the blockhouse, as was their usual procedure, Nichols and Superata hopped into Nichols's car and sped to the opposite end of the track where the sled would come to rest. They wanted to be able to get to Stapp as quickly as they could when it was all over. They weren't talking about their concerns—to each other or to Stapp—but they'd all worked together long enough that none of them had to say it.

The trusted Northrop crew prepares the test subject for the Big Run on the morning of December 10, 1954. Stapp described himself as "grim and without purpose." (Photo courtesy of U.S. Air Force)

This was the moment Stapp dreaded most, the final piece of waiting. He avoided the word "fear," but in his own written record of the event, he offered a view of his state of mind that was the closest thing to it: "Subject had considerable apprehension and uneasiness with cold sweat of axillae and palms."

From Stapp's view the rocket sled rails ran to the edge of the earth. The sky was suddenly deep blue; the remnants of the morning clouds seemed to have evaporated in mere seconds. He could see mirage lakes forming and shimmering in the pink stretch of desert before him.

As T minus one minute approached, Stapp felt suddenly light-headed, as if he was suffocating. He tried to speed up his breathing, gulping for air. He pressed himself into the padded seat back, chomped down on the bite block, and began breathing as hard and fast as he could. The countdown reached single digits. When he heard *"Three!"* he pulled hard on the lanyard that triggered the onboard camera.

At the same instant he felt the thunder of the T-33 come screaming up behind him, Stapp clearly heard the dry pop of the igniters. Precisely 0.067

As the nine rockets fire, Joe Kittinger in a T-33 above hits his mark at 350 miles per hour. Stapp and Sonic Wind *would, from a dead stop, pull away from the jet and outrace it to the brakes. (Photo courtesy of U.S. Air Force)*

Stapp on Sonic Wind: *faster than a speeding bullet. The man operating the camera that got this shot dove for a trackside foxhole at about this point. (Photo courtesy of U.S. Air Force)*

seconds later, the seat slammed him in the back. A 35-foot river of flame shot rearward as nine rockets fired simultaneously, delivering 45,000 pounds of bone-humming thrust in an instant. The g-forces ground him into the seat. His vision blurred and quickly went black.

Five seconds later, Stapp hit rocket burnout at a top speed of 639 miles per hour, moving faster at that moment than a .45 caliber bullet shot from a pistol. Joe Kittinger and the photographer in the jet above—they'd hit their mark perfectly and were even with Stapp as the rockets fired—were astonished to watch Stapp in *Sonic Wind* below pulling away from them.

As the rockets went silent, the terrible pressure in Stapp's back vanished and he had a floating sensation for a sliver of a moment as he sailed into the brakes.

One very brave cameraman who was positioned between the rails at the end of the track got footage of the sled bearing down on him. At the last instant, he dove for a nearby foxhole that had been carved out for just this purpose.

The sled's frontmost scoop cracked the first of the dams and Stapp hit the water. In a microsecond, all momentum was reversed. His organs and fluids surged forward and Stapp's visual field went from black to yellow in a stroboscopic flash. The pusher section of the sled detached and was left behind. The forward section continued through the second dam. The last residue of breath was crushed from Stapp's lungs and a terrible needle-like pain hit him in his gut. Then the pain all seemed to shoot into his face as the sheer force attempted to gouge the eyeballs from their sockets. He described the sensation as "somewhat like the extraction of a molar without anesthetic." The world went from pink to red in "a salmon-colored blur of scintillating patterns."

It was a few moments before Stapp realized it was over. He was no longer moving, having come to rest a mere 20 feet short of the end of the track. He was vaguely aware of the crackling roar of the T-33 diminishing and the fountain of water from the brakes spattering all around him. He was trying to breathe, sucking on a wind that wasn't there. He couldn't seem to pull anything into his lungs and felt himself slumping numbly forward. Then he became aware that hands were all over him, one of them with a knife slashing at the straps. It was hard for him to make out what the voices around him were saying. Somebody pulled the bite block out of his mouth, while somebody else freed the chest straps. Stapp gasped but still couldn't seem to get any air.

They pulled him out of the seat and off the sled. He was on his back on the ground and they were talking to him. Stapp could hear the reassuring voice of George Nichols. Nichols had insisted on parking his car within 20 feet of the edge of the track and had been the first one to reach the sled. He got there so fast the cascade of falling water soaked him.

They got the helmet off and David Simons stuck an ammonia-soaked sponge under Stapp's nose—with no effect. Stapp could feel fingers and thumbs trying to press an oxygen mask onto his face, but he waved it away and put his fingers to his eyes and pushed the lids open. He could feel the eyeballs bulging against the orbit bones.

Stapp turned in the direction of Nichols's voice and asked: "What do my eyes look like?"

Years afterward, George Nichols teared up as he described the scene that day. "I looked at his face and it was just . . . horrible. I had nightmares for months because all it looked like was just total blood." Nichols recalled the conversation:

Stapp asked again: "What do my eyes look like?"

"It's all just blood," Nichols told him.

"Does it look like anything's leaning on the eyeball, on the lens?" Stapp asked.

"Nope," Nichols said. "It's just blood."

"Maybe the retinas are still in place," Stapp said. He found himself wondering if any of the chimps had seen this same sizzling pink as they'd come to following a high-speed run.

One reporter claimed he heard Stapp say: "I can't see." Nichols denied it: "He was very calm. He didn't *say* he couldn't see. But his eyes were all red and I knew he wasn't seeing anything."

By this point the news photographers had made their way to the sled and Stapp could hear their cameras chattering. He could still feel a pain somewhere inside him. Intestines or kidneys? He wasn't yet sure.

Major Simons had a blood pressure cuff on Stapp's arm and a stethoscope on his chest. He later reported that "the subject's face began to turn cyanotic." Stapp was turning blue. They lifted him onto a stretcher in order to carry him to the ambulance, but Stapp protested. He told them he wanted to see if he could walk.

Nichols and the rest of them helped Stapp up and supported him, but after a couple of staggering steps, he allowed them to lay him back on the stretcher and slide him inside the ambulance. Throughout, Stapp kept his eyelids pinched open in case his sight returned.

PART III

ABOVE THE ARMSTRONG LINE

16

THE CONFERENCE

> These deaths, for the most part, occur because the motorcar manufacturers make no provision whatsoever for the control of the occupants when they must decelerate rapidly.
>
> —*Dr. Horace E. Campbell, Surgery,*
> **Journal of the American College of Surgeons**

WHEN NICHOLS AND SUPERATA and the rest of the Northrop men arrived at the hospital about four hours after the Big Run, Stapp's eyes were still bloody lumps. They all cringed when they saw him.

"He said he could see shapes by then," Nichols said, "but I asked him and he couldn't tell me how many of us were standing there beside the bed."

Stapp later said he had seen a flash—just a quick flash—of blue sky in the split second before they'd loaded him into the ambulance.

The men returned on the morning of the 11th and found Stapp somewhat improved. "He could count us by that point," Nichols said, "but he couldn't really identify us. He didn't know who was who. But by the end of the day, he was able to recognize and identify people."

Stapp had survived his foray across the onset-of-injury borderline, and his doctors were at least tentatively encouraged. His eyesight was clearly improving, and tests had found no serious internal injuries, which is not to say the patient wasn't suffering. Stapp had severe bruising on his shoulders and upper arms, and his torso was flecked with small blood blisters. What amazed the hospital staff was Stapp's energy and humor. It was as if the Big Run had charged him up with a long-lasting shot of adrenaline.

On his third day in the hospital, Stapp invited photographers in. He had two massive shiners, and his eyeballs were still a frightening sight, but everyone seemed relieved to see the fastest man on earth sitting up in bed, grinning and cracking jokes. Stapp berated reporters who suggested that he'd been knocked unconscious, taking it as a personal insult.

"I've never been unconscious after a sled run in my life!" he told them. He did his best to shrug the whole thing off. "They brought me to the hospital and I had two lunches because I hadn't had any breakfast." Only days after the Big Run, Stapp—still in the grip of the most intense case of survivor's euphoria he'd ever experienced—was eagerly talking about his next one. In fact, he said, he'd begun planning for a run he wanted to make using the high-speed track at Hurricane Mesa in Utah, where the sled would run off the edge of a cliff and fall beneath a big drogue parachute. For his next deceleration test, he told the press, Northrop was already building his new sled. It would be called *Sonic Wind II*, and aboard it he intended to achieve a speed

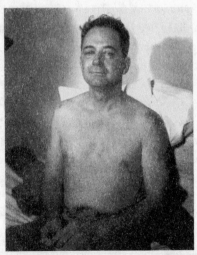

In the hospital two days after the Big Run. As he waited for his vision to improve, Stapp was thinking ahead to a 1,000 miles per hour ride. (Photo courtesy of U.S. Air Force)

in excess of 1,000 miles per hour. The reporters ate it up. The *El Paso Times* proclaimed: "STAPP OUT TO BREAK SOUND BARRIER."

The doctors remained concerned. The ophthalmologist who examined Stapp and treated him following the Big Run described his observations: "Retinal vascular spasm was evidently present, followed by retinal edema and serous exudates and, within four hours, by fogging of vision in both eyes. The next day, dimness remained in the right eye, with paracentral and peripheral scotomata. The vision in the left eye had cleared, although a peripheral hemorrhage producing scotomata existed." The same physician noted: "One might possibly expect, after 29 experiments, brain damage such as is found in a 'punch drunk' pugilist. The contrary is the case, as any one can testify after a few minutes' conversation with Col. Stapp."

The Holloman team's final report on the Big Run told the story. The deceleration Stapp had experienced was judged to be the force equivalent of an open-seat ejection from a jet traveling at 1,800 miles per hour at an altitude of 36,000 feet. The astonishing thing was that he emerged not only alive but fully conscious. "I felt like a fly riding the nose of a bullet," he said. It was a testament to the team's ingenuity and painstaking obsession with detail, as well as to Stapp's grit. It also established, to a greater degree than ever before, that airmen could withstand severe punishment previously thought disastrous. From a near supersonic top speed of 639 miles per hour to zero in 1.37 seconds. Mach .9—measured at 937 feet per second—to a full stop in an instant. As *Sonic Wind* hit the water brakes, Stapp had been exposed to 46.2 g's—a force of almost four tons—and had walked away, albeit only a couple of steps and with some assistance. An Air Force general compared the jolt Stapp had endured with that "an auto driver would experience were he to crash into a solid brick wall at 120 miles per hour."

Word of Stapp's latest achievement traveled almost as fast as he had. He got a letter from his father on the day after Christmas: "Now I have read the daily papers and to-night heard your voice on the radio, friends and neighbors have talked about you and the phone has been ringing too." Charles included a note from his second wife, Pearl: "It was exciting to hear about you and also to hear your voice on the radio. But it gives me cold chills up and down my spine to think of what you are doing . . . I surely do wish you would be satisfied now and not go on another even worse test! This last one just seems terrible to me!"

William Bloom had traveled from Los Angeles to Alamogordo by train to witness the Big Run, and had gotten a chance to visit Stapp in the hospital only briefly before returning to Hollywood to work on the movie. Bloom brought with him fresh reels of rocket sled footage that he hoped might form the visual centerpiece of the film. Just before New Year's, Stapp—still suffering from blurry vision in one eye—drove his Cadillac across the desert to California. He did not see Bloom or Frank Cockrell or anyone at 20th Century Fox. Instead, he headed for San Diego.

When Celia Richards's parents had returned her to Lancaster after practically kidnapping her, she'd written to Stapp, desperately seeking advice. He'd encouraged her to apply to college and suggested San Diego State. He wired her the money for tuition, and had made a brief visit to help her enroll there just in time to begin the fall term. She declared a major in German and a minor in Spanish, both of which Stapp said he'd be available to help her with.

When Stapp arrived in the Cadillac, Richards found his appearance and behavior worrisome. The bruising around his eyes was still prominent, and the eyes themselves were still flecked with blood from the ruptured capillaries. She'd commented that the only thing missing in her life was music, and Stapp insisted on buying her the latest hi-fi equipment and spent nearly an entire day attempting to connect it and set it up in her apartment. She was surprised to see him fumbling with what should have been—for someone of

Celia Richards, the beautiful redhead from Kerosene Flats. Stapp never forgot her. (Photo courtesy of Celia Richards Wilson)

his capabilities—a simple wiring task. She thought he seemed confused and unsteady, and she attributed it to what he'd suffered on the Big Run.

Before returning to Holloman, Stapp once again proposed marriage. She listened to his proposal, but again she said no. She was immensely grateful for everything he had done, she said, but while she was tempted she knew that marriage would be a mistake and she held firm. As Stapp drove away, she wondered whether she should have tried to stop him and whether he ought to be behind the wheel at all.

. . .

A still groggy but jubilant Stapp had called Air Force headquarters from his hospital bed late on the afternoon of the Big Run to report a mission accomplished. Yet despite ambitious plans for a spectacular 1,000-miles-per-hour rocket sled run at Hurricane Mesa, General Flickinger issued orders that any future human experiments on the high-speed tracks would require his specific approval in advance, and it was made clear—if not explicit—that there were no plans to grant such approval in Stapp's case. There would be no appeals. Stapp was now considered too valuable an asset for the Air Force to risk him on a rocket sled. As always, Stapp was aggravated by what he viewed as paranoid preventive discipline, and he complained bitterly. Flickinger's caution was vindicated, however—though Stapp would never have admitted it—when a high-speed chimpanzee run on January 25, 1955, ended in disaster. At 640 miles per hour, one of the steel slippers coupling the decelerator to the rail fractured and split apart. The sled hurtled off the end of the track, tumbling end over end like a one-ton bowling pin across the desert floor for 1,600 feet, decapitating the subject.

If the dream of Hurricane Mesa was dead and the days of human testing on the rocket sleds were coming to an end, Stapp would need to begin looking for his test cases in the real world. He had been waiting for several years now for the first supersonic bailout survivor. On February 26, just a month after the disastrous chimp run at Holloman, a North American Aviation test pilot named George Smith became the first man to survive such an ejection. Smith lost control of his F-100A Super Sabre off the coast of Newport, California, and punched out at a speed of Mach 1.05 and an altitude of about 6,500 feet. The impact knocked him unconscious. Luckily, he came down under a

damaged parachute canopy a few yards from a Coast Guard boat on a training mission and was rescued.

Smith didn't come to for five days, and his memories of the event would
always be spotty, but based on eyewitness accounts and aircraft debris
recovered from Newport Bay by Navy divers, Stapp—who was brought in for
the investigation—was able to piece together the ordeal. Smith's helmet had
been ripped off by the windblast almost instantly. Stapp estimated that he
may have sustained as much as 60 g's, and concluded that so much air had
been blown into Smith's stomach as he'd exited the plane that he floated like
a cork when he hit the water.

Stapp, believing that internal injuries were likely, recommended exploratory abdominal surgery. Hospital personnel demurred. Stapp insisted that
they cut the patient open, but the attending surgeon at the small civilian
facility in Newport Beach refused. Three days later, with jaundice setting in,
emergency exploratory surgery discovered a tear in Smith's bowel and significant liver damage.

"So," Stapp explained later, "they had to cut things apart and rejoin them."
Smith recovered and was able to leave the hospital after three months. The
ejection system had saved the pilot's life, but engineers such as Red Lombard
went back to work on better ways to keep helmets secured to pilots' heads.
Smith—who lost 65 pounds and much of his liver function as a result of the
accident—was back flying shortly after his release from the hospital, a living
testament to the latest USAF survival systems.

Not long after the Smith ejection, experimentation targeting a variety of
problems related to tolerance to mechanical force was being conducted at
Holloman under Stapp's command. The challenge of how to rescue pilots in
distress, especially at high speeds and extreme altitudes, was still the driving force behind the research. Despite important advances in safety systems, some 23 percent of pilot ejections still resulted in fatalities, and Stapp
remained unsatisfied. With the exception of Flickinger's—in Stapp's mind—
overly intrusive concern for his own personal safety, he was operating now
with practically free rein, and for the first time was able to watch his prodigious ideas transformed directly into practical hardware. The rocket sled track
was always the most prominent Aero Med Field Lab venue—NASA astronauts
orbiting the earth in the years ahead would report spotting the gleaming Hol-

loman rails from space—but the rocket-powered runs were expensive as well as dangerous, and Stapp had come to believe that there might be easier and cheaper ways to gather some of the complex impact data he needed.

One of Stapp's designs, built by Northrop to his specifications, was called the Crash Restraint Demonstrator. It consisted of an 800-pound aluminum chair-sled on a 15-foot section of track bolted to a concrete platform inside one of the Aero Medical buildings. A motorized winch pulling an elastic shock cord drew the chair back like a slingshot, and a cock-and-release lever fired the assembly. Once released, the chair shot forward until it hit a set of highly controllable spring-tensioned brake pads. The Bopper, as Stapp's slingshot came to be known, could deliver a full 30 g's. This valuable piece of test gear would soon be in demand not only by the Air Force but eventually by NASA's Projects Mercury, Gemini, and Apollo. Another of Stapp's impact devices was called the Swing Seat. It was basically an aircraft seat suspended by a cable that was lifted, released, and allowed to smack into steel pilings.

Preparations for a 1957 test of an early bandolier-style seatbelt on the Bopper at Holloman Air Force Base. Anesthetized chimpanzee subjects provided crucial data that helped define safe limits for human tests. (Photo courtesy of U.S. Air Force)

If he could dream up a way to subject men to deceleration or impact punishment, his team found a way to build it, and Stapp insisted on trying out all of these devices personally before turning them over to volunteers.

Stapp's most versatile and successful Holloman brainchild was the Daisy Track, a 120-foot test track constructed of tubular steel rails and designed to fire a sled powered by compressed air. Stapp named it after the Daisy air rifle he'd owned as a teenager in Texas. It was the first track apparatus built specifically and exclusively to study human tolerance to impact rather than deceleration forces, and it was used for a wide range of aeromedical experiments, including the simulation of aircraft crashes. A sled on the Daisy Track could achieve a full 100 miles per hour—and deliver 200 g's—in a distance of a mere 40 feet, and could be stopped in just 5 inches by a perforated water-filled cylinder at the terminal end. Despite Stapp's banishment from the high-speed rocket sled, he served—anonymously and without seeking permission—as subject for a number of lower-profile Daisy Track experiments.

The Daisy Track may have been Stapp's most valuable invention. In a distance of just 40 feet, the device could deliver up to 200 g's to a test subject. Thousands of experiments were conducted with this apparatus over the years, many of them on behalf of NASA. The Daisy Track was nearly dismantled and sold for scrap in the 1990s. It is preserved today at the New Mexico Museum of Space History. (Photo courtesy of U.S. Air Force)

Stapp's biodynamics research continued in full force at Holloman throughout the 1950s and beyond, and the Daisy Track was in almost constant demand by the Air Force and the Army—and later by NASA and, for a while, by the automobile industry and the Department of Transportation. By the time of its decommissioning in 1970, the Daisy Track had been the venue for more than 5,000 human experiments. One of the wildest would occur on May 16, 1958, when Air Force Captain Eli "Lack" Beeding—like Stapp, a graduate of Baylor University—inadvertently absorbed a scary 83 g's for one-tenth of a second in a backward-facing run. He said it felt like Ted Williams had whacked him in the spine with a Louisville Slugger. Though he went into shock briefly following the run, Beeding apparently suffered nothing more serious than a sore back. Stapp said, "We put him in traction and gave him a six-pack of beer and he was fine." It was, however, the most severe deceleration ever experienced by a human volunteer at any research institution. There was no question in the minds of the test team that Beeding would have been killed if he'd made that run facing forward.

This near-disastrous experience taught researchers a valuable physics lesson they'd failed to predict with the mathematical models they'd constructed: namely, that increasing g-force levels, or plateaus, are accompanied by quantum increases in mechanical force. Up to 35 g's, they found a 3:2 ratio of g-levels measured on a subject's chest as compared to the vehicle itself. In other words, if an accelerometer on a speeding sled measured 40 g's, an accelerometer on the chest of a man riding on that sled would measure 60 g's. Stapp had learned this way back in his Muroc days. But, mysteriously, once the vehicle is exposed to a rate higher than 40 g's, the ratio goes rather spectacularly to 2:1, so that a man on that same sled experiences 80 g's. This is what happened to Captain Beeding, and it was confirmed by subsequent tests with dummies and black bears (selected because bears have a recurved pelvis similar to the human pelvis). Stapp never tired of reminding his men of the dangers inherent in human testing. He liked to say that the Daisy Track was like a contraceptive: "For if either fails, no money-back guarantee is going to do the slightest good."

Stapp's successes with the Daisy Track and the Bopper didn't mean he was quite done with high-speed rocket sleds. His studies had finally concluded that *Sonic Wind* was too heavy and insufficiently aerodynamic ever to achieve supersonic velocities, and so he authorized Northrop to speed up the design of *Sonic Wind II*. Most of the testing Stapp envisioned for it was related to windblast, which required a sleek, super-lightweight, ejectable windshield. But the tests didn't go well. They had all kinds of problems with the new windshield, and the rockets were misfiring and even exploding. Stapp was never satisfied enough with the new sled to seek authorization for any human runs.

All rocket sled testing at Holloman would be suspended in March of 1957 while the track was extended to a length of 35,000 feet, and *Sonic Wind II* was transported to California for use at the Naval Ordnance Test Station at China Lake.

.   .   .

Though he'd made it clear before ever arriving at Holloman that ground vehicle crash research would be one of his priorities there, he'd managed little demonstrable progress by this point. Stapp's ground vehicle research plans had never received the support he had hoped for from the Air Force, and it galled him. The Department of Defense had ruled that all of its ground vehicle research work would be conducted at Holloman—and then completely failed to fund it. As Dr. David Bushnell, Holloman's historian, wrote: "The presence of such a task at an aeromedical research institution . . . has caused much raising of eyebrows in some quarters." And yet, Bushnell added, "There was good reason to undertake such a program at Holloman's Aeromedical Field Laboratory in particular, in view of the extensive background of Colonel Stapp and his co-workers in the study of impact forces."

Stapp had been studying military death-by-accident statistics for some time, and had already documented the fact that the Air Force was losing more men in ground vehicle crashes in and around its bases than it was in airplane accidents. He'd spent many hours attempting to calculate the monetary losses associated with car crashes, and the totals ran easily into the millions of dollars. He had also been making inquiries in the automotive community, exploring the possibility of a relationship with the auto industry, trying to find some way to provide a transfer of his data and his methods to civilian safety engineers, who might be able to put them into play as Detroit released its new model designs each year.

Finally, thanks to the publicity generated by the rocket sled runs, it looked as if he might be on the verge of a breakthrough. Back on December 14, 1954, as Stapp lay in the Holloman base hospital recuperating from the Big Run, he'd received a letter from Don Blanchard of the Society of Automotive Engineers (SAE). Blanchard had written: "Recently the Society appointed a Committee to study the Seat Belt Proposals and related matters, and this Committee is most interested in having any information about the test work you are planning which you are in a position to disclose." This was precisely the opening Stapp had been waiting for, and he pounced on the opportunity. After months of communication with headquarters of the ARDC, Holloman Air Force Base, the Aero Medical Field Laboratory and other stakeholders—and after paying personal visits to researchers and automakers in the company of the chief of the Holloman Biodynamics Branch—he managed to arrange clearances and approval for a program of field demonstrations at Holloman and a conference on ground vehicle crash protection.

On May 17, 1955, Lieutenant Colonel Stapp hosted a unique gathering of some two dozen individuals—including members of the SAE's Seat Belt Committee, along with a group of civilian doctors and scientists that he'd selected—for a program of presentations, demonstrations, and open discussion on matters related to the protection of ground vehicle occupants involved in crashes. In addition to SAE representatives and a handful of Stapp's Air Force colleagues, organizations represented at the conference that spring included the Ford Motor Company, General Motors, American Motors, Chrysler, Studebaker–Packard, the American Medical Association, the Cornell Aeronautical Laboratory, the American College of Surgeons, the Mayo Clinic, the Harvard School of Public Health, and the University of Minnesota. This seminal gathering of minds—a hugely significant event in the history of the American automobile—was billed as the first annual Automotive Crash Research Field Demonstration and Conference.

Both Don Blanchard, on behalf of the SAE, and the public information officer at Holloman agreed that the press should be excluded. Auto safety was a sensitive subject with far-reaching financial, legal, and political ramifications. The public was aware that accident rates were rising—the 1955 fatality rate was up more than 6 percent over the previous year—and the auto industry was extremely sensitive about negative perceptions of its

products. It was important to the individuals associated with the auto companies to be able to speak freely and without attribution. If Ford and GM
were going to swap ideas about safety system design, it would have to be
strictly off the record.

Early in the morning on the day the conference was scheduled to begin,
attendees were ferried by bus out to the rocket sled track to witness the
maiden run of *Sonic Wind II*. The experiment was a high-speed test with a
chimpanzee in the forward-facing configuration, and it went off like clockwork. The chimp was unharmed, but the violence of the whole business
got everyone's attention. After lunch back at Holloman, Lieutenant Colonel
Stapp conducted a crash test using a sedan with four dummies, two of which
were restrained with a webbing-and-straps system and two of which were
completely unencumbered. Stapp had the sedan, with its top cut away to
enable observation, towed at 40 miles per hour into a 70,000-pound timber
barrier. It was Stapp's graphic example of what was and was not happening
on the nation's streets and highways: the restrained dummies suffered only
very minor damage, while the other two were sent sailing headlong into the
barrier. Stapp then took his visitors to one of the Aero Med Lab buildings and
fired up the Bopper to show them what he'd learned about different types of
harnesses with a variety of body positions.

That afternoon John Moore of the Cornell Crash Injury Research facility
gave a presentation on statistical evaluation of lap belts in cars, and Dr. Dun-
can McKeever, an orthopedic surgeon from Houston who'd himself survived
a 60-mile-per-hour crash thanks to a custom lap belt he'd installed in his car,
shared some thoughts about his own experience. The evening was devoted to
a wide-ranging discussion of the problems faced by those interested in incor-
porating safety devices into America's automobiles.

One important figure familiar to most of the attendees was not present at that first conference, though he would be there for the next one:
Hugh DeHaven. If in subsequent years John Paul Stapp would sometimes be
referred to as the father of automobile crash protection, Hugh DeHaven must
be considered its grandfather. DeHaven was a pilot and engineer—largely
self-trained—who'd established what he called the Crash Injury Research
(CIR) project at Cornell University in 1942. Five years before Stapp arrived
at Muroc, DeHaven had published a paper that anticipated the results of the

rocket sled experiments: "The human body can tolerate and expend a force of two hundred times the force of gravity for brief intervals during which the force acts in transverse relation to the long axis of the body. It is reasonable to assume that structural provisions to reduce impact and distribute pressure can enhance survival and modify injury within wide limits in aircraft and automobile accidents."

Stapp and DeHaven had already been corresponding for years by the time of the Holloman conference, and DeHaven's ideas were invoked often during the evening's discussion. DeHaven's most important concept was that of the "second collision." There were two impacts involved in any vehicle crash, DeHaven proposed. The first was the collision of the vehicle itself with an external object; the second was the collision of the vehicle's occupants with the interior surfaces of the vehicle. These were related but quite separate events, and thinking of them like that led to the notion of the need to "package" vehicle occupants in the same way that fragile objects are packaged for shipment. DeHaven had staged demonstrations in which he dropped eggs from a ladder onto a thin foam rubber pad to illustrate his contention that, if properly packaged, people riding inside a speeding vehicle might survive extreme impact.

DeHaven, like Stapp, had encountered his share of opposition over the years. Beginning in 1951, first General Motors and then Ford had begun to inquire about DeHaven's research. The Crash Injury Research group at Cornell had published a number of statements critical of the industry, and GM was particularly concerned about their effect on public opinion. By the time of the Holloman conference in 1955, which kicked off what would be three years of serious and often controversial work on car crashes at Holloman, the world of automotive safety was already highly charged. Lieutenant Colonel Stapp, as the principal player in this drama who was associated neither with the industry nor a university that received funding and support from the industry, was now on the automotive community's radar.

Meanwhile, recognition of Stapp's biodynamics work on the rocket sleds continued. In August of 1955, he was awarded the Defense Department's Cheney Award, given to an officer or enlisted airman for "an act of valor, extreme fortitude, or self-sacrifice in a humanitarian interest." Ed Rees, the writer who'd been assigned the Stapp cover story *Time* would publish in its

September 12 edition that year, traveled with Stapp, George Nichols, Jake Superata, and Stapp's brother Celso—with whom Stapp was still in the process of repairing his relationship—to the award ceremony in Washington, DC. There, Rees found himself fascinated by Celso, who'd just returned from his own two-week trip to the Far East, where he'd conducted inspections of armed forces hospital facilities. In a dispatch to his editors, Rees noted that the younger sibling continued to argue against further rocket sled rides, and quoted Celso as saying: "I have argued with him as a brother, a doctor and as an officer. He listens, as he always does when anyone argues with him, but he feels that he must continue his work and my arguments have no effect . . . I tell him what he already knows, that he gets hurt a little bit on every ride. And he tells me right back, 'Okay, I know I'll get hurt, but I take inventory of my injuries *after* a ride, not before.' This is not bravado with him. He's not a reckless man and he's never done a foolhardy thing in his life. It's just that he is completely devoted to research and there is no room in him for fear."

For his own part, John Paul spent most of the flight to Washington absorbed in newspapers and correspondence. He did take a break to chat with Rees about an invention he'd been working on: a nonlethal "stun gun" that police could use without fear of killing offenders or innocent bystanders. Potential practical implications of his work on mechanical force and impact seem to have been constantly on his mind. Stapp also told Rees about a new idea he'd had that involved studying the mechanics of head trauma. He'd proposed a project ("Measurement of Forces and Effects of Head Impacts Fortuitously Incurred in Boxing as a Sport") to the University of Michigan, he said, in which he would recruit members of the school's boxing team for his research. The proposal apparently caused, in Stapp's words, "mass hysteria at Ann Arbor. I don't know what got them so excited," he told Rees. "I'm limited to studying boxers because this is the only socially acceptable form of mayhem; I can't go around hitting people on their heads with mallets to learn the phenomenon of knockout."

At the award presentation the next day, Rees commented that "there was enough brass at the ceremony to plate the first two floors of the Pentagon." Just prior to the ceremony, Stapp—who paid little attention to such things— was handed a shoeshine kit and sent to a bathroom to touch up his shoes.

Upon his return to Holloman following the Cheney Award, Stapp was right back into field work by August of 1955. His building obsession with automotive safety, bolstered by the enthusiasm of the engineers and doctors who'd made the trip to Holloman for the conference in March, had only intensified. Work done by Hugh DeHaven and others had shown that rigid steering wheels were the cause of many driver deaths, and the Ford Motor Company was already working on a collapsible—or energy-absorbing—model that would give way in a severe impact. It was a promising concept and Stapp eagerly offered his facilities at Holloman for testing. He hoped that by establishing an ongoing relationship with the auto industry he might build a conduit through which to move his own ideas from the Tularosa Basin onto the factory floors of Detroit.

With the assistance of Aero Med Field Lab personnel, Stapp and Ford engineers worked side by side to conduct a series of crash tests using the Holloman Swing Seat. The test subjects were anesthetized hogs. At impacts of 20 miles per hour, tests with both conventional and the experimental energy-absorbing steering wheels showed a significant reduction in injury with the new design. Stapp knew they'd gotten some valuable results, and he knew that his own cooperation was crucial because Ford's safety engineers had been prohibited from running certain kinds of tests at their own facilities. While the company would eventually conduct its own limited crash tests using hogs, Ford discouraged experiments involving animals. Yet without live test subjects, these kinds of advances in crash protection could never, in Stapp's estimation, be properly vetted. It was a vexing problem that would haunt impact research for decades.

Not all of Stapp's work on automotive safety, however, was done at Holloman. In December, Stapp would sign the first of a series of contracts with civilian research organizations that required the cooperation of field laboratory personnel and equipment. That first contract was with the University of Minnesota for the design of a hydraulic bumper that would reduce the force of impact on vehicle occupants and a superstructure for open-top vehicles that could offer protection during a rollover. The work at Minnesota was conceived and managed by Professor James J. Ryan, not coincidentally a participant in the Holloman conference that spring.

The Minnesota tests were a rousing success. When coupled with the use of lap belts, the new bumpers were shown to be capable of offering protection from serious injury during a "second collision" impact with a solid barrier at speeds up to 30 miles per hour. It was a practical application of DeHaven's notion of packaging the human being like an egg in a carton. Ryan's tests also proved that a loop or bar of strong metal tubing extending above a weapons carrier or a convertible could keep occupants from being crushed during rollovers at 40 miles per hour.

Shortly before Christmas, Stapp inked another deal, this time with the Institute of Transportation and Traffic Engineering at UCLA where researchers, under the direction of Dr. Derwyn Severy, would attempt to duplicate and validate much of the auto safety work done at Holloman. This cooperation between the military, private industry, and academic institutions— imagined and orchestrated by Lieutenant Colonel Stapp, with at best only shaky support from the Air Force—previewed a promising new paradigm for advancing the cause of vehicle safety.

. . .

The *Time* issue hit the newsstands in the second week of September 1955, with a cover painting by Boris Artzybashoff, who'd previously done dozens of *Time* cover portraits including those of Harry Truman, Ernest Hemingway, and Adolf Hitler. This one depicted Stapp's helmeted head atop *Sonic Wind* with fountains from the water brakes arcing above like a pair of giant angel's wings. The expression on Stapp's face was serene and steadfast. The caption:"SPACE SURGEON STAPP." The story ended with a quote: "'The human body,' says Colonel Stapp, 'comes in only two shapes and three colors. I don't expect there will be any changes, so what we learn about it now will serve us for a long time to come.'"

Stapp was generally pleased with the article, with one exception: he was annoyed that it made no mention of his colleagues Nichols and Superata. He would be annoyed again later that year when *Time* selected Harlow Curtice, the president of General Motors, as its Man of the Year for 1955. If an auto manufacturer deserved to be singled out, Stapp thought the Ford Motor Company a better choice based on its record-breaking charitable contributions to higher education and hospitals. He wrote: "I consider Mr. Curtice a little too mal-a-droit (badly to the right, in the literal French) a choice."

Not long after the *Time* article appeared, Stapp was in Washington lobbying for Aero Med Lab funding when he was informed he'd need to make a quick trip to Los Angeles to attend a safety conference. When he arrived—Stapp strolled along Hollywood Boulevard with General Leighton Davis, who'd accompanied him on the flight, both of them in uniform, believing he was on his way to deliver a speech on impact research—a smiling man with a microphone accosted him. They were standing in front of the El Capitan Theater, and Stapp's interrogator was Ralph Edwards, host of the television program *This Is Your Life*. Lights flashed on and a cameraman stepped from the shadows.

Edwards moved in front of Stapp and a sheepish General Davis, who was in on the blindside, and asked: "You're known as the fastest man on earth, aren't you?"

Stapp, understanding immediately that he'd been lured into something and appearing slightly peeved—on the plane he'd spent his time writing a speech for a conference that didn't exist—answered: "Yes, but not fast enough to get away from you, was I?"

Stapp and Davis were escorted inside the theater for the show that would be dedicated to the life of John Paul Stapp. As he was ushered through a curtain and onto the set—the show was filmed in front of an audience—Stapp saw Edwards excusing General Davis, whose job had been to get Stapp to Hollywood for the show. As Davis turned to leave, Stapp made a half-hearted attempt to tag along behind and escape into the night, but was stopped short by a production assistant and coaxed back onto the stage.

After a brief recounting of Stapp's heroics on the Big Run, Edwards asked: "What went through your mind before the final run?"

The audience was hushed, and Stapp answered quickly. "Isn't there a better way to make a living?" But then he went on: "I've had a good life," he said, "more than I deserve. I thought: if this is it . . . well, all right."

Then Edwards began to bring his surprise guests onto the stage. First came the Reverend Charles Stapp, mustachioed and white-headed now in his seventy-fourth year, speaking in the pompous cadence of the missionary preacher. "Paul was always very adventurous," he offered. He told the well-worn story of his eldest son sawing off the coconut tree limb on which he was sitting.

"Not the last time I've been out on a limb," John Paul quipped. Stapp was smiling nervously by this point, clearly uncomfortable in the spotlight.

Edwards then trooped out, in succession, Stapp's Uncle Hansford and Aunt Verna from Burnet, a friend from San Marcos Academy, and a woman who'd been a beneficiary of the Curbstone Clinic at Muroc, who told Edwards: "We idolized him!" Then came Celso Stapp, looking—as Ed Rees of *Time* had noted—much like John Paul, but "kangarooish" with his prominent pot belly, and finally the ever-dependable George Nichols and Jake Superata. Of them all, Stapp seemed most grateful to see these last two guests. He told Edwards: "I'm the only doctor to put his life in the hands of engineers twenty-nine times and live to tell about it."

Nichols was honest about his reaction to Stapp's rocket sled rides. "I don't enjoy these runs," he said.

This Is Your Life gave Stapp a gold watch, along with something that was of much greater value to him. It was the only thing that seemed to get his juices going during the whole affair: a certificate for $1,000 worth of RCA Victor record albums. All he could think of was Lovers Lane and his hi-fi set, and more symphonies than he could listen to in a year. Stapp beamed at the audience while his family and friends whomped him on the back.

Stapp had now entered the mainstream of American celebrity, and despite his own penchant for showmanship and drama, anyone viewing the *This Is Your Life* episode could see that Stapp was conflicted about this new level of attention. He would later say to an editor at *The Reader's Digest*: "Television is a rather unappealing prospect for me, personally, but apparently a necessity for the Air Force." Yet even before leaving the El Capitan that night, Stapp was already scheming about how he might be able to make use of the spotlight that had been trained on him.

17

SPACE SCIENTISTS IN HOLLYWOOD

It's the biggest piece of miscasting I've ever seen. He doesn't
fit the real life role he plays.

—*William Bloom, USAF interview*

AFTER THE TV SHOW AIRED, Stapp was approached by several publishing
houses interested in his story. The most enthusiastic inquiry came from
editor William E. Buckley at Henry Holt and Company in New York. Buckley
offered an advance for Stapp's story in his own words, but Stapp waffled. He
doubted he'd have the time to write given his current duties and pursuits, he
said, though he maintained that he did intend to put his thoughts on paper
as soon as he got the opportunity. Buckley was understanding but persistent.

"I certainly won't give up," he wrote to Stapp, "and I know you won't. I
look forward to progress reports from time to time, and the first draft as
soon as you have written it. There's a great book there, and I want to see us
publish it."

Stapp had thrived in the late months of 1955 despite a broken heart
that continued to pine for Celia Richards. His eyes mostly healed, able to
employ the full range of his intellectual skills, his creative impulses largely
unchecked, Stapp could indulge the urge to push forward on all fronts at
once. Friends and colleagues wondered how he kept up the pace. Some of
them urged him—always unsuccessfully—to slow down.

Then on March 9, 1956, the fastest man in the world was cited by local police for doing 40 miles per hour in a 20-miles-per-hour zone. The incident made the front page of the *Alamogordo Daily News* and was subsequently picked up by papers around the world. The consensus seemed to be that mad scientists such as Stapp were far too absorbed in deep thought to notice trivialities such as speed limits. But Stapp's celebrity, along with his continuing willingness to attend to the medical needs of friends and strangers alike, always free of charge—the Curbstone Clinic had never gone out of business—gave him certain advantages. It seemed like he'd helped everybody in town at some time or other. When he appeared before the justice of the peace on the speeding charge, the judge had the ticket reissued to a fictitious "Captain Ray Darr," and proceeded to pay the $12.50 fine out of his own pocket.

Four days after the speeding ticket, Stapp was in Hollywood for the premiere of *On the Threshold of Space*, starring Guy Madison, John Hodiak, Virginia Leith, and Dean Jagger. The film had actually been previewed for

Stapp, the inspiration for 20th Century Fox's On the Threshold of Space *and the lead consultant on the project, with some of the film's stars at a special preview event at Holloman. Though the Air Force brass was generally pleased by the favorable publicity, dissenters suggested that Stapp was mainly out for his own glory. (Photo courtesy of Wilford Stapp)*

Stapp and selected guests at the Holloman base theater a few days earlier. The garish posters and lobby cards proclaimed: "THEY SOAR TO GLORY! THE FLYING SCIENTISTS OF THE U.S. AIR FORCE! THE MOST DANGEROUS FRONTIER OF THEM ALL LIES JUST 17 MILES FROM YOUR DOORSTEP, STRAIGHT UP!"

The plot revolved around a band of brave men—the characters contained elements of real-life Holloman personnel such as David Simons and Joe Kittinger—who were battling life-threatening dangers as they paved the way for a glorious future of space travel. Except for the spectacular rocket sled footage that 20th Century Fox got from Holloman, nothing much had gone right on the set. The character that writer Frank Cockrell had based on Stapp, Major Ward Thomas, was played by forty-one-year-old leading man John Hodiak. In the middle of production the previous October, Hodiak had suffered a fatal heart attack one morning while preparing to shoot one of his primary scenes. Cockrell had been forced to rejig the plot in order to keep Hodiak's already-completed scenes in the picture. At one point producer William Bloom had complained to Stapp: "We're on the rewrite trail again. Oh, how I'd like to see this on the screen awreddy!" Things became such a mess that Darryl Zanuck fired Bloom before the movie was released.

The end product can charitably be described as only marginally substandard B-movie fare: cheesy, disjointed, and cliché-ridden. The *New York Times* went relatively easy on it, while noting that Cockrell and team "have not fashioned inspired personal dramas." In fairness, however, the film was prescient about the high-altitude balloon programs that would blossom in the years immediately following. It was also on the leading edge of a new "scientist as hero" theme in popular culture that would last for another decade at least. Newspapers panned the film and moviegoers ignored it. Stapp tried to hide his disappointment in his continuing correspondence with Bloom and Cockrell, but as time went on he would judge *On the Threshold of Space* to be "perhaps the worst movie ever made" and complained to a friend that he was "thoroughly disgusted with the bungled up job they did with the story . . ."

The Pentagon brass, on the other hand, were generally pleased with the heroic light in which Air Force men had been portrayed on the silver screen, though their delight did not always extend to the film's informal sponsor, Lieutenant Colonel Stapp. Brigadier General Leighton Davis gave Stapp an official letter of commendation that read: "Such films as 'On the Threshold of

Space' do an excellent job of telling the story of ARDC research projects and in doing so gain favorable public relations," but others were not so complimentary. While in California that March, Stapp had given a talk to the Press and Union League Club in San Francisco. Prior to his appearance, an overeager Fox press agent had supplied the club with a backgrounder on the movie that included some erroneous information about Stapp. The press release forwarded by the club to the *San Francisco Chronicle*, a version of which appeared on the next day's front page, confused the film's plot with details of Stapp's real-world accomplishments. The release claimed that he had personally made a balloon flight to 110,000 feet as well as a bailout from 62,000 feet. It was a genuine embarrassment to both Stapp and the Press and Union League Club's chairman, who apologized profusely.

Even worse, however, on April 3, 1956, Stapp received a handwritten letter of admonishment from Colonel A. M. Henderson at USAF headquarters. Henderson had been skeptical of the Hollywood crew from the start, he said, and suspected Stapp of seeking to exploit the publicity and attention to further his own ambitions. The letter invoked the names of generals at ARDC whom Henderson claimed were furious with Stapp. "I personally feel," Henderson wrote, "that you have been amply rewarded for your sled runs without gathering two years of personal publicity. Isn't it time to curb it and get on with more important projects?" He went on: "As a senior USAF officer I must voice my personal opinion and that of many others that you are only hurting yourself, the medical service, and your publicity has become a joke."

Stapp's reply was curt. He explained what had happened in San Francisco. Alluding to Colonel Henderson's imminent retirement, he wrote: "The most charitable view toward the attitude you expressed in your letter of 3 April is that you were emotionally disturbed by the anxieties attendant to leaving the service."

Henderson, who later apologized to Stapp, wasn't the only high-ranking Air Force figure bothered by Stapp's extracurricular activities. Brigadier General Marvin Demler, deputy commander of research and development at ARDC, issued a formal objection to Stapp's many public appearances. Demler argued that Stapp's frequent travel and participation in events unrelated to his official duties—many of these were Air Force events such as awards ceremonies; most others were hosted by professional societies that had invited

Stapp to speak or to consult—had caused "dissipation of his time." Demler listed sixty-two appearances away from Holloman in the first eight months of 1956, and joined Henderson in suggesting that surely enough was enough.

Membership in the Stamp Out Stapp club seemed to swell in direct proportion to Stapp's burgeoning celebrity, and new club members were coming increasingly from the higher ranks of the Air Force. When Stapp applied to headquarters that summer for a reclassification of his ground vehicle crash study project—he wanted a stand-alone program that would receive its own funding separate from aircraft crash studies—he was promptly and unceremoniously denied.

Stapp appealed the decision, and forwarded to headquarters a long and passionate defense of his auto safety research. In a report titled "Justification for Car Crash Project," he tried a fresh approach based on a metaphor: "In the era dating from 166 B. P. (Before Penicillin), the number one problem of manhours lost in hospitalization in the military services related to venereal disease." Despite years of attempting to reduce VD rates through preventive measures, behavior modification techniques, and military discipline—all of which had been ineffective—within fifteen years of the introduction of penicillin the disease went from a major problem to a medical inconvenience. "The analogy is obvious," Stapp insisted. "All preventive angles have been applied to the utmost in the vehicle accident business . . . with what results?" The new number one cause of man-hours lost due to hospitalization was the car crash. Safety systems built into new cars could be the penicillin. Restraint systems and overall safety design held the promise of a huge new class of survivable accidents. The solution was within their grasp. The Air Force could not afford, Stapp argued, to turn its back on this chance to attack the epidemic ravaging not only its own ranks but those of the entire nation.

The response from the mahogany desks at ARDC headquarters, drafted by Colonel Charles Roadman, was clear: "This headquarters must nonconcur in the subject project as written. It is our belief that the emphasis should be on accident prevention and not on survival in accidents."

Dejected, Stapp took time out that June from his battles with Air Force brass to return to the family alma mater, Baylor University, to deliver the commencement address to the 1956 graduating class and to accept an honorary doctor of science degree. He told the school he considered it the single greatest honor of his life to that point. He asked, as a favor, that Baylor invite the ailing Reverend Charles Stapp to be present. His wish was granted and not only Charles, but all three of John Paul's brothers—along with aunts, uncles, cousins, and in-laws—were present in Waco for the event.

Stapp's commencement speech, titled "Science as a Social Force," was short on practical advice for the graduates. Instead, it was a philosophical meditation on the role of science in the modern world. In it, he made a subtle point, one that undergirded most of the research that had been his life's work. Science, he suggested, has a larger purpose than simply to offer explanations and interpretations of the workings of the world. The truly valuable work is to connect the lessons of science to the world for the betterment of mankind. "Consider," he challenged his audience, "the effect of the invention of the mariner's compass on the science of navigation, and its application in exploration and discovery. With it Columbus unlocked a new world for immigration and economic expansion. This was not the most important contribution of his famous voyage in 1492. The most important thing he did was to get back where he started. He invented the guaranteed round trip." It's one thing to travel to parts unknown, but quite another to return home to one's family safe and sound; one thing to go hundreds of miles per hour, and another to survive the stop.

The Stapp brothers, reunited briefly at Baylor, had followed the lead of their faithfully correspondent father and had continued to keep in regular touch over the years through the mails. Their exchanges were occasionally in Portuguese or Latin, usually in English, and very occasionally in German. Robert Stapp, who came between John Paul and Celso in age, had kept the loosest orbit. It would be unfair to call Robert the black sheep, but he was clearly, in Wilford Stapp's words, "the brother who had a hard time." The only Stapp not to embrace the sciences in one form or another, Robert had gravitated toward the arts and humanities, earning from his famous elder sibling the unkind sobriquet "No-Go Van Gogh." While at Baylor, Robert had designed stage sets for the theater department, and subsequently studied architecture for two years at the University of Texas, where he also tried his hand at creative writing. Those who knew him in those days remember him as sickly and tormented by migraine headaches. Robert taught history as well as theater and art at small colleges in Arkansas and Kentucky, and had several modest exhibitions of his own paintings. But he never found the professional success his brothers enjoyed.

Once the family dispersed following the Baylor ceremony, the rest of the summer months of 1956 were for John Paul a period of seemingly nonstop activity. He travelled on Air Force business that May to Oslo, Copenhagen, Brussels, and Farnborough, England, taking a few days off that spring to visit his brother's family in El Paso, where Day Stapp threw a party in her brother-in-law's honor. It was at the party that John Paul met Lillian Rose Marie Lanese. She was a stunning brunette, fourteen years Stapp's junior, with brown eyes and an almost regal carriage. She had a self-confidence and poise—some read it as haughtiness—that distinguished her from the crowd. She was a ballerina and had the air of an aristocrat. Born in Brooklyn to a policeman father and a French mother, she had started dancing at age fourteen at George Balanchine's School of American Ballet in Manhattan. Not much later, she joined the Ballets Russes, and—while still a teenager—toured

Lillian Lanese as a soloist in the American Ballet Theatre's production of Helen of Troy, *New York City, 1950. Her legs were legendary. (Photo courtesy of Wilford Stapp)*

Europe with the Ballet Theatre. She went on to dance as a soloist all over the world, also appearing in a handful of Broadway shows and television specials. Edward Denby, poet and dance critic for the *New York Herald-Tribune*, had written: "She is a dancer of the highest quality . . . the most magnificent legs of the season." She had come to West Texas for a couple of weeks to visit a friend from the Ballets Russes who'd settled in El Paso.

Unbeknownst to Stapp, the meeting had actually been arranged by Day—one of Celso and Day's daughters had briefly taken ballet lessons from Lillian's friend—who was forever in matchmaking mode when her brother-in-law was around. Lillian, who knew all about John Paul from the magazines and television shows, told Day after the party that Stapp had been "older and more tired-looking" than she'd expected. Still, there was some kind of spark between these two highly successful, high-profile people. She found him brilliant and witty, and he found her uncommon grace and ease refreshing. Just the idea of an accomplished ballerina, an artist of the highest ideal and the classic physical manifestation of the eternally unattainable, excited Stapp.

Lillian returned home to the Upper West Side of Manhattan, but she and Stapp would maintain a correspondence. She wrote him: "I hope you won't mind me placing your picture amongst my favorite people."

The promise of this relationship may have energized Stapp, because the pace of his work seemed only to increase from that point on. Undeterred by the less-than-positive response his auto safety studies had met at Air Force headquarters, Stapp convened the second Car Crash Conference in September 1956. By now, most everyone involved in car safety research had heard about Stapp's conference, and the attendance this time was nearly double that of the inaugural meeting. Automobile accident rates in the United States remained alarmingly high—nearly 38,000 Americans would die on roads and highways that year—and the public was becoming concerned for its own safety. For a number of the engineers, doctors, and scientists who gathered at Holloman, crash protection was coming to feel like a crusade. The open discussions, which had been mostly reserved and polite in 1955, were becoming, as the second conference moved through Stapp's agenda, intensely passionate. They often led to, in Stapp's words, "heavy arguments." He relished them.

In spring of the following year, Stapp wrote to a friend of his, a brigadier general at the Office of the Surgeon General, about the car crash program at

Holloman and the actionable results it was producing. "I think there is no question we can justify lap belts, head level seat backs, locking doors, and thin spun aluminum instrument panels, etc., for absorbing energy of impacts of heads and knees... Drastic changes in automobile frame specs, etc. require car crash experiment justification." Stapp also complained about the lack of a formal mechanism by which the Air Force could effectively share this research with the auto industry. Over time, the conference could help enlighten the car makers as their engineers returned to the corporate labs in Detroit having been exposed to the latest discoveries, but there were also managers and lawyers involved. It was an inefficient path to enlightenment.

Stapp asked the ARDC to help him find a way to collaborate directly with Detroit, to find a way inside and show them how to save the lives of their customers. Who could argue with that? Who could oppose it?

As it turned out, the conference had a quite unintended side effect: it galvanized new chapters of the SOS club. In the words of Joel Eastman, historian of early automotive safety efforts: "The conferences served to make Stapp's unorthodox research—especially the use of human volunteers for automobile crashes—visible and subject to criticism. Perhaps influenced by the lobbying of some of the automobile manufacturers, Congressional committees overseeing the military budget threatened to cut off all funds for aeromedical research unless the automobile crash testing was ended."

The Air Force was listening.

18

SPACE SCIENTISTS IN SPACE

> Victory awaits him who has everything in order, luck,
> people call it.
>
> —*Roald Amundsen,* The South Pole

E VEN THROUGH THEIR DISAGREEMENTS and the garden variety of intra-family offenses over the years, the Stapps had remained connected and ultimately supportive of one another. So it was a blow when, on October 26, 1956, the Reverend Charles Stapp died in Atlanta at the age of seventy-five. Though Charles had been stricken with both respiratory problems and a severe kidney infection that put him in the hospital for several weeks prior to his death, the news seemed to take his sons by surprise. With both Charles and Mary Louise now gone, the Stapp men all reported feeling cast adrift. John Paul wrote an unusual first-person obituary that was given to the news-papers in the Texas Hill Country, avoiding the language of the missionary while yet managing to capture a sense of the struggle that had defined his father's steadfast life: "To me death is the last friend that gives us a way out of all the accumulated pain and disability that life has to offer after all the good we can accomplish has been done. It is unfortunate that we are taught to fear it. In it all living things become equal. Through it we live again more perfectly. It is our bridge to eternity... With so much evil rampant in the world, it can ill afford to lose such good men. Such lives and examples are the foundation of civilization."

While Stapp mourned his father, Major David Simons's Holloman-based radiation experiments using small animal subjects and high-altitude balloons were shifting into high gear. By mid-1955 Simons had conducted dozens of research flights and what he'd learned was that fears about the dangers of primary cosmic rays on living tissue were mostly unfounded. Once he'd gotten the technical and logistical problems of launch, recovery, and capsule environment worked out, Simons was bringing the mice, guinea pigs, and monkeys back to earth unharmed. Relatively brief radiation exposures looked to be tolerable for human pilots or—one day—for astronauts. This was very good news for enthusiasts of space exploration. But one more test was required before they could say with confidence that space was safe for man. A human pilot would have to make the trip.

Stapp had been following the Holloman biology flights closely, and he had become frustrated with their limitations. "The animals did nothing up there but breathe, eat, and defecate," he observed. "They didn't talk on the radio, shift around in a 180-pound mass, fidget in a pressure suit, or try to grab scientific observations out of those saucer-sized portholes." So in August of 1955, shortly before making his trip to Washington to collect the Cheney Award, Stapp had summoned Major Simons to his office and asked if he was ready to begin working on the real thing: a manned research flight. The question caught Simons off guard. There were some calculations he'd need to do to spec out the cooling requirements and the breathing system, but he told Stapp that he didn't see any reason why it wouldn't be possible. Stapp immediately authorized the planning necessary to develop a capsule capable of carrying a human pilot into the upper stratosphere and returning him safely to earth, and he told Simons that he expected him to volunteer to make the flight himself. Nothing drove attention to detail like knowing it was your own life on the line.

Simons was thrilled—and not merely for the opportunity to experience a trip into near space, but for the possibility of doing unprecedented research above earth atmosphere. It would be the culmination of all his work with the V-2s and the years of animal flights. But it was also, Simons would admit, a rare chance for a scientist and doctor to get out ahead of the fighter jocks and test pilots. It was a motivation Stapp understood well.

The Air Force's lack of enthusiasm for space research seems only to have hardened a stubborn John Paul Stapp's desire to help clear the path to earth orbit and beyond. His work at Wright-Patterson ten years earlier on the liquid-oxygen breathing system had given him exposure—literally and painfully—to the challenges of high-altitude survival. The prospect of contributing to the development of the systems and protocols needed to protect human beings during space travel, the need to determine whether it would even be possible to offer adequate protection, had a strong appeal. In Stapp's mind, it was simply another angle on the problem of returning the "pathetically vulnerable" human traveler to his starting place in one piece, of the guaranteed round trip. The space biology work offered Stapp a fresh outlet for his missionary hatred, shifting at least some of his focus from the specter of crippling mechanical force and impact to the near-vacuum and sub-freezing environment of the deep stratosphere.

Near the end of 1955, just after returning from an exhausting month-long trip to Japan, Korea, and the Philippines as a guest of the Fifth Far East Air Force, Stapp met with Simons and German immigrant engineer and entrepreneur Otto Christian Winzen to discuss what it would take to raise a pressurized vessel with a man aboard to an altitude of 100,000 feet. Back in 1947, Winzen—then working for General Mills—had built some of the world's first polyethylene balloons, pioneering the process for reliably heat-sealing long gores of wispy-thin plastic extrusions and enabling the construction of massive envelopes that were both extremely light and extremely strong. The Air Force had used Winzen's balloons for some of its aerial spy programs. When Winzen left General Mills in 1949 to form his own company, Winzen Research of Bloomington, Minnesota, Simons began using Winzen balloons for his animal research flights at Holloman. Stapp offered Winzen a contract for both a balloon and a pressurized gondola suitable for the manned flight they were envisioning, and thus Project Manhigh—America's first space program—was born. At the time of Stapp's conversations with David Simons about a manned flight to 100,000 feet, the formation of NASA was still three years distant. Nobody was yet quite sure if human beings could really handle a prolonged journey beyond the atmosphere.

These were the pioneer days of space research, and though this sort of work was supposed to be a detriment to an Air Force man's career, you couldn't prove it by John Paul Stapp. Three days after New Year's, 1957, Stapp received a short note from his old nemesis at Edwards Air Force Base, Major General Albert Boyd. Boyd, now at ARDC, wrote to inform Stapp that the Air Force had approved his promotion to full colonel. Boyd's battles with Stapp had never been personal; while still at Edwards, Boyd had written a formal commendation for Stapp, saying, "I commend him highly as an officer and gentleman and as a scientist." Stapp was ecstatic, and immediately booked the Holloman officers' club for a celebratory cocktail party. "If you could only imagine the shock power—pleasant, of course," he replied to Boyd. "Expressed as g's, it would just about double anything I have experienced to date." The promotion, Stapp knew, would boost his credibility and the credibility of some of the unconventional research he wanted to conduct, but it would also pull him further and further away from the day-to-day project problem-solving that he loved—and threaten to make him a full-time manager.

Stapp's managerial talents would be tested with David Simons. On one hand, Stapp had great professional respect for Simons as a medical researcher. The two men had important things in common. They both labored within a world of military affairs that would always place doctors and scientists on a shelf below the great warriors; they were both possessed of an intelligence and creativity that was never valued as highly as they believed it should be; and they were both unconventional men who succeeded, from time to time, in raising the ire of their superiors. An officer friend of Stapp's who had made a visit to the ARDC offices in Baltimore even before Stapp arrived to take charge of the Holloman field lab reported that Simons was already a cause of concern at headquarters. "They are quite worried about Dave," the letter said. "Apparently people have raised hell everywhere he has gone. Flickinger seemed to think you were the man for the job if anyone was. So if you do have difficulties, Flickinger is already aware of the problem and from the conversation is prepared to remove him if worst comes to worst."

When an Air Force general who'd received some unfavorable reports about Simons contacted Stapp for his evaluation, Stapp replied: "The consistency of Major Simons' performance in an assignment where he has found an outlet for his urge to do research begins to look like excellent possibilities for a Nobel Prize someday." Stapp was supportive of Simons during the long months when Simons was chasing balloon payloads around the upper

Midwest, regularly checking in on Simons's family—they rented a house just down the street on Lovers Lane—and of course providing Curbstone Clinic medical services when required. One of Simons's children was born prematurely in the summer of 1955 during a balloon research trip, and Stapp provided updates on the home front by letter: "Your son, Stevie, now weighing four pounds eleven ounces, is to be brought in from Beaumont today or tomorrow by ambulance, and turned over to Libby. She is holding up pretty well under the strain."

Stapp understood that Simons's success would be his own. The articulation of Project Manhigh's mission emerged from a series of lengthy planning sessions Stapp held with Simons and Otto Winzen. They agreed initially on a flight of at least twenty-four hours' duration to an altitude somewhere around 115,000 feet. Stapp insisted from the beginning that the project pursue original research objectives that did not duplicate studies being done elsewhere—the Navy had its own high-altitude balloon program—and that any emphasis on altitude records be proactively minimized. "We will not," he announced, "have track meets in the sky."

The gondola developed at the Winzen plant was a cylindrical aluminum-alloy capsule eight feet tall and about three feet wide constructed in three sections that were clamped together. There were six glassed portholes to provide visibility for the pilot. The internal volume was approximately 50 cubic feet, meaning the pilot compartment was extremely cramped: a man could not stand fully upright inside and had very limited lateral space. It was small enough that the heat generated by the onboard electrical components made a heating system redundant even in the brutal cold of the stratosphere. The interior was packed tight, with three batteries beneath the pilot compartment floor, a drinking water reservoir, a liquid oxygen supply and converter, radios and cameras, a spot photometer, a fire extinguisher, an electrical control panel, an instrument panel, a lighting system, and a thermostat. It was, in the words of a future Manhigh pilot, "the smallest cabin a man could occupy and still do anything."

Winzen's balloon, on the other hand, was all about size. It was a billowing leviathan: 238 feet tall and 200 feet wide fully inflated. It could hold a volume of three million cubic feet of helium, capable of lifting a payload of two and a half tons.

The Manhigh system was highly experimental, and fundamentally different from the small aluminum globes that had been used for the animal flights. Stapp quickly became convinced he needed some sort of test flight to qualify both the balloon and the gondola before a twenty-four-hour research flight could be authorized. Too many things could go wrong. Murphy's Law still held, and so Stapp—who already had Baltimore breathing down his neck about the riskiness of the whole thing—began considering test pilot candidates, men with the technical skills to give the Manhigh capsule a good workout and who possessed the ability to handle an emergency involving either the survival systems or the balloon.

The first candidate Stapp thought of was Captain Joe Kittinger. Having used Kittinger so effectively for both the zero-gravity research and for the photographic mission during the Big Run, there was no one on the base Stapp trusted more completely. Kittinger was, in Stapp's estimation, "the sharpest pilot I'd ever met." Lieutenant Colonel Oakley Baron was the director of Holloman's Flight Test Division, and—while he had a hard time understanding why Kittinger would want to ride balloons when he had an array of the latest jet fighters at his disposal—he signed the transfer papers.

Soon after, Stapp prescribed an exacting test regimen that both Kittinger and Simons would have to complete before they'd be cleared for a stratospheric flight in the Winzen system. Manhigh pilots would be required to prove that they could endure isolation inside a small compartment for twenty-four hours: a claustrophobia test, as Stapp referred to it. They'd have to submit to exposure to a near-vacuum in a decompression chamber, obtain a balloon pilot's license, and complete parachute training to prepare them for an emergency. Throughout the months of 1956, both Kittinger and Simons pushed themselves until they had satisfied Stapp's prerequisites. Yet they did so not as companions chasing a common goal, but as rivals.

Joe Kittinger and David Simons were two very different souls. Kittinger had openly ridiculed Simons for the nausea he'd experienced on the zero-g flights, seeing him as a fussy lab scientist desperately trying to impersonate a pilot. Simons, for his part, looked down on Kittinger as an unsophisticated thrill-seeker whom he suspected was motivated largely by the desire to make a world-record flight and to beat Simons to the edge of space. Stapp attempted to assuage Simons's bitterness about the assignment of Kittinger to make the Manhigh test flight. But the way David Simons saw it, Kittinger was simply there to steal his thunder.

Despite this simmering feud, preparations for the test flight shifted into high gear during the early spring of 1957. Joe Kittinger spent long hours at the Winzen plant in Minnesota, conferring with the mechanical engineers and, in the best test pilot tradition, learning everything he could about the craft to which he was going to entrust his life. His job was to put the Manhigh system through its paces and to illuminate any glitches in hardware or procedures so they could be corrected prior to Simons's prolonged research flight, but part of that job would be simply staying alive. Stapp harangued both of his Manhigh pilots about the hellish environment they aspired to visit: the freezing, suffocating near-vacuum that would render a man unconscious in seconds and destroy his brain in less than a minute. Their bottom-line objective, he told them, was survival.

Since Stapp had invented and proven the procedure of breathing pure oxygen prior to high-altitude flights to alleviate symptoms of the bends, the pre-flight breathing ritual would be carefully monitored: 100 percent oxygen for a full two hours. The Manhigh team had settled on a novel three-gas cabin atmosphere: 60 percent oxygen, 20 percent helium, and 20 percent nitrogen. The mixture alleviated the fire danger of pure oxygen and reduced the problematic nitrogen content of earth atmosphere. This work would prefigure much of the design and planning that would be required for high-altitude jet flight and rocket travel.

On June 2, 1957, Captain Kittinger completed his pre-breathing and was sealed into the Winzen capsule in preparation for a predawn launch from Fleming Field in St. Paul, Minnesota. He wore an Air Force standard-issue partial-pressure suit and sat sealed in the tiny compartment as the ground crew completed inflation of the massive balloon. Then, at about 6:30 a.m., with Stapp and Simons and Otto Winzen all looking on, *Manhigh* was launched.

It took Kittinger just a couple of hours to reach 96,000 feet, a deadly altitude if anything were to go wrong with the pressure suit—more than six miles beyond what's known as the Armstrong Line (discovered by Harry Armstrong and John Heim in 1935 at Wright Field), the height at which gases in the bloodstream begin to come out of solution and "boil."

It was at once both an astonishing voyage and unprecedented vantage.

Captain Kittinger was able to gaze out and see the earth stretching away for hundreds of miles in all directions. A handful of jet pilots had reached these heights, but only briefly and at blurry, bone-shaking speeds. This was something altogether different, a silent floating platform with an airspeed of zero that provided an opportunity to carefully survey a significant stretch of the planet's geography. Unfortunately, due to mechanical failure of the onboard VHF radio transmitter, Kittinger was unable to communicate by voice with his team and was forced to resort to Morse code. Stapp and Simons, meanwhile, had followed the balloon's path in a Kaman H-43 helicopter.

Not long after Kittinger reached peak altitude, the Manhigh team discovered a more serious problem. The carbon dioxide levels inside the gondola were rising. Stapp had intended to monitor Kittinger's heartbeat and respiration on a designated telemetry channel on the radio, but that channel was now being used for the Morse code messages. Simple readings that should have taken seconds were requiring many minutes for Kittinger to transmit with the telegraph-style key. When Simons asked for Kittinger to report his capsule oxygen supply level, the report came back—in Morse code taps—HALF. The supply should have been nearly full. Simons requested a repeat reading. Kittinger's reply came: "DOUBLE CHECKED. HALF FULL. WHAT GIVES."

Simons tried to contact the Winzen plant in Bloomington for an answer, but the chopper was now too far away for a clear signal. Thinking quickly, Stapp ordered the helicopter pilot to put down near a rural gas station. Using the station telephone, Simons got hold of Otto Winzen. By this point, only 40 percent of *Manhigh*'s oxygen supply remained. Based on the fact that Kittinger's cabin pressure was holding steady, Winzen concluded that there was no leak in the capsule. The problem had to lie elsewhere.

At 8:54 a.m., with the helicopter airborne again, Stapp advised Kittinger to begin his descent immediately. The intent had been to leave the gondola at altitude for several hours, but they were out of options. The onboard readings at that point showed that the oxygen supply was down to only one-tenth of what Kittinger had left the ground with, and he was still 18 miles up. The situation was fast becoming an emergency. If Kittinger started down now, he might be able to get back into a breathable environment before his oxygen ran out or he blacked out from CO_2 poisoning.

Then came, through the dry clicks of the telegraph key, Kittinger's response to Stapp's directive that he descend: "COME UP AND GET ME."

The "breakaway phenomenon" was a psychological syndrome that some Air Force psychologists speculated had the potential to affect high-altitude pilots, lulling them into a false sense of security and creating a hallucinatory sensation of detachment from the earth. The theory was that this mysterious phenomenon might lure pilots ever higher even in the face of imminent danger—lure them to their deaths. Stapp was skeptical, but David Simons had gone on record with concern about breakaway and his concern was coupled with his distrust of Kittinger's motives. Following the young captain's delight in his parachute training in the months leading up to the flight, Simons had become worried that Kittinger might manufacture a crisis that would "require" him to make a skydive from record heights, putting the future of the entire program at risk.

For the first time, doubt crept into John Paul Stapp's mind about Joe Kittinger. Stapp dismissed Simons's breakaway fears, but he couldn't dismiss the notion that an excess of carbon dioxide might have clouded his favorite pilot's judgment. As the chopper sped south on a course to intercept the balloon's path, Stapp grabbed the microphone and issued an unconditional order.

"Captain," he said sternly, "if you do not start your descent immediately, we will cut you away from the balloon and bring you down."

The clicking from Kittinger came at once: "VALVING GAS." He was on his way down. But Kittinger's race against his oxygen supply was going to be a nail-biter. The sun's heat had warmed the helium inside the great balloon, and the system was now in a superheated state—it was trying to rise at the same time the pilot was trying to bring it down. It took Kittinger nearly three hours to return to the troposphere, and by that time his onboard oxygen was spent. Greatly relieved to be able to vent cold air into the capsule and refresh himself, Kittinger landed at three minutes before 1 p.m. on the muddy bank of Indian Creek near Weaver, Minnesota, about 80 miles southeast of Minneapolis. The gondola touched down, wobbled for a moment, and then tipped gently over into the shallow water.

The helicopter landed in a nearby clearing moments later, and David Simons brushed past Stapp—who was, as always, in full uniform with hat and necktie knotted tight, looking out of place in the scrubby woods—and was the first one to reach the capsule. Finding the small turret window on the gondola open, Simons peered inside, where he saw the impish face of Joe Kittinger grinning back at him.

Once he'd managed to climb out of *Manhigh* and wriggle out of the partial-pressure suit, Kittinger explained to Stapp and Simons that he'd been joking. By the time he'd received Stapp's order to descend, he had, he said, already begun valving gas to set up his descent: "I'd already been working my butt off for an hour valving." He was a professional test pilot and he had no death wish, he assured them, and he had certainly not been stricken by any breakaway phenomenon. It had just been good old smart-ass Joe Kittinger, having some fun.

Stapp laughed it off. It reminded him, he said, of something he might have done himself. He was not necessarily happy with the excess drama that Captain Kittinger had manufactured, but neither was he upset. The same could not be said for David Simons. He took Kittinger's cavalier attitude personally. Manhigh was *his* project and he was not going to stand by and let some redneck captain derail it.

They all climbed aboard the helicopter and headed back to Bloomington, in silence, for a debriefing. The oxygen supply problem, it was quickly determined, had been caused by a valve that a technician had installed backward and the VHF radio malfunction was traced to a faulty transmitter knob. It was a perfect expression of Murphy's Law, though both Stapp and Kittinger wondered how accidental the snafu had really been. Years later, Stapp told an interviewer: "I suspected sabotage by somebody. Because Kittinger could not only please but could irritate." Regardless, it was validation of Stapp's insistence on conducting a shakeout flight of the Manhigh system. Human testing, Stapp would say, is the ultimate testing. There was no substitute for it.

Following the debriefing, Stapp addressed the team. They should all be proud of themselves, he told them. Test flights were designed to uncover problems. They had just participated in a historic flight above 98 percent of the earth's atmosphere, the edge of space, and had returned their test subject safely.

In the days following the flight, Major Simons was still seething. Kittinger had gotten the glory and, in spite of scaring the hell out of everyone in the process, had come out smelling like a rose. Simons only hoped that Stapp would be successful in securing the additional funds they'd need to complete the full-scale research flight. They were targeting the month of August, and from Simons's perspective, August was a very long way away.

Stapp, in the meantime, had other projects that required his attention. Renewed focus on the Holloman rocket sled work developed in the summer of 1957 following the appearance of an article by Colonel Stapp in the *American Journal of Surgery* titled "Human Tolerance to Deceleration." Stapp's office received hundreds of requests for copies, although not all the attention was welcome. That June, ARDC headquarters issued word of new regulations being prepared under the title "Testing on High Speed Tracks." The directive forwarded to Holloman staff explained that track testing was to be considered a "suspect area" of research. Stapp had no doubt that any restrictions being considered were aimed squarely at him—and specifically at his ground vehicle program.

In the meantime, Holloman engineers were hard at work building an extension to the rocket sled track that Stapp had long envisioned and, at the same time, replacing the original rail sections with continuous-weld track that would provide improved strength and require minimal maintenance. The new track, which was months away from completion in the spring of 1957, would eventually extend 35,000 feet—more than six and a half miles. It would accommodate much higher sled velocities and would allow Stapp to complete high-speed windblast tests without the need to stop the sled so quickly. They could hit Mach 1-plus speeds and simply let the sled coast to a stop.

There was still a great deal of work to be done on tolerance to windblast. Pilots ejecting from supersonic aircraft needed proper restraints to prevent limbs and head from flailing, but they also needed protective garments capable of withstanding the punishment of gale-force-plus wind. With this in mind, Stapp was unwilling to wait until the new track at Holloman was complete. In late June, just three weeks following the flight of *Manhigh*, Stapp joined his new Biodynamics Branch chief, Captain John D. Mosely, at the Navy's 22,000-foot supersonic research track in China Lake, California, some 75 miles north of Edwards Air Force Base. Mosely had been at China Lake for months running windblast tolerance tests by the time Stapp arrived, and had already orchestrated a run that reached Mach 1.7 (approximately 1,350 miles per hour) with an anesthetized chimpanzee named Tom II wearing a special flight suit and one of Red Lombard's custom helmets. This was the first test to use a single 100,000-pound-thrust rocket they called the Mega-boom Booster and the *Sonic Wind II* sled. But the test had gone poorly and the rocket sled's headrest collapsed even before supersonic speed was reached, snapping the chimp's neck, while the huge Megaboom rocket vibrated free of the sled and flipped off the track like a TinkerToy. To make matters worse, the latest protective flight suit—they had already experimented with weave cotton and heavyweight denim, both unsuccessfully—turned out to offer nowhere near the protection required. Mosely discovered severely burned tissue on the chimp's body caused by the horrific windblast. An autopsy also found hemorrhages in the lungs, heart, trachea, esophagus, and eyes.

On June 27, with Stapp in attendance, Mosely and team conducted another test. This one also achieved Mach 1.7, and a windblast force of 3,500 pounds per square foot, with results that were no more satisfactory. The beefed-up flight suit worn by the chimp subject, this one a 110-pound male named Ichabod, was ripped open by the windblast. The autopsy noted second- and third-degree friction burns over 40 percent of the carcass. On the same June 27 run, a bio-acoustics team from Wright-Patterson had attached to the sled three guinea pigs: two restrained only with nylon net-ting, and the third in a metal container with 1" × 2" perforations. In a staff briefing, Stapp reported that all three guinea pigs had simply been obliter-ated by the windblast and had "vanished into thin air." Clearly, supersonic wind was nothing to be trifled with. Stapp observed: "Now you understand why hurricanes can tear whole towns apart."

After experimenting with a variety of fabrics, Mosely and Stapp even-tually found a material that could stand up to severe windblast: 12.5-ounce Dacron sailcloth over a thin layer of aluminum to reflect heat, which they immediately recommended be used for flight suits for all high-speed mis-sions. A Red Lombard suit design using the sailcloth would later be recom-mended to NASA with a testament from Stapp: "Dr. Lombard's work was so exceptional . . . that a recent test has indicated beyond doubt that man would survive with full protection from craft even at speeds several times that of present supersonic planes."

Somehow, while working with Mosely during the summer of 1957, Stapp found time to complete the requirements for a private pilot's license: forty hours of flight time, including ten solo cross-country hours, along with two written examinations and a physical. He joined the Holloman Aero Club, and he told friends he intended to try and fly as many different models of light planes as possible, and eventually to obtain free-balloon and sailplane licenses. Stapp and Duke Gildenberg even went partners on a 1948 Cessna 120. Not surprisingly, though, the demands on Stapp's time made it impossible for him to pursue the pilot's life he'd always admired. His logbook remained mostly empty.

.   .   .

On August 5, 1957, Colonel Stapp appeared before the United States House of Representatives at the invitation of Representative Kenneth Roberts, a Democrat from Alabama. Roberts chaired the Special Subcommittee on Traffic Safety of the Committee on Interstate and Foreign Commerce. While influential members of the Senate had been trying for months to catalyze action on auto safety, the first hearings on the subject would be conducted in the House. Roberts, who'd survived being shot on the House floor during a bizarre attack by Puerto Rican nationalists two years previously, was an early consumer safety advocate. Roberts had heard Stapp address an audience and suspected he would make a powerful witness.

The Roberts hearings were intended to investigate the crashworthiness of automobile seatbelts, and Stapp was grateful to have a chance to testify. He was also anxious to take advantage of the opportunity to publicly challenge those in the federal government who were opposed to the idea of using Air Force resources to study car crashes. He knew there had been grumblings in the House Appropriations Committee about his car crash work at Holloman. He also knew that the moment was ripe: the American public was rightfully concerned about the continuing carnage on its roadways and a clamor was building for somebody to do something about it. Just the previous week, an op-ed piece in the *New York Times* had suggested that even as consumers were demanding safer cars, the automobile industry was turning its focus away from safety in favor of the latest sexy design innovations.

Stapp had also seen the news stories that appeared only a few weeks prior to his appearance before the Roberts committee claiming that the Air Force was wasting taxpayer money by crashing perfectly good automobiles. One of those stories quoted UCLA researcher Derwyn Severy as saying that a lap belt alone would have little life-saving value in a high-speed collision. Stapp's critics used these reports to bolster their opposition to seatbelts and to Stapp's work in general.

Colonel Stapp marched into the fray with guns blazing. He arrived on Capitol Hill in full uniform—this time with shoes polished and shined—armed with a slideshow and a thick satchel full of background materials. There were two main points he was determined to make, the first of which he knew would be controversial. Living test subjects, he told the congressmen, are not a luxury but a requirement in the quest to determine the injurious or lethal effects of mechanical force. Neither Oscar Eightball nor Sierra Sam would ever be good enough. "I have never found a successful technique," he told the congressmen, "for doing an autopsy on a dummy." Point number two: it was vitally important to continue live testing because Stapp could now say with certainty that human subjects protected with nothing more than a lap belt were capable of enduring the force of 27 g's for one-tenth of a second, which is equivalent to a car crash impact. If we can learn how to properly restrain and protect drivers and passengers, Stapp claimed, and put them in vehicles designed to mitigate crash forces, we can keep a great many of them alive in accidents previously assumed to be unsurvivable. It is possible to walk away. "I know," Stapp said, "because I did."

There was nothing at all incongruous, he said, about the Air Force studying car crashes: "I have done autopsies on aircrew members who died in airplane crashes. I have also performed autopsies on aircrew members who died in car crashes. The only conclusion I could come to was that they were just as dead after a car crash as they were after an airplane crash." He showed the panel an article from the *El Paso Times* about five Air Force men from Walker Air Force Base who had died in a three-car pileup in New Mexico just that spring. He showed them his statistics indicating that the number one cause of hospitalization for Air Force personnel was injuries sustained in ground vehicle accidents. He told them that seatbelts alone could save 125 USAF lives a year and half a million dollars in direct costs alone. If seatbelts were installed in *all* vehicles and were used by *all* citizens, he said, the country could save 20,000 lives a year.

Like Kenneth Roberts, who had opened his hearings by announcing that "we are not out looking for villains, and we are not out to destroy an industry," Stapp was careful not to appear to attack the automobile companies directly. That wasn't a battle he was prepared to fight. In fact, he praised the cooperation he'd gotten from the Society of Automotive Engineers, most of whose members worked for the Big Three automakers. He mentioned a generous gift of testing equipment that General Motors had made to Stapp's team at Holloman, yet he made it clear to the committee that the safety engineers at the car companies would never be allowed to perform the kinds of tests required to figure out precisely how to implement even the most obvious safety systems.

He'd been invited to consult on safety tests at Ford earlier that year, he explained, and while there he had convinced the engineers that anesthetized hogs were the perfect test subjects for certain kinds of impact studies. Stapp had even located a source, a Michigan pig farm, and had arranged to purchase a test specimen for twenty dollars. When the Ford legal team heard about it, however, they immediately put a stop to the proposed tests. "It seems to be a touchy subject at all the motor companies," Stapp observed. With the cooperation of the Ford engineers, Stapp moved some of the testing program to Holloman, where he could classify it and operate outside the purview of Ford's or anyone else's lawyers.

"I make no special plea about doing animal experimentation," Stapp told the subcommittee, "because I know that animal lovers are very fond of telling stories about how an animal would lay down his life for his master, and I think that when the Air Force purchases a hog to experiment with, that hog is being given the opportunity to lay down its life for all of its masters." Stapp went on: "We consider them [animal test subjects] to be good soldiers, and we give them the very best care that we possibly can, and we do our experiments in strict accordance with the American Medical Association's rules for animal experimentation . . . And with a limited number of animal experiments, we can assure the safety of a practically unlimited number of human experiments."

Stapp proceeded to give Roberts and his subcommittee a fairly technical tutorial in biodynamics, with charts demonstrating the effects of multiplying g-forces and how rate of onset is calculated. He concluded by noting that Air Force scientists were doing valuable work that was not—and perhaps could not—be done elsewhere, and furthermore that this data was being shared not only with the SAE (and through them with the automakers), but also with the Armed Forces Epidemiology Board, the Cornell Crash Injury Research Institute, and the American Medical Association.

The next phase of the work before them, Stapp told the panel, was to continue with more sophisticated animal research, and then to move into detailed impact experiments with human volunteers. The nation needed this next generation of data. Keeping people alive in rapid transit is, he said, "our most serious preventive medicine problem in the United States today."

Some of the congressmen were a bit overwhelmed by the technical nature of some of Stapp's presentation, but most seemed to support the general thrust of his arguments. In follow-up questioning, however, Chairman Roberts noted that in a recent debate on a defense appropriations bill one of his House colleagues had referred disparagingly to the Air Force's "crashing of cars" and had demanded that funds for such work be cut. Roberts asked Stapp, for the record, how expensive the auto crash program had really been, whether it was true that he was purchasing and destroying perfectly good cars, and finally whether his vehicle research had produced any tangible results to date.

The cars he'd used in his crash studies at Holloman, Stapp explained, had all been obtained at essentially no cost from salvage yards and had monetary value only as scrap metal. He also explained that the same equipment and personnel used to study ground vehicle crashes were used to study aircraft crashes. There was marginal incremental cost in either manpower or hardware needed to continue work on the Air Force's "foremost cause of hospitalization and the second-most cause of death." As to the question of results, Stapp mentioned that MATS, the Military Air Transport Service, had reversed passenger seating (from forward-facing to aft-facing) in all transport aircraft, reducing death and injury rates from crashes by two-thirds. He also mentioned that crash harnesses for aircrews in bombers and fighters had been redesigned to comply with strength requirements indicated by Stapp's research. What he wanted, he said again, was to achieve similar results with cars.

Representative Walter Rogers of Texas asked Stapp, seemingly in passing, if he had installed seatbelts in his own car. "I have 9,000-pound-test lap belts in my car," he answered, "and I could not afford to be caught dead without them."

In the days following Stapp's testimony, Rogers summoned spokespersons for each of the Big Three automakers to address the question of seatbelts. Ford was up first, and Alex Haynes, Director of Advanced Product Study, asked to have entered into the record a statement from Robert McNamara, Group Vice President at Ford, extolling the value of seatbelts and estimating that crash fatalities could be halved if all drivers and passengers had and used seatbelts. Yet, under questioning, Haynes was unwilling to commit his company to installing such devices as standard equipment in its cars. Ford's policy was that seatbelts should continue to be optional, dealer add-ons. The problem with that approach, as Stapp and others had been arguing, was quality control. In 1956 there were more than a hundred companies manufacturing seatbelts, most of them with no experience in the design and test of restraint systems. State testing laboratories in California had shown that many of the after-market belts were shoddy and would be useless in a crash.

General Motors' Vice President of Engineering, Charles Chayne, like McNamara, praised the potential value of seatbelts. He explained, however, that GM believed "public acceptance and demand might be extremely small." The company's own testing, he said, suggested that seatbelts might in fact create more injuries than they would prevent. The most Chayne would admit was that seatbelts *might* be beneficial "under many accident conditions." He then proceeded to argue that since most vehicle occupants would probably refuse to use seatbelts even if they were made available as standard equipment, the belts would raise costs for all consumers while benefitting only the few who chose to wear them. The many, therefore, would be asked to subsidize safety for the few. It was a particularly legalistic line of reasoning.

Roberts asked Chayne if saving the lives of those who chose to wear seatbelts would "be worth the additional cost." Surprisingly, Chayne admitted that from General Motors' perspective, "it might not be." Samuel Friedel of Maryland pointed out that the auto companies had installed other safety devices such as turn signals and windshield wipers and headlights, and wondered why seatbelts were so different. He asked Chayne if GM would at least consider installing seatbelts in all their new models as a public service, since Stapp and other experts had testified so convincingly to their life-saving potential. Chayne responded flatly: "I am not in a position to discuss any of the features of our 1958 automobiles."

Ray Haeusler, an automotive safety engineer at Chrysler, offered much the same story. The Big Three automakers all praised the work of Stapp and his colleagues, and none had any objections to voluntary guidelines, but all three made it clear that they would vigorously oppose any effort by the federal government to define standard equipment in new cars. Such mandates would be a declaration of war and it was clear that the companies' formidable legal teams—GM was one of the largest, and arguably one of the most powerful, corporations in the world at that time, and its legal assets were legendary—were poised to fight such efforts. As Charles Chayne had put it, from the point of view of the carmakers, the real problem with automotive safety was "one of education," and had less to do with the cars and trucks themselves.

Following his hearings, and citing Stapp's testimony as a primary basis for the committee's conclusions, Kenneth Roberts issued a formal report to the 85th Congress on automobile seatbelts. The report went fairly easy on the auto industry, but was firm in its dismissal of industry criticisms of seatbelts: "It is the opinion of the Subcommittee on Traffic Safety," Roberts wrote, "after having listened to many witnesses, having asked them many questions and having carefully studied all the data presented, that seat belts, properly manufactured and properly installed, are a valuable safety device and careful consideration for their use should be given by the motoring public."

The public seemed perfectly willing to consider using seatbelts. But would the car companies ever install them?

19

THE MOST SIGNIFICANT APPARATUS

My body tingled with the mixture of apprehension and
excitement that is born of the finality of commitment to an
unknown. It was a pleasant sensation, the same sort of feeling
I suppose that forces other men to demand the chance to make
of themselves fools or heroes.

—*David Simons*, **Man High**

STAPP RETURNED FROM WASHINGTON in the summer of 1957 to find preparations for the full-scale Manhigh mission underway. Following new
chamber tests of the capsule and the survival systems, David Simons supervised the installation of a five-inch telescope and the development of some
twenty-five onboard experiments. Beyond the science agenda, Stapp had
insisted that Manhigh focus on the ability of a human being to tolerate the
flight conditions and to negotiate any number of tasks that might eventually
be required of an astronaut. Simons added experiments in astrochromatics:
they color-coded the operational systems and used pastel hues on the capsule interior to create more pleasing surroundings for the pilot. But the most
significant of all the experimental apparatus, Stapp pointed out, was Simons
himself. The project team had taped a section of photo film to his right forearm to capture the precise location of cosmic radiation particle hits. They'd
inscribed the location of the film plate in permanent hash-mark tattoos on

Simons's arm so they could revisit that same patch of skin in the months and years to come for the purpose of monitoring effects of the radiation.

Stapp, comfortably in his element, appointed himself flight surgeon for the research mission, but he acquiesced to Simons's insistence on one modification to Project Manhigh. Joe Kittinger, originally slated to serve as coordinator of air support operations, was replaced by the commander of Holloman's 6580th Field Maintenance Squadron. No more intramural rivalry. Kittinger had enjoyed the Manhigh assignment, but he'd burned his bridges with Simons, and he figured he was probably done with high-altitude balloons forever. He went back to his first love: airplanes.

With team and mission defined, the site selected by Stapp and Simons for the *Manhigh II* launch was the abandoned Portsmouth iron pit-mine near the town of Crosby, Minnesota. They had needed to find the perfect spot. This balloon would be significantly larger than the one used for Kittinger's flight, likely the largest balloon ever constructed to that point, and the wind protection provided by the cliff walls surrounding the 425-foot-deep pit would offer the best chance for a clean launch. August 16 was the original target date, but cloud cover caused the flight to be delayed twice. Major Simons was finally sealed into the capsule late on the night of the 18th to begin the pre-breathing ritual. After a final equipment check, *Manhigh II* and its pilot were loaded onto the bed of a pickup truck and driven down into the Crosby mine. The Holloman-based balloon launch crew was getting quite good at handling these big balloons, but there were always new wrinkles with each flight, and things inevitably went wrong. Stapp had always taken operational problems in stride, viewing them as opportunities, as lessons to be learned. On the *Manhigh II* launch, a restraining band had gotten snagged on the plastic envelope 40 feet above the ground and had to be cut away by Otto Winzen's wife, Vera, with the aid of a hook and ladder truck. Vera had supervised the balloon construction back at the plant and refused to let anyone else near the wispy envelope with a pair of scissors. The liftoff finally occurred at 9:33 in the morning. By that point, Simons had already been sealed inside the capsule for eleven hours.

Stapp and other members of the ground crew crowded into a van that had been outfitted as a communications station and headed out in an attempt to

Manhigh II *with David Simons aboard clears the cliffs of the Minnesota iron mine on its journey to the edge of space. At peak altitude, the helium bubble would expand and the polyethylene balloon would round out—in Simons's words, "like a basketball being dribbled in slow motion in an upside-down world." (Photo courtesy of U.S. Air Force)*

chase the flight path of *Manhigh II*. Stapp had done the best he could with the funds he'd been given, but they had never really been enough and at one point he'd had to convince Winzen Research to invest company funds in the project, or the research flight might never have left the ground at all. They had no choice but to be resourceful. Cabin air conditioning was provided by a can of water, a fan, and a hose. As the water evaporated at altitude, a fan blew the cool vapor into the capsule. Their equipment may have been low-tech, but the Manhigh team had never compromised its lofty goals. In Stapp's view, they were conducting nothing less than a rehearsal for manned space flight. If they could bring David Simons back safely following a full day and night beyond earth atmosphere, they would go a long way toward debunking the myth that space conditions were inherently untenable for man.

As the balloon gained altitude, the van carrying Stapp and the communications team headed west out of Crosby on State Highway 210, tracking it at a rate of 20 miles per hour. For the next two hours, the ascent went flawlessly. *Manhigh II* reached the balloon's equilibrium—the point at which the helium gas had expanded to completely fill the polyethylene envelope and the system's lifting capacity equaled the weight of the payload—at 101,500 feet. Simons's radio dispatches to the crew were enthralling. He described the balloon as "gently bouncing like a basketball being dribbled in slow motion in an upside-down world." He reported being able to see 400 miles, and to discern the curvature of the earth. For a while he became obsessed with the color gradient of the sky: "The color has intensified a bit . . . the pink or reddish lowest band has now disintegrated into a clear salmon red pink . . . bottom area grading through a distinct yellow into the luminous light blue so distinctive of a high-altitude sunrise which very quickly shades off into the black of night and a star-studded sky." It was almost as if Simons was anxious to surrender to the breakaway rapture. "This cloud layer is so solid," he said, though his words could not be heard on the ground at this point due to radio problems, "it gives one a feeling of being in heaven above the rest of the world where you can look down over the earth and see the poor faltering mortals."

In spite of the awe-inspiring vistas, the flight of *Manhigh II* was never a joyride. Trapped in the seat inside the tiny capsule, Simons suffered as the daytime hours wore on. Exhaustion set in, and his knees and his back began to ache. To make matters worse, the VHF radio the team relied on to transmit vital signs back down to Stapp had ceased to work at all. To his credit, Simons insisted on remaining aloft through the night as planned, as he knew Stapp would expect him to. In fact, Stapp was on the radio urging him on: "As far as we're concerned, things are going so well that I see no reason why you shouldn't stay up during the night." As dusk settled on the upper Midwest and the gondola's westward swing slowed, the communications van pulled off the road near Fargo to wait. At one point that evening, Stapp thought he noticed a slight slurring in Simons's speech, and he immediately got on the radio to ask when Simons had last eaten. As usual, Stapp's diagnostic suspicions were on the money. Simons admitted that he'd simply forgotten to eat anything since that morning, and he quickly downed a chocolate bar to try and boost his energy and concentration. Stapp had put Simons on a low-fat diet in the days leading up to the launch to reduce the chance of painful gas expansion in the intestines, but Simons had reported some gastric issues nonetheless. "Still have a little nervous diarrhea which is part of the business." The astronaut's role would not always be a glamorous one.

The radio van wasn't the only vehicle tracking the flight of *Manhigh II*. Two airplanes and a helicopter were airborne as well. Wright-Patterson had sent a C-47 that carried a camera crew from the 1352nd Motion Picture Squadron. Simons had occasionally lost payloads on his unmanned radiation research flights. The Air Force was definitely not going to lose this one.

As the night wore on, the Manhigh flight delivered plenty of surprises. The gas in the balloon cooled and contracted with nightfall and, as it gave up altitude, Simons found himself in the 70,000-foot range floating just above a huge bank of thunderheads with lightning strikes crackling all around him. Meteorologists had had no idea that storms extended so far into the stratosphere. Simons was probably the first man to witness and was certainly the first to photograph a thunderstorm from above.

Manhigh II also had its fair share of problems, including communications trouble throughout the flight. The radio van kept jockeying as best it could to get positioned more advantageously. The interior of the capsule was far too hot at times, with temperatures in the mid-80s. Because the electrical supply was limited, Stapp had had to instruct Simons to shut off the fan on the bootleg air conditioning system to conserve power. Condensation in the cabin shorted out a photo flash unit during the night. The chemicals in the air regeneration system lost their effectiveness as the mission ground on, and—as with Kittinger's flight—the carbon dioxide levels climbed dangerously. Simons was forced to use an oxygen mask just to breathe.

Some of these complications were undoubtedly a result of the extended duration of the flight. The objective had been twenty-four hours aloft, but the descent took far longer than planned. By the time Simons touched down in a South Dakota alfalfa field at 5:32 the following afternoon, he'd been in the air for thirty-two hours and ten minutes, and had been sealed inside the capsule for nearly forty-four hours. One of the first to reach him following the landing was flight surgeon Stapp, in full uniform as always. As "the most significant apparatus," before Simons could even be congratulated by the rest of the team, Stapp stuck him with a needle to draw a blood sample. Otto Winzen observed for the benefit of reporters covering the flight that Simons had been inside his craft longer than Lindbergh had been in the *Spirit of St. Louis* on his solo trip. Simons lost 17 pounds in a day and a half. Stapp was animated as he addressed a press conference the next day, and made a point of singing his project officer's praises: "He was sort of a one-man band . . . Dave could sit in a gondola, handle 20 emergencies, and not die once."

The flight of *Manhigh II* was a *Life* magazine cover story. Doubleday offered Simons a book deal. An op-ed piece in the *New York Times* ran under the headline "First Space Man," and declared that Simons "made the trip for all of us." For a brief period, Simons surfed in the wake of Stapp's scientist-as-action-hero celebrity. Both Simons and Kittinger were awarded the Distinguished Flying Cross by the Air Force. Yet for all of the fawning press and the mission's achievements, Stapp was unable to convert the attention into solid support for a follow-up flight—perhaps with the new two-man gondola design Winzen Research was already proposing. "Human performance in an environment equivalent to space is now known to be possible," Stapp announced, making sure that everyone understood what they'd done and why it mattered. Nevertheless, space research continued to get a cold shoulder from the United States Air Force. One aviation writer labeled this attitude "a huge scientific-political boycott of the future," and that was exactly the way Stapp felt about it. He continued to make what began to seem like a quixotic argument for not only a renewal but an expansion of Manhigh. Nobody, however, seemed very interested.

Meanwhile, though it continued to be a time of almost frantic activity for John Paul Stapp, he'd somehow managed to maintain his long-distance relationship with Lillian Lanese. Except for an interlude during a quick trip to New York for an appearance on the television show *To Tell the Truth*, Stapp's only contact with Lillian during this period was epistolary. He had begun tucking short poems into his letters, which, she said, "certainly makes one's heart feel warmer." She was working on a series of specials for Canadian television, but she promised she'd arrange another visit to El Paso as soon as her schedule allowed, even hinting that as her career wound down she was considering leaving the hustle and crowds of New York for good, and that El Paso might be a good landing spot.

That move came in September, and Stapp began arranging regular trips to see her whenever he could spare a few hours. Lili joined the staff of her friend's ballet school and began to acquaint herself with some of the new skills she'd need for a life in the wide open spaces. "I will try to remember all the do's and don'ts of driving," she wrote Stapp. "I could not help but write you after you left El Paso Sunday evening to say how very fond I have become of you, Paul."

Unfortunately, Stapp's opportunities to court the ballerina whose eye he'd managed to catch were rare and brief. The formal promotion to full colonel had come in April, and the scope of his duties continued to expand. While he retained command of the Aero Medical Field Lab at Holloman, which gave him responsibility for a wide array of programs, dozens of personnel, and a roster of civilian contractors, he also participated in front-line research at the project level when he could. That was where, he had once told William Bloom, the real work got done. Where the lives got saved.

Out on the front line, the ejection seats, parachutes, and other emergency escape technology that had been the focus of so much of Stapp's work had already proven their life-saving value for pilots many times over. They would prove it again in late September 1957, serendipitously this time for Manhigh pilot Joe Kittinger. On a routine test flight out of Holloman, his F-100 Super Sabre had caught fire seconds after takeoff. At an altitude of only 800 feet, the airplane pitched nose-up and stalled. Kittinger punched out, though he was by most estimates too low for a standard parachute opening. Luckily, the ejection seat fired perfectly, and his canopy deployed and bloomed just a few yards above the ground. Kittinger walked away crowing his trademark fighter-pilot motto: "No sweat." In truth, it was a near miracle. Stapp, immensely heartened that all the hard work that had gone into qualifying the escape technology had now saved the life of a valuable colleague and friend, named the young captain a member of the Survivors Club.

· · ·

All eyes turned back to space on October 4, 1957, when blockbuster news reports announced that the Soviet Union had managed to put a satellite—a

184-pound orb about twice the size of a basketball—into elliptical earth orbit. About once every ninety minutes, *Sputnik* completed another loop around the globe. It was a stunning coup for the Soviets, and it generated a level of anxiety and turmoil in the United States that is hard to exaggerate. *Sputnik* was widely seen as a frontal attack on America's cultural core. Decrying both the symbolism and the technical achievement of the incident, a prominent senator issued a call to arms that summed up the mood in Washington: "What is at stake is nothing less than our survival!" A poll of American citizens revealed that the vast majority favored immediately raising the national debt ceiling and hiking federal taxes in order to step up America's entry into the space race. Among the beneficiaries of this new attitude toward the formerly ridiculed supporters of space research were John Paul Stapp, David Simons, and Otto Winzen, accompanied by the Air Force's new darling: Project Manhigh.

No longer would stratospheric research programs need to plead, hat in hand, for funding. Now the Air Force brass were summoning Stapp to high-level strategy sessions in Washington and asking how they could help, how much money he could use, and how long it would take him to get the next Manhigh flight up. Not that there weren't plenty of strong opinions about exactly how space research ought to be conducted. As Major Simons said following a private meeting with General Flickinger at a conference in Barcelona: "The people who had opposed me earlier were suddenly telling me how to do it."

Planning for a third Manhigh flight began only a few days after *Sputnik* rocked the world. A "fact sheet" created by the Air Force redefined the Manhigh mission: "The purpose of the *Manhigh III* flight ... is to continue research into the field of space medicine. It will involve the investigation of human reactions under space equivalent conditions with special emphasis on the psycho-physiological area of interactions." Colonel Stapp announced that he wanted to focus special attention on the pilot selection process. He wanted to identify the profile of the astronaut, and then devise testing protocols that would allow specialists to evaluate and measure candidates. There was a range of qualities they would be looking for, and the most important of these in Stapp's mind were psychological. He wanted someone physically fit and intellectually advanced, but the nature of a space mission also required someone who could remain calm under pressure, who could handle complex problem-solving in a stressful environment, and who possessed a healthy blend of confidence and caution. Another new wrinkle would be the formation of a "panel of experts" that would direct the Manhigh pilot during the mission. This panel would include subject-matter experts in space medicine, physiology, meteorology, astronomy, atmospheric physics, and other relevant disciplines. It was the model that would be adopted first by the Air Force's secret space project, Man In Space Soonest (MISS)—designed around a capsule much like Manhigh's—and eventually by its civilian counterpart and usurper, NASA.

Stapp cleared Simons to begin preparations for the flight of *Manhigh III*. He made it clear they were not going to rush it. They were going to operate—from a human factors perspective—a full dress rehearsal for the launch of a manned orbital flight. There were thousands of details to be worked out and Congress was still months away from passing the National Aeronautics and Space Act of 1958, which would authorize the formation of America's civilian space agency. In 1957, as meteorologist Duke Gildenberg said of the Holloman balloon team: "We *were* the space program."

Late that year, Simons and Stapp traveled to Huntsville, Alabama, to the Redstone Arsenal, an Army rocket development facility, at the invitation of Wernher von Braun to discuss the Army's Project Adam. Von Braun, perhaps the world's most notable evangelist for space travel, wanted to employ a Redstone rocket to propel a Manhigh-like capsule to a height of 150 miles, and he told Stapp he wanted to offer the design and fabrication contracts to Winzen Research. He suggested that the Adam system be used as a vehicle to help establish a lunar military base. Simons was particularly excited about the prospects of an intra-service effort he wanted to call Project Man Very High. However, only a few months later, orders came directly from ARDC headquarters forbidding Air Force officers from collaborating with the Army on von Braun's plans. "The Air Force officially decided not to support *Adam*," wrote one historian, "because it feared the ballistic capsule project might take resources away from the service's X-15 rocket plane."

Simultaneous with the Air Force's decision not to cooperate with the Army's rocket program, Stapp received a directive forbidding him to participate in any additional committee work. There were those at the mahogany desks in Baltimore and Washington, DC, who remained suspicious of Stapp's activities, and who continued to attempt to curtail the exposure and publicity surrounding his work at Holloman. ARDC headquarters informed Stapp that he would not receive approval for a European trip he'd planned for the fall. Stapp wrote to a friend at Allied Powers Supreme Headquarters in Europe: "Puns fail me! That should tell you how depressing this is."

In the meantime, that November, in between quick trips down to El Paso to see Lillian, Colonel Stapp hosted the third annual Car Crash Conference at Holloman, and this one was the best-attended yet, with some eighty-one participants who included doctors and engineers from the auto companies, and also representatives from police departments, insurance companies, the National Safety Council, the Veterans' Administration, the U.S. Public Health Service, and the United States Congress. The star of the show that year was Professor James "Crash" Ryan from the University of Minnesota. Ryan was a mechanical engineer and inventor with a flair for the dramatic. With the assistance of Stapp's team, Ryan staged a number of live tests—one of which involved Ryan driving his own hydraulic bumper-equipped Ford into a solid barrier at 20 miles per hour—simultaneously demonstrating energy-absorbing bumpers and collapsible steering wheels. Months earlier, Ryan had

Professor James "Crash" Ryan hits a solid barrier at 20 miles per hour in a demonstration of his hydraulic bumper at the Third Car Crash Conference. The conference was focused not on preventing accidents but on helping people survive them. (Photo courtesy of U.S. Air Force)

crashed the same car into an iron barricade at 40 miles per hour with no damage to the dummy passengers or the structure of the car.

Stapp arranged for all of the conference attendees to personally experience a 6-g deceleration in the Bopper. The conference was coming to be regarded as the premier event for automotive safety specialists and word was leaking out about the latest advances in restraint systems and other technical breakthroughs that many thought should be getting more attention in Detroit.

The auto industry's approach to safety had traditionally been to focus the blame for what Stapp referred to as "this black plague of automobile deaths and injuries" on everything but the automobile itself. One study of the American automobile industry described Detroit's longstanding attitude: "The industry argued that automobiles almost never cause accidents and that it was not normal for one to be involved in a crash—thus, there was no obligation to design a car with this possibility in mind ... The solution to the safety problem lay in improving the driver and the highway."

A General Motors vice president stated: "I am convinced that more progress can be made in traffic safety by emphasizing the relations between the driver, the signaling system, and the road, than by undue emphasis on a crash proof car." Another GM VP put it in even starker terms: "We feel our cars are quite safe and reliable ... If the drivers do everything they should, there wouldn't be any accidents, would there?"

Stapp's Car Crash Conference, however, was pointedly uninterested in the problem of preventing car crashes. The conference began with the practical assumption that crashes were going to happen. The only thing Stapp and his colleagues concerned themselves with was how to protect human beings when the inevitable occurred.

Stapp's was not the first national auto safety conference. The Automotive Safety Foundation had held meetings as early as 1937 to address what had then been referred to as "the accident problem," and the foundation's president had said publicly that "the driver was where the focus of attention should be." The foundation, along with the National Safety Council, both of which were supported financially and otherwise by the automakers, adopted the institutional views of the industry, and excluded individuals and projects focused on the car.

Back in 1936, President Franklin Roosevelt convened an Accident Prevention Conference that reached what at that time was a fairly radical conclusion: that the solution to the proliferation of traffic accident death and injury must, in fact, lie with the cars themselves. The APC urged automakers to begin voluntary efforts to incorporate safety design into their new models, predicting that "a growingly impatient public" would eventually demand it. In the end, FDR's conference failed to significantly influence the thinking or behavior of those who made the public's cars.

Two decades later, President Eisenhower convened his own White House Conference on Highway Safety, appointing as chairman Harlow Curtice, president of General Motors. The conference's advisory board included vice presidents of both GM and Ford. Three years after, Curtice summed up his own and the board's determination in testimony before the House of Representatives: "I have reached the firm conclusion that driver education offers the most fertile field by far for making substantial progress toward our objective." The implication was that there would be no great need for an emphasis on safety engineering in Detroit if only federal and state governments would get busy instructing the nation's drivers how not to crash. Forget lap belts and collapsible steering wheels: the solution to traffic accidents was to be found in the classroom.

Even as the American public was becoming alarmed by "the accident problem," influential forces were hard at work attempting to divert that attention, and certainly to forestall any restrictions or regulation on the design of new cars. Twenty years earlier, Paul Hoffman of the Studebaker Corporation, who was also chairman of the National Automobile Chamber of Commerce's Traffic Planning and Safety Committee, had come to the conclusion that traffic accidents were becoming a deterrent to the sales of new cars, and his committee had been instrumental in promoting the highway safety movement as an antidote. Years later, Hoffman admitted that the committee had made a conscious choice not to investigate the cars themselves. He confessed that since, by one account, "he and his associates were selling automobiles, they did not want to admit that they were not safe—but rather attention was focused on some of the other, easily identifiable problems, such as dangerous intersections, bad driving, glaring headlights, and inadequate driver training."

Hoffman and the expanding coterie of industry insiders who were engaged in debate on the accident problem without reference to the cars themselves would come to be known in the circles Stapp moved in as "the safety establishment." It was not a term of endearment. The Stapp Car Crash Conference became the first of its kind to buck the trend and to ignore the official position of the industry.

A few weeks after the conference, on December 18, 1957, Stapp finally acquiesced to editor William Buckley and signed a publishing contract with Henry Holt and Company for an autobiography to be titled *Nothing Ordinary: A Chronicle of Space-Age Man*—or, as Stapp referred to it, "this confounded book"—for which he received a check for $1,500 as an advance against royalties. The manuscript was to be delivered on May 1 of the following year.

First, however, the would-be author had more pressing issues to attend to. To the surprise of almost everyone who knew him, Colonel John Paul Stapp married Lillian Lanese at his brother Celso's house in El Paso, in a small ceremony two days before Christmas, 1957, with Major General Leigh-

General Leighton Davis gives away the bride at the wedding of John Paul Stapp and Lillian Lanese in El Paso. Though Stapp would later complain that he'd been pressured into the union by his brother and sister-in-law, the couple remained married for forty-two years. (Photo courtesy of Wilford Stapp)

ton Davis—Stapp's commander at Holloman—and Wilford and Margaret Stapp in attendance. Wilford never believed it was a marriage based on passion, but was rather the pragmatic union of two mature, accomplished individuals who were ready for companionship: "Not a burning romance, but just a darn good match." He would even go on to describe it as an arranged marriage, concocted primarily by Day Stapp. John Paul himself complained to a friend many years later that he had been railroaded. "To the end of his days he was very disgruntled about being pushed or forced into that marriage sooner than he would have wanted."

Whatever the truth of its origin or the motives of the couple, the marriage suited both parties. Lili provided Stapp with an attractive wife who'd achieved her own stature in the world, who'd accomplished big things and made a name for herself, and he'd given her something of at least equal value in return. A friend who knew them both observed: "He made her laugh!" Lili moved into the house on Lovers Lane and began immediately to put her own artistic stamp on the longtime bachelor quarters, taking over maintenance of the garden and gradually acquainting herself, to her great disappointment, with the duties of an Air Force colonel's wife.

20

KICKED UPSTAIRS

Ryan? Stapp here. The worth of a scientist can be measured by
the degree to which he's willing to do battle with the savages
and cannibals of management. Good day.

—John Paul Stapp, complete transcript
of telephone conversation with the author

B Y THE TIME OTTO WINZEN received the contract from Colonel Stapp for the
Manhigh III gondola—a completely redesigned craft, a foot taller than its
predecessor, with a rocket-age turret machined from stainless steel—Stapp
had already gotten word that he was going to be reassigned. The way Stapp
saw it, he was being "more or less suspended." He'd hurried to ARDC headquarters to see if he could get the decision reconsidered. "General Flickinger
told me that I could have a choice of working at his headquarters or being
the chief of the Aero Medical Laboratory at Wright-Patterson. I made it clear
that I didn't want either job." A desk at headquarters in Baltimore was a hor-
rifying prospect. So, after some reflection, Stapp opted for the lesser evil and
agreed to return to Dayton.

The move surely solved one problem at ARDC, conveniently separating
Stapp from the controversial ground vehicle testing program. A proposal
Stapp had made the previous summer that would have established a Direc-
torate of Biomedical Sciences at Holloman to continue the study of restraint
systems had been summarily dismissed by Brigadier General Don Ostrander

with the comment: "I do not feel we should re-organize just because John Stapp got promoted." Coincidentally, at almost exactly the same time, the Air Force began the process of installing seatbelts in all of its ground vehicles nationwide in compliance with a mandate for which Stapp had long been advocating. Some base commanders embraced the measure and went further, requiring seatbelts even in privately owned vehicles kept on base. Others complied only grudgingly.

Colonel Stapp arrived back in Dayton on April 20, 1958, exactly five years from the day he'd left for his tenure in New Mexico. On the 23rd, he received a second invitation to testify before Representative Kenneth Roberts's subcommittee in the House of Representatives investigating research needs in the area of traffic safety. Stapp had been in the Capitol only the week before, when he had publicly taken on a prominent Army crash injury researcher over the issue of aft-facing seats in airplanes. Stapp used his congressional appearance to bolster support for the inclusion of seatbelts in new car models. He also, however, offered an impassioned plea for increased federal funding for car crash research using human volunteers, citing the "difficulty in terms of Air Force monetary support," complaining that "support is dwindling to nothing." The human volunteers part was crucially important, he explained. "That is the ultimate measure of the adequacy of a protective device and certainly a way of certifying it for human use more realistically than can be done with dummies, animals, or similar simulators."

Even though he felt he'd been exiled from Holloman, he vowed to continue his crusade to protect the crash program there. Congressman Roberts had told Stapp, well before his House testimony, that he would "be displeased if the Air Force drops the automobile crash program." That seemed promising. Nevertheless, the House Appropriations Committee would shortly vote to eliminate all military funding for auto safety research—a slap in the face to Stapp that would effectively kill the Holloman program along with the various projects it had operated through contractors. Stapp suspected that lobbying efforts on behalf of the auto industry were a major factor in, if not the principal impetus for, the decision. At any rate, with Stapp in Ohio and the funding for the Holloman work reallocated, the Car Crash Conference, the nation's only independent forum on auto safety, appeared all but defunct. As a result, the cause of coordinated crash research in the United States was thrown, at least temporarily, into neutral.

Meanwhile, Stapp shuttled between Wright-Patterson and Holloman so that he could remain involved in the planning for *Manhigh III*. The task of identifying a new Manhigh pilot—and, by extension, helping to define the profile of the astronaut—began in earnest in the early weeks of 1958. Stapp and Simons had worked out an exhaustive selection process. It included an interview to determine a candidate's motivations and probe the depth of his scientific background; a four-day medical evaluation; time in a decompression chamber and in a "hot box" with temperatures of 155 degrees Fahrenheit with 85 percent humidity; a twenty-four-hour claustrophobia test; a daylong series of interviews with clinical psychologists and psychiatrists; isolation in a soundproof, lightproof sensory deprivation chamber; stress tests that included grueling centrifuge rides; and a session with the candidate's feet submerged to the calves in ice water.

David Simons, who had received a promotion to lieutenant colonel following his triumph with the Manhigh research flight, selected the first two pilot candidates himself: Otto Winzen, who had been auditioning for the job since the earliest days of Manhigh, and Captain Lack Beeding, the Baylor grad who would narrowly escape severe injury on the Daisy Track later that spring. Both men, however, were eliminated for unspecified reasons following an intense battery of psychiatric and psychological evaluations. Another candidate, Captain Grover Schock, was disqualified following a ballooning accident that left him with numerous broken bones and severe internal injuries. A veteran parachutist was considered until the medical exam revealed an elevated blood cholesterol level. But as the weeks went on, a capable young pilot and engineer who had been brought in by Colonel Stapp to undergo the selection process for the purpose of establishing baseline measurements for the other candidates began to turn some heads himself.

South Carolinian Clifton "Demi" McClure—he got his nickname because he had been, he explained, "the first Democrat born on the night Franklin Roosevelt was elected President"—desperately wanted to be part of the effort to put human beings in space, and he saw Project Manhigh as his ticket. He knew that space research was generally unpopular even at Holloman, and he determined to take advantage of that fact as he got familiar with the Balloon Branch operations on the far edge of the base. "At this time no decent, intelligent, sane Air Force officer would find his hat over there in that damn place . . . They wouldn't volunteer for any of the flights, and they were scared to death to be associated with it. They used to make fun of you." While no one had yet suggested that McClure be considered as a Manhigh pilot candidate, it was a perfect opportunity. The stress tests actually sounded entertaining to him and McClure approached them as if it were all a game. "I started getting myself evaluated along with the rest of these people," he explained. "And I beat 'em every time!"

The word about McClure spread quickly through the Manhigh operation. It wasn't merely his ability to handle difficult or unpleasant situations with aplomb, but he seemed to possess a rare blend of practical know-how and formal education. "I've never met anybody any smarter," Lieutenant Colonel Simons would say.

Simultaneously, at Wright-Patterson, Stapp announced that he would be changing his organization's name, infuriating the outgoing Aero Medical Laboratory commander and his staff. It would henceforth be known as the Aerospace Medical Laboratory. Colonel Stapp was determined to shake things up. It was one of the prerogatives of rank and he intended to exercise it. At the same time, he approached his new assignment with trepidation. "I just hadn't been a paper shuffler," he said. "I had been in the field working. It was not an easy transition." Later that summer, he would write to a friend about his new life as a colonel in charge of the aeromedical assets at Wright-Patterson: "I really miss Holloman and the carefree productive days there. Here I feel like I'm doing rather mediocrely a job that almost any flight surgeon could do much better. I like to produce more than I care for just overseeing."

Wright-Patterson was also a difficult transition for Lili Stapp. She had always been a loner, and her background as a dancer had not endeared her to Air Force society. Lili refused to join the officers' wives club at Wright-Patterson—by all accounts one of the snobbiest of such clubs in the Air Force, always eager to punish social missteps—and had little interest in hosting dinner parties or attending public events. This would be a black mark against Colonel Stapp in the upper echelons of the Air Force, but Lili could not bring herself to play the game.

For his part—though he was no longer formally associated with Project Manhigh and had been forbidden to work with the Army on its plans for space flight—Stapp remained deeply engaged with the problem of how to protect human beings beyond earth atmosphere. Even at his desk job at Wright, Stapp did plenty of thinking about pilot bailout from aircraft, and now a new set of problems presented itself. Would it be possible to rescue the pilot of an X-15 at extreme altitudes or astronauts during suborbital missions? Would an ejection seat system be viable in the upper stratosphere? Could a man tolerate a fall from such heights? These questions led Stapp fairly quickly to imagine a program designed to study the biophysics of super-high-altitude bailout using a human volunteer as a subject.

As it turned out, a similar program already existed, or had existed. The Air Force's Operation High Dive had been founded four years earlier to investigate one of the enduring riddles of free fall: a rapidly falling body's tendency to twirl, spinning like a platter, faster and faster, reaching deadly rates of 200 revolutions per minute. The phenomenon was known in the parachuting world as "flat spin." Skydivers had learned how to use subtle body-control techniques to arrest flat spin in the years following World War II, but such techniques required extensive training. What Operation High Dive had explored over a period of three years was a method that a pilot with no parachute training might use to free-fall safely from extreme altitudes. High Dive had used balloons, outfitted with cameras, from which they dropped anthropomorphic dummies. But the project was disbanded in 1957 after failing to make any significant progress. After reviewing the technical reports, Colonel Stapp was convinced that an answer still might be found.

Stapp began to think about the qualifications of the individual he'd need to lead such a project. He'd want an experienced test pilot with good aviator instincts, someone with some parachuting experience, someone who could handle a high-altitude balloon. Most of all, however, he needed someone with the guts and creativity to persevere in the face of the inevitable naysayers he knew would come out of the woodwork during a project like this. One name leapt to the top of the list: Captain Joe Kittinger. The only downside was that Kittinger had never led a research project of any kind. He'd been trained originally as a fighter pilot and didn't have a lot of experience with—nor probably, Stapp guessed, much patience with—the niggling details of management and organizational discipline that would be required.

Nevertheless, Colonel Stapp invited Joe Kittinger to join him at Wright-Patterson in the summer of 1958 and named him director of a new emergency escape research project Stapp had decided to call Excelsior—a Latin word meaning "ever higher" or "ever upward." *Excelsior* is also the title of nineteenth-century Italian composer Romualdo Marenco's best-known ballet, written in 1881, which was intended as a tribute to scientific and industrial progress. Could Lili have introduced it to Stapp? Did they listen to the score on Stapp's ever-evolving hi-fi system? The ballet's final act closes with an apotheosis of the genius of the human race: Science, Progress, Brotherhood, and Love celebrate the glory of the present and the hope for the greater glory of the future.

Stapp asked Kittinger, whom he'd assigned to the Biophysics Division in the Escape Section of the Aerospace Medical Lab, to initiate Project Excelsior with a study of the records of Operation High Dive—which had been conducted at Holloman while they'd both been there—and to identify an approach to safe high-altitude bailout that might avoid the failures of the earlier effort. Kittinger quickly settled on two key strategies that he felt could be game changers: first, he wanted to design a completely new kind of parachute that could automatically control the jumper's body position; second, he wanted a human test subject rather than a dummy, for reasons he hardly needed to explain to Colonel Stapp. Kittinger also desired access to one of the Air Force's rocket planes for the experiments he was envisioning.

Captain Kittinger presented all of this during a meeting in Stapp's office at Wright-Patterson. One of the reasons Kittinger had been so eager to follow Stapp to Dayton was that he saw Stapp as a leader who was willing to make tough decisions rather than dithering forever while he gauged the political winds that were always swirling about the Air Force. "That was the great thing about Stapp," Kittinger said. "It was what set him apart from other senior officers. He didn't care where an idea came from or how many feathers it might ruffle. He was interested in one thing: results."

The meeting ended with Stapp granting Kittinger full approval to proceed with Project Excelsior, though he made it clear there was no chance they'd be able to get access to high-altitude jets, which left balloons as the only option for getting a test subject up into the stratosphere. Kittinger hated the idea of using balloons because balloons have no airspeed, and parachutes don't open without an airstream. "The balloon complicated the whole damn thing," he complained. In order to simulate the airstream that an ejecting pilot or suborbital astronaut would encounter, the subject would need to delay any activities until he'd built up sufficient speed—which created another set of challenges for the parachute design.

Meanwhile, by late summer, Demi McClure had officially been named the pilot for the *Manhigh III* mission. Had they in fact found the prototype for America's astronauts? Stapp certainly thought so. The Air Force introduced McClure, a ceramics engineer with a master's degree from Clemson and a hotshot pilot, to the press to see how he would handle the public relations aspect of the job. He was charming and disarming. "I think the least we can do is to approach the problem without being afraid of the supreme sacrifice," he told reporters when asked if he harbored any fears about the Manhigh mission. "I'm afraid I'm not afraid. Anything of this sort has its serious side, but I think it's also going to be quite a bit of fun." McClure had a lot of Joe Kittinger in him. Perhaps predictably, then, as summer waned, he began to clash with his project officer, David Simons.

With Stapp out of the picture—at least in terms of the day-to-day operation of Manhigh—Simons struggled at times to maintain his grip. There were a thousand details, and Simons was never a very willing delegator. To complicate matters, during the run-up to the third flight, Simons fell in love with Otto Winzen's attractive young wife, Vera. The daughter of a Minneapolis society photographer, Vera Habrecht Winzen had been instrumental in the business of Winzen Research. In fact, she had invented several of the novel processes that made the development of giant polyethylene balloons possible. With her interests split between science and the arts, Vera had taught herself enough about finance to keep the company afloat through some thin times—something Otto had never paid much attention to.

More or less simultaneously, Vera divorced Otto and David Simons divorced his wife, Libby. It disturbed many of those connected with the project, especially those who'd become close with the Simons family in Alamogordo—including Colonel Stapp and Lili, who lived practically next

From left to right: Otto Winzen, Vera Winzen (later Vera Simons),
David Simons, and Stapp pose in front of the Manhigh *capsule.*
(Photo courtesy of U.S. Air Force)

door. "It really bothered Paul a lot," Wilford Stapp would say. "Because Simons and Libby had four children. And he just walked off and left the family, and I think everybody was just down on him. And I know that Paul and Lili kept up with Libby after that. I think that was the breaking point." Simons and Vera married two years later, in 1960. The entire drama certainly did nothing to lessen the challenges facing Manhigh.

As it turned out, one of the biggest challenges was Colonel David Simons's project management. Cooling for the gondola, during the period after the pilot was sealed inside and before the balloon reached the strato-sphere, was provided by a 30-pound package of dry ice that was placed inside the cap on top of the craft. It had worked quite well for the first two flights. But *Manhigh III* left the ground without the dry ice cap being replenished in the minutes prior to launch, as protocol mandated. This was a particularly critical error because McClure—having been forced to repack a parachute that had popped open after he had been sealed inside the capsule—was per-spiring heavily and creating more interior heat than they'd accounted for.

Somehow, an adequate supply of dry ice had been unavailable, and the decision was made to launch without it.

Was it Murphy's Law? Or Stapp's Ironical Paradox? The consensus was that someone just forgot to pick up the dry ice that morning. Simple as that. According to McClure: "Colonel Simons *volunteered* to pick up the CO_2. And he *forgot* it!"

Without the dry ice, the balloon lifted off from a Holloman runway at 6:51 a.m. on October 8, 1958, and it reached equilibrium at 99,700 feet at precisely 10 a.m. McClure remained at that altitude for nearly three more hours, conducting experiments, taking measurements, and making a series of observations that included watching a ground-launched missile destroy a target at 80,000 feet. Throughout, however, both the telemetry and the pilot's radio transmissions indicated that all was far from well. McClure was having trouble getting water through a rubber drinking tube. At one point he reported feeling exhausted. The panel of experts, airborne in a C-47, asked for some readings, which is when they discovered that the interior of the capsule had heated up to 96 degrees Fahrenheit.

McClure was asked to supply a reading of his body temperature, and he reported it as 101 degrees. By 1:50 p.m., McClure's thermometer was reading 103.4. The onboard air conditioning system had failed. Colonel Rufus Hessberg, aboard the C-47, ordered McClure to begin his descent. McClure pleaded for more time. He still wanted to spend the night aloft, a notion that was quickly rejected by the panel.

By the time the balloon had descended to 85,000 feet, McClure's temperature was a shocking 105.2. At this point, the C-47 and the ground crew lost radio contact with *Manhigh III*. Several of those on the C-47 were convinced they were going to lose the pilot; others feared they already had. Duke Gildenberg said it was like a scene from a Hollywood disaster movie. To make matters even more dire, nightfall was looming and the balloon was on course to come down in the jagged San Andres Mountains west of Holloman. It was a worst-case scenario.

It was full dark by the time the gondola touched down in a rocky gully thousands of feet above the valley floor. A helicopter managed to put down within a few hundred yards, and the rescue crew scrambled and clawed its way up toward the landing site. When they made it, they found that McClure had climbed out of the gondola and was pulling his helmet off. When he saw them, he flashed a grin. He refused to allow himself to be carried back down to the helicopter. "I don't know what all the fuss is about," he said. "You don't have to treat me like an invalid."

A medic got McClure to sit down on a rock ledge long enough for a brief examination. His pulse rate was 180 beats per minute, and his body temperature was 108.5. It was astonishing that he was still conscious. It could be that the project's pilot selection process had produced one of the few individuals who could have survived such an ordeal.

The mix-up with the dry ice cap and the subsequent overheating of the *Manhigh III* capsule had very nearly led to a tragedy that might have dealt a serious blow to the continuation of American space research—at least in the short term. While it had all occurred after Colonel Stapp had left Holloman, he resisted the notion of publicly criticizing David Simons or anyone else connected with the management of the program. Simons's nemesis, Captain Joe Kittinger, on the other hand, was not so restrained. "Neither McClure nor I had much respect for Simons," Kittinger said. "It was absolutely ridiculous that they launched that thing without the dry-ice cap. Poor Demi ended up getting almost killed. Stapp told me one time he thought that maybe Simons didn't want McClure to beat his record. I can tell you this: if either Stapp or I'd been there it would never have happened. That cap would've been on there. I'd have made sure of it."

21

THE SURVIVORS CLUB

I am not a daredevil. Stapp was not a daredevil.

—*Joe Kittinger, interview with the author*

THIRTY DAYS AFTER the launch of *Sputnik*—and the awakening of the American public to its apparent second-banana role in the dawn of the space age—the Russian rocket men went one better and shot *Sputnik II* into orbit. This was not only a bigger and more technically impressive spacecraft, it was a biological flight. Aboard the 1,118-pound craft was a small, female mixed-breed dog named Laika (Russian for "barker"). Laika died early in the mission due to capsule overheating, but the achievement brought the already high level of paranoia in the American aerospace community to a fever pitch.

In January of 1959, NASA quickly arranged to begin shuttling candidates for its astronaut corps to Dayton for evaluation by Colonel Stapp, the only man in the U.S. who'd actually developed and run a formal selection process for space travelers. The new national space agency would need to rely on help from both private contractors and existing military programs. NASA's Space Task Group had initially considered a number of profiles for the role of astronaut: pilot, athlete, explorer, daredevil, mountaineer, deep-sea diver. Stapp was on record as preferring "either an engineer with medical training or a medical man with engineering training." He eventually came to agree with President Eisenhower that test pilots represented the most promising pool of potential astronauts, in large part because they would be the easiest to train, and he set to work modifying the Manhigh protocols to match the long list of prerequisites NASA officials had issued. The initial group of 137 applicants was winnowed to just thirty-seven after preliminary tests in Washington, DC, and at the Lovelace Clinic in Albuquerque, where some of the testing for *Manhigh III* had been conducted.

At Wright-Patterson, candidates were sealed into capsules for extended periods, during which they had to demonstrate tolerance to sudden pressure drops and massive temperature swings. The Air Force exposed them to ear-splitting noise, ran them on centrifuges, and locked them into ink-black isolation chambers. They were strapped to "shake tables" that vibrated intensely, and they were subjected to invasive medical exams, personality evaluations, and psychiatric interviews. Stapp gave them complex tasks, both physical and mental, to perform while they were made to endure all manner of unpleasant conditions. In Stapp's mind, one of the most revealing procedures was one they called the Frustration Test. Candidates were instructed to press a series of buttons in response to lights that flashed on and off, but the results they thought they were trying to achieve were in reality impossible. The idea was to monitor the candidates' ability to control their emotions when presented with their own failure to complete a task. An emphasis on psychological endurance was one of Stapp's major contributions to the Mercury selection process, and he judged the value of the physiological stress tests by their ability to reveal psychological reactions, though these tests greatly annoyed the pilot subjects. Interestingly, the German aeromedical community that had done so much work on physiological response to flight conditions in the 1940s had, at Adolf Hitler's personal directive, largely ignored the mental dimension of pilot capability. The Führer apparently believed such concerns to be "Jewish science."

After a week of intensive evaluation, the Air Force settled on twelve finalists and presented them to NASA officials as the best of the best. NASA quickly trimmed the list of twelve to just seven, over Stapp's protests. "This is a monumental blunder that NASA committed," insisted Stapp, who'd given them a dozen finalists for a reason. "They picked seven just because they had seven Mercury flights projected. I just couldn't understand it. Why didn't they want some spares?" As the Mercury program progressed, Stapp's objections proved to be justified. "They paid dearly for it," he said, "because the seven were a pretty firm unit knowing that they had no spares to worry about. When they wanted something, they just resolutely banded together . . . and they'd get it." Stapp was involved in a number of Project Mercury planning efforts, but was often frustrated that the civilians in control of the program seemed to distrust and devalue the Air Force's contribution. It was, he thought, a textbook example of the NIH syndrome: Not Invented Here. In Stapp's estimation, much of the blame for this attitude should be assigned to contractors such as McDonnell-Douglas. "But the NASA people themselves," he was at pains to point out, "I got along with extremely well . . . I was being consulted all the time from Houston."

During this period, in addition to his work for NASA on astronaut selection, Stapp was preparing white papers for the agency on a number of other aspects of space travel planning, some of which appear to have been influential. In "The 'G' Spectrum in Space Flight Dynamics," Stapp covered $+Gx$ and $-Gx$ tolerances, and went on to document some of his own ideas on rocket launch logistics, propulsion systems, astronaut performance, the predicted long-term effects of zero-gravity exposure, and the effect of exposure to massive vibration. Though there was much more to penetrating the heavens than mere rocketry—"To conquer space," Stapp wrote, "man must first conquer his own limitations"—he believed that the work done at Edwards and Holloman and Wright-Patterson had put the beginnings of the space program on a good footing.

Stapp enjoyed his informal association with NASA, often working on agency assignments after hours and on weekends. He also quickly became an inspirational spokesman for the U.S. space effort, one who was able to effectively weave together both the romance and the technology of space exploration in a way that sometimes left audiences spellbound. A trip to address the Travis County Medical Society in Austin in March of 1959 offers an example. Stapp told the overflow crowd, which that night included his old professor and mentor Dr. E. J. Lund, that based on his examination of the astronaut candidate corps, the problem was no longer with the ability of the human being to withstand the hazards. "All we need," he said confidently, "is the vehicle that will bring him back." He explained that the Air Force and NASA, working together, had determined how to counteract the three main dangers: radiation—in Stapp terms, "cosmic bullets"—heat from the buildup of ionic pressure against the spacecraft's surface, and turbulence. "Man is ready to go!" he insisted almost defiantly, throwing down the gauntlet before the engineers. "We're just waiting for the ship." He got a standing ovation. The *Dallas Times Herald* ran it on the front page the next day: "AF Doctor Claims Man Ready for Moon Flight."

In the meantime, both competitors in the race to the moon had critical decisions to make about the individuals who would make the journey. NASA's astronaut selection process differed in two important demographic respects from that adopted by the Soviet Union. First, the initial Soviet space travelers were nearly a generation younger than the American test pilots of Project Mercury. Yuri Gagarin was twenty-seven when he made his orbital flight; John Glenn was forty-one. The Soviets wanted maximum shelf life for their astronauts. Given the expense of selecting and training someone for a space mission, they wanted to optimize the investment by starting with younger candidates. Stapp agreed with the Soviets on this, and would have preferred a younger Mercury corps. But on the second point, Stapp and his Soviet counterparts differed.

The Soviets were also determined to launch the first woman into orbit, and as part of their early space program trained five female cosmonauts for orbital missions. John Paul Stapp, however, was a staunch opponent of women in space, offering three principal arguments for his position: 1) "Physiologically, women are considerably less efficient than men." He claimed that given equal weight, height, and age, females had proven in testing to be about 85 percent as efficient in relevant physical performance as men. 2) "Psychologically, there is reason to believe that they are not equipped by nature to take the emotional stresses peculiar to the conditions of space flight." While acknowledging that there was little data to back up such a claim, he considered exposing women to the hazards and stresses of space flight to be irresponsible. And, 3) "Economically, the cost of putting a woman in space is prohibitive . . . strictly a luxury we can ill afford." The inclusion of women, he thought, did not justify the expense of alterations in test methodologies and equipment. Stapp was not an outlier on this subject within the American space community. Both James Webb, who would become the NASA admin-

istrator to oversee the race to the moon, and President Dwight Eisenhower
viewed female cosmonauts as cynical propaganda stunts on the part of the
Soviet government.

One staunch opponent of this view, unsurprisingly to any who knew her,
was Lili Stapp—and she was not shy about expressing her opinion. Her years
of strenuous training in the practice of a demanding and highly disciplined
physical art—an art, she pointed out, that required her to take flight dozens
of times in a single performance—were, she believed, ample evidence that
women were well up to the challenges and stresses of taking a seat in a capsule and riding it into space.

Russian skydiver Valentina Tereshkova became the first woman in
space in 1963. Her performance and those of other women cosmonauts
finally settled the argument. As the years went by, Lili Stapp would remind
her famously visionary husband of this fact. He was not, even in matters on
which he was an acknowledged expert, infallible.

· · ·

Stapp had, meanwhile, been elected president of the American Rocket Society in January, and on the pleasant morning of April 24, 1959, departed
Dayton in the back seat of a T-33 trainer en route to Denver, where he was
scheduled to address a society meeting that same evening.

Joe Kittinger had originally been scheduled to fly Stapp that morning.
When he was available, Kittinger was always Stapp's first choice. But Joe had
had to cancel a few days earlier. That's when Stapp contacted a colleague of
Joe's, thirty-six-year-old Captain Harry Davis, a Wright-based fighter pilot
who'd served with distinction in Germany in the early 1950s.

When Stapp realized that Davis—who seemed distracted by an argument he'd had with his wife before the flight—intended to try and make it
all the way to Denver without refueling, Stapp began trying to convince him
to turn back. "He went absolutely haywire," Stapp reported. "Preoccupied,
fatigued brains under stress suffer deterioration of the highest functions
first—thinking may be rational, analysis of events correct up to the point of
judgment, then there is a gradually enlarging breach with reality, a hypnotic
intrusion of wishful thinking." Apparently Davis was convinced he could
glide safely into Buckley Field outside Denver.

As the plane descended through 10,000 feet, running on fumes, Stapp convinced Davis they would both need to eject. Buckley was still 17 miles out. "I finally talked my way out of the plane at an altitude of 400 feet above the terrain," Stapp said. This was ridiculously low. Stapp ejected in the vicinity of Watkins, Colorado, doing four quick reverse somersaults in the ejection seat before separating. He described the sequence of events surrounding his first parachuting experience: "I was in the air just eight seconds after my parachute deployed before landing in 20-knot gusts that dragged me to the brink of a rocky gulley before I was able to release myself from the harness. The ejection seat and completely automatic equipment I spent 15 years helping develop functioned with absolute precision and saved my life." Later, saying that he'd been too occupied with staying alive to be terrified, he elaborated: "I was in the air just long enough to watch the chute deploy, then started looking down and before I saw the ground I hit it . . . still with a horizontal component from the 135-knot aircraft speed, I snubbed down my heels and rolled on rump, shoulders and head, then flopped back."

Back in Dayton, word got out that Colonel Stapp had gone down. According to testimony from the tower at Buckley Field, Captain Davis had finally ejected—but at a height of only 100 feet above the ground. He was killed on impact. The plane, with no fuel to ignite, came down in a wheat field and was later discovered intact with only minor damage. Television news in Dayton reported that Stapp was presumed dead, and Lili spent a couple of hours believing the worst. Stapp called her as soon as he got to the hospital in Colorado, and he asked her to pay a visit to Captain Davis's wife, which she did that evening. Stapp signed himself out of the hospital once the doctors had gotten x-rays of his sprained neck and fashioned a cast on his foot to protect a chipped ankle bone.

"FASTEST MAN ON EARTH BAILS OUT IN CRASH!" The papers covered the story, but perfunctorily. The incident might have gotten more attention had it not been for Stapp's discreet efforts to suppress it. Harry Davis was one of the few African American Air Force officers flying fighters at Wright-Patterson in 1959, and Stapp was concerned that the crash might be exploited by those who would argue that black pilots weren't sufficiently capable or disciplined. He'd grown impatient with racial intolerance in America. Two years earlier, he'd railed to Lili about Arkansas governor Orval Faubus calling the National

Guard out to stop black students from entering Little Rock Central High School. But the main thing was that Davis had been a damn good pilot. Stapp turned down several opportunities to publicly explain the details of the crash and the events that led up to it. *The Reader's Digest* offered him $2,500. Stapp explained in a letter to an editor at Henry Holt: "It would grieve and embarrass a widow and 3 children who would perceive it as an attack on their race as well as on their father. With a bit more time, particularly if Mrs. Davis marries again, I can tell the story without opening her wounds." Stapp told friends years later that in fact she never had been able to forgive his survival, and continued to blame him for her husband's death.

Even as Stapp recuperated from his hard landing in Colorado and got around with a painful limp for some weeks following, he received an anonymous letter on stationery from the Brown Palace Hotel in Denver. It read: "So – we lost a good pilot bringing you out here so you could receive one more 'honor' so dear to your warped mentality – But you got out of it – Congratulations!"

Ignoring the SOS club, Stapp, during his recuperation in the summer of 1959, set about reviving the Car Crash Conference. Because he no longer had the access to the Holloman test facilities or the funding that would allow him to host the event, Stapp got Professor James Ryan to agree to organize the gathering under the auspices of the University of Minnesota's Extension Division. They billed it as the Fourth Stapp Automotive Crash and Field Demonstration Conference. Ryan staged rollover demos using dummy test subjects and rented a crane to raise a car 58 feet in the air and drop it onto a wooden pallet. The three-day conference drew seventy-one registrants from thirteen states. Stapp only hoped, he said, that he could convince the Air Force to resume car crash testing some day.

. . .

Jetting back and forth between Holloman and Wright-Patterson, Joe Kittinger labored throughout the spring and summer of 1959 on Project Excelsior. He assembled an all-star team of Air Force engineers and civilian contractors, and worked them day and night. Perhaps the most important contributor was Francis Beaupre, a tough-talking, cigar-chomping civilian parachute designer who'd been exploring some novel ideas with a parachute team at Wright.

The parachute dated back to the mid-1770s, when the Montgolfier brothers of France—inventors also of the hot-air balloon—perfected their design by dropping sheep from towers beneath small fabric canopies that resembled umbrellas. In the first decade of the twentieth century, an American inventor named Charles Broadwick figured out how to fold a silk chute into a backpack in such a way that it could be released after jumping. Deploying a parachute in the stratosphere, however, created all kinds of new and exotic problems that the Excelsior team would have to solve in order to qualify emergency escape systems for high-altitude flight.

Luckily, Beaupre quickly came up with an ingenious design for a multistage parachute: shortly after jumping, an 18-inch pilot chute would deploy and that chute would pull from a backpack a 5-foot drogue chute that was intended to stabilize a jumper in free fall without really slowing his descent. It effected a sort of modified free fall. The Excelsior team believed it was the key to arresting the flat spin that had plagued Project High Dive, and it was exactly the kind of breakthrough that Kittinger had been counting on.

One of Stapp's requirements for Excelsior, as it had been for Manhigh, was that backup balloon pilots be trained and ready to assume a seat in the gondola should that become necessary. Which is why Captain Kittinger lifted off from Holloman on a training mission on the evening of May 20, 1959, with captains Dan Fulgham and William Kaufman aboard. The three men suffered a hard landing early the next morning about 10 miles from the town of Roswell, and they all tumbled out of the gondola onto the desert floor. Unfortunately for Fulgham, his helmet got pinned for a moment between the lip of the gondola and the ground. Blood vessels in Fulgham's head burst, and by the time the balloonists' chase helicopter got to them, his face and scalp had swollen grotesquely. According to Kittinger, Fulgham looked like a "big blob." Years later, Kittinger described the incident: "We could see his head just expand . . . The guy's head was the size of a basketball. You could barely see his nose."

The doctors at the hospital did not believe the injuries were life-threatening, but they wrapped Fulgham's head in bandages to inhibit further swelling. Kittinger, meanwhile, who'd sustained a deep laceration on his own forehead, was concerned about the imminent arrival of Air Force crash investigators whom he feared might ground them all. Because Project Excelsior was still in ramp-up mode and still regarded skeptically at Air Force headquarters, Kittinger was justifiably worried that bad publicity might derail the progress he and his team had made. Once his forehead was sewn up and bandaged, he put in a call to Colonel Stapp in Dayton, who verified the identities of the goverment men for hospital personnel and ordered Kittinger to get his team back to Holloman as rapidly as possible.

At that point Kittinger began barking orders, apparently intimidating Glenn Dennis, a Roswell mortician in attendance at the hospital. Kittinger took custody of Fulgham—looking now every bit the part of the alien invader with his swollen head wrapped in bandages—and escorted him to the Holloman chase helicopter, which took off before any investigators could arrive, leaving a long-running mystery in its rotor wash. Mr. Dennis later described a creature with an oversize head that he'd seen walk into the hospital under its own power, surrounded by heavy security. Dennis told reporters that a belligerent, redheaded Air Force captain had accosted him and told him: "You did not see anything. There was no crash here. You don't go into town making any rumors that you saw anything or that there was any crash . . . you could get in a lot of trouble." When Dennis protested, he claimed the captain threatened him: "Somebody will be picking your bones out of the sand." Glenn Dennis began telling anyone who would listen that he'd witnessed the government whisking away a captured alien. The Roswell papers ran a few stories, but the incident otherwise got little more attention than had the space alien stories from a few years earlier following the Project Mogul balloon crash.

Fulgham's cranial hematoma subsided after a few days and he made a full recovery. The incident was covered up to avoid any embarrassment for the Holloman crew and to deflect attention from Project Excelsior—which only exacerbated the suspicions of Dennis and others who believed they'd been privy to a conspiracy.

Back in Dayton, the redheaded Captain Kittinger's own biggest headache during Excelsior's design and test phase was the chief of the Parachute Branch at Wright-Patterson, a longtime Aero Med Lab member and Francis Beaupre's boss. The chief, believing the multistage approach was too complicated to be practical, issued a requirement that in order to qualify the new design for a live test, the team would need to complete thirty-five consecutive test drops without a single glitch of any kind. It was an onerous demand in light of the project's meager budget, and Kittinger complained loudly to the branch chief—to no avail. Finally, after thirty-two near-perfect parachute tests, the team experienced its first malfunction, and Kittinger immediately asked for a meeting with Colonel Stapp. Stapp summoned both Kittinger and the branch chief to his office and listened to both sets of arguments. When he'd heard enough, Stapp turned to Kittinger and posed a straightforward question. Would the parachute work?

When Kittinger told him it would, Stapp stood up and announced that the meeting was over. Project Excelsior would continue toward its goal of a jump from the stratosphere. No more dummy tests of the new chute design would be required. Though Stapp could no longer afford the luxury of rolling up his sleeves and working alongside his project teams, he could at least knock down walls to free up his best people to do what they did best. "Stapp went to bat for us on that," Kittinger said.

The team went back to work on the project's survival systems, as well as on its gondola and balloon. The gondola was a far simpler affair than the one used on Project Manhigh. To Kittinger, it made no sense to use a pressurized capsule since it would need to be depressurized at altitude prior to jumping. He reasoned that if a problem developed in his partial-pressure suit or his helmet, he wanted to discover it on the way up rather than deep in the stratosphere. The other thing was, an unpressurized gondola would be much cheaper—and the project needed to cut expense wherever it could safely do so. Kittinger's ultimate plan was to complete a jump from 100,000 feet, but he wanted to do it in stages: first, a test flight to somewhere in the vicinity of 60,000 feet; then another flight to about 75,000 feet; and finally the big one: up where the Manhigh flights had gone.

By late fall, the team was ready to go, but before final clearance to proceed could be obtained, Systems Command (the new designation for what had been the ARDC) in Baltimore asked for a briefing.

"The gun was loaded against Stapp," Kittinger said. The young captain got the distinct impression that some of the generals were looking for an excuse to run Stapp out of the Air Force. At headquarters, before the commander, General Bernard Schriever, and several other generals seated around a conference table, Kittinger laid out the project's objectives, described the preparation that had gone into achieving them, and explained exactly how they intended to accomplish the mission. He took a few questions. When he'd heard enough, Schriever turned to Stapp. Kittinger remembered the exchange.

"Colonel," Schriever asked. "Will you approve this?"

Stapp answered without hesitation: "Yes sir."

"Do you understand the ramifications of this flight?" Schriever asked.

"Yes sir," Stapp said again.

It was clear to Kittinger that Stapp was going out on a limb for him— perhaps even putting his own career in jeopardy.

General Schriever asked them both to step out of the room. Ten minutes later, the supplicants were ushered back into the conference room and informed that a stratospheric jump had been tentatively approved. But just a single jump. If, afterward, Excelsior still wanted to move forward, both Kittinger and Stapp would have to return for another hearing.

"I felt obligated to Colonel Stapp to make certain everything was right," Kittinger confessed. "I felt very responsible to him."

Stapp, however, had a dilemma with regard to his favorite pilot, one that had nothing to do with balloons or parachutes. When NASA had announced that it would begin taking applications from test pilots for Project Mercury, Kittinger had immediately gone to Stapp seeking his advice. He knew that Stapp was going to be involved in the selection committee, and he wanted to know if Stapp would support him as an astronaut candidate. Stapp wanted to, but with Project Excelsior in full swing, he needed Kittinger in Dayton. There was also another argument against NASA—at least at that point— from a pilot's perspective. "When the Mercury program started," Kittinger explained, "those guys were just going to be along for the ride. They were just living organisms."

In the end, Kittinger agreed with Stapp that he could make a greater contribution with Excelsior, helping to develop a workable escape system for astronauts and high-altitude jet pilots alike. "I never regretted it," Kittinger said. "Ever."

With Kittinger firmly aboard, the Excelsior team continued working toward its first manned launch. Stapp could only follow the flight from his office at Wright-Patterson through intermittent reports. Excelsior was precisely the kind of innovative effort he had always championed, and he badly wanted to be on the ground with his men, but his new duties—committees, meetings, and strategy sessions—made it impossible.

Early on the morning of November 16, at a makeshift launch site on the outskirts of Truth or Consequences, New Mexico, Kittinger completed his pre-breathing ritual to denitrogenize his blood and climbed into the open gondola to await his return trip to the stratosphere. The balloon, again constructed by the reliable Winzen Research, rose at a rate of 1,200 feet per minute, but the view the pilot was expecting was marred by the intense glare of the rising sun. For much of the ascent, Kittinger found himself all but blinded. He also had problems with the faceplate of his helmet fogging up with condensation. Even more alarming, however, was the feeling that his helmet was tugging upward on his neck.

Kittinger remained so focused on his helmet problem that he missed his planned jump altitude at 60,000 feet. He kept rising. By the time he'd completed his jump preparations, he was at nearly 76,000 feet, some three miles higher than he'd wanted to be.

Then, in his haste to exit the gondola, Kittinger inadvertently tripped the timer for the multistage parachute ahead of schedule, causing the pilot chute to deploy only seconds after he jumped. Because he had not yet fallen long enough to build up sufficient airspeed to fill the 18-inch pilot canopy, the little chute whipped about like a rag before wrapping around Kittinger's neck. A few seconds later, he entered the dreaded flat spin that had been the project's enemy from the beginning. Falling through the stratosphere at hundreds of miles per hour, Kittinger did what the dummies on Project High Dive had done a few years earlier. He twirled like a leaf. "All of a sudden I started turning violently to the left and I couldn't stop it . . . I couldn't pull my arms in, which gives you some idea of what the centrifugal force was. I fought, but I passed out." Only a specially rigged reserve chute set to open automatically at a predetermined altitude saved his life.

"I landed pretty hard," he said. "I was really in pretty bad shape."

The flight and jump from *Excelsior* looked, on the face of it, to have been a catastrophe. After opposing more extensive testing of the multistage parachute and promising the Air Force brass that he knew what he was doing, it had all come apart on Joe Kittinger. Following the flight postmortem,

Kittinger—with his confidence in his team's system unshaken—returned to Dayton to explain to Colonel Stapp what had happened. They knew without a doubt how to solve the problem that had caused the foul-up with the timer, he said.

Within days, both Kittinger and Stapp were back in Baltimore, where Kittinger took a page from Stapp's approach during the Battle of Muroc, arguing that the test flight had done exactly what tests are intended to do: reveal critical flaws in the system and enable the team to improve it. "This jump proves why we're there," he remembered telling Schriever and the other generals. "This justifies the research we're doing and what we're trying to prevent. And by God, we just showed you what happens when a body free falls from these altitudes."

Kittinger told the generals that he wanted another shot. Once again, Colonel Stapp backed him up. He told the brass that Kittinger was the finest pilot he'd ever encountered and that if Joe was convinced he could make Excelsior a success, then the Air Force would be a bunch of fools not to let him try. They'd invested too much time and energy to quit now. They were too close to solving the problem of high-altitude escape.

Schriever's approval came the following day, and on December 11, Kittinger and the Excelsior crew returned to Truth or Consequences and executed a flawless flight to and jump from 74,700 feet. This time, the multistage parachute system performed beautifully. The whole thing had been simple, Kittinger told a reporter a few days later, "like riding a motorcycle." *Excelsior II* was an unqualified triumph, paving the way for preparations for the project's pièce de resistance: an epic jump from above 100,000 feet. But Kittinger knew he would have to get everything right this time. If something went awry and he failed to survive *Excelsior III*, the rerpurcussions wouldn't stop there.

"If I'd been killed," he said, "they would have buried me—but they would have hounded Dr. Stapp forever. They'd never have left him alone."

Stapp had placed his bet on Excelsior, and having done so, turned to other things. During the project's test phase, he had continued typing away on the manuscript of *Nothing Ordinary*, even soliciting the help of Ed Rees, the journalist who'd written the cover story for *Time* back in 1955. Still, he'd been unable to submit enough suitable material to please his editors. Henry

Holt and Company was eager to cash in on the scientist-as-hero moment, and Stapp was anxious to oblige, but the demands of his Air Force work and the sheer breadth of his own interests and ambition conspired against them both. "Autobiographical writing is not easy," Stapp admitted, "even with as interesting material as I have at my disposal." He struggled to keep his narrative from veering into highly technical detail. "Another difficulty," he wrote the publisher, "is related to the recounting of things that actually happened, as they happened, gruesome things like picking up the pieces of a pilot and performing an autopsy on him in order to try to reconstruct the events of an aircraft crash and to bring home to the reader that crashes are deadly violence that can happen to anyone—not exempting the reader, with his obdurate refusal of reality, and the need of research to develop protection from the Kinetic Plague of the steel age. One Air Force public relations officer who looked at this chapter of the book called me up to say that it had kept him from sleeping, and accused me of sensationalizing bloodshed." Stapp must have begun to sense that his style and approach to the telling of his own story were not what the popular press wanted for its readers. Finally, in January 1960, the publisher pulled the plug. "I am sorry to say," a senior Holt editor wrote Stapp, "we have no alternative but to suggest that the agreement be cancelled and that you return to us the $1,500 advance that was paid for the book."

By April 15, Stapp had managed to pay back $1,000 of the advance, and made good on the remaining $500 by the end of that same month. The experience had been a bitter disappointment for him, but the truth is that he had never really had the luxury of time that would have allowed him to focus for long on any book project. Between his duties to the Air Force, his work for NASA, near-constant domestic and international travel, his never-ending list of speaking engagements—he made 202 appearances on behalf of the American Rocket Society in 1959 alone—John Paul Stapp continued to be an extremely busy man. His marriage to Lili survived this whirlwind because, in Stapp's words, "I was married to a lady who had been a ballerina and she knew all about traveling." He took her on some of his trips, but Lili admitted that she actually enjoyed the solitude of her time alone. She had never been able to find much of interest in the world of aeromedical research or military politics.

At about this time, Stapp heard the first of an incident that had occurred shortly before he'd first met Lili—and it came as a blow. In his obstetrical clinic in El Paso, Celso Stapp had performed an abortion for Lillian Lanese. John Paul Stapp was a lifelong opponent of abortion—which was at that time an illegal operation in the United States—except in cases of medical necessity. Stapp felt not only betrayed, but robbed. He would have wanted to begin his own family if he'd been able and deeply regretted not having had the chance to raise Lili's child. "One of the things that I thought was missing in his life," Wilford Stapp observed, "was children. I think it was his greatest disappointment."

"I would have raised it," Stapp said. "I would have raised it gladly."

The knowledge of what had happened, made worse surely by the fact that the secret had been kept from him by his wife and his brother for well over a year, seems to have affected his relationship with both. He confided in a friend that he was not sure he would ever be able to forgive his brother.

22

EXIT FROM VALLEY FORGE

As an individual leader of men, Stapp was unexcelled. People who worked for him admired him greatly. But Stapp was not highly regarded for his management skills by the Air Force. He was not a 'nuts and bolts' commander.

—*Richard Chandler, interview with the author*

THE YEAR 1959 seems to have been tough on John Paul and Lili Stapp's relationship. His schedule had him gone most of the time, and even when he was home he was consumed with work. The return to Dayton had been no less difficult from the perspective of Stapp's Air Force command. Though he had not asked for nor particularly enjoyed his role as a senior officer at Wright-Patterson, he had tried to do it with the same passion he brought to his work at Holloman. Since arriving from New Mexico in 1958—and mindful of General Flickinger's order to reduce his travel and speaking commitments—he'd been reorganizing the original Aero Medical Laboratory. He'd presided over a modernization of the research methodologies and equipment in an attempt to reinvigorate a staff that had, in his estimation, degenerated into merely "burnishing trivia in airplane hardware." He transferred anyone who declined to cooperate fully, and created positions for new skill sets that he hoped could open fresh inroads into bioastronautics, bionics, and closed-system human ecology—even when this work threatened to overlap the charter of NASA.

Then, early in 1960, General Flickinger began suggesting that it was time for Stapp to move again, perhaps to assist in the development of a new life sciences division to be headquartered in Washington, DC. Stapp protested that his work in Dayton remained unfinished.

That February, Colonel Stapp made a final, impassioned attempt to convince his boss to give him more time to complete his mission. "Washington's birthday brings to mind the Revolutionary War," he wrote, "which was won by taking calculated risks, calculated losses, and going on in spite of them to victory; and Valley Forge, a bloody winter when everything looked very grim for George and his cause. I wouldn't begin to compare our situation to that, of course, except that our cause is much less in jeopardy and our risks less fraught with disaster—but we can still take example from Father George's courage and resourcefulness, his perseverance and patience."

He pleaded for more time. Even a few more months, he suggested, would be enough to finish the job of bringing the Air Force's greatest research and development group kicking and screaming into the space age.

"Later I can go enjoy life in Texas," he told Flickinger, "but I can't walk out on Valley Forge." Though their relationship—like most of Stapp's with his Air Force superiors—had escalated into a pitched battle of wills, John Paul Stapp and Don Flickinger were more alike than either of them might have wanted to admit. They were both men of science with an uncommon talent for innovation and original thought. Flickinger had fought hard to get the Air Force oversight responsibility for the American exploration of space; the Man In Space Soonest concept had been his baby. But they were both stubborn as Hill Country goats. "He and I locked horns considerably," Stapp said.

A few weeks after his final request for an extension at Wright-Patterson, Colonel Stapp was reassigned to Brooks Air Force Base in San Antonio and named Deputy Chief Scientist of the Aerospace Medical Division there. Valley Forge would have to find its way without him. It would be Stapp's last command position with the United States Air Force. He was about to turn fifty as he prepared for his return to Texas, and he couldn't help but wonder whether his best days were now behind him. His biggest single regret was the failure to make more progress on automobile safety.

As he took up residence at Brooks, Stapp noted, perhaps predictably, "I wasn't very well liked by the management." He claimed that his lecture slides were occasionally stolen when he sent them to be copied, and that a superior officer confiscated an original set of plans for the Nazi sled track at Tempelhof which he'd acquired years earlier at Wright Field. Nevertheless, it turned out to be an extraordinarily productive time for Stapp's work on problems related to aerospace medicine and—unexpectedly—on ground vehicle safety as well. His contributions were rarely anymore in the field, on the front lines of research, but had moved into the realm of policy, influence, persuasion, and even inspiration. Blowing sand and rocket engines had given way to steel desks, telephones, and hermetic conference rooms. Though Stapp actually spent little extended time in his office at Brooks—instead, he doubled down on his already prodigious travel regimen, roaming the world preaching the gospel of safety and survival.

In the summer of 1960, he addressed medical and flight safety conferences in his native Brazil and made several other stops across South America and Europe—one to attend a vehicle safety conference in Sweden, where he made important connections with members of the Volvo automotive engineering team. Throughout that same year, he spoke to dozens of domestic organizations in nearly every state in the Union on the future of U.S. space exploration. His prescience is evident in an interview he gave to the *Detroit Free Press*: "So far as space research is concerned, we are in the same position as the alchemists of old who promised princes they'd change base metals to gold . . . We don't promise gold but knowledge from our space research. And from the technical developments should come a wealth of by-products."

Stapp was not only part of a national effort to assure America of its readiness for the challenge of *Sputnik*, he became a leading figure in it. In dozens of speeches across the country, he described how astronauts would eat freeze-dried foods. "The food then can be reconstituted with water obtained by refining and purifying the human waste product." He explained that the astronaut might also obtain water by concentrating the sun's rays onto rocks and boiling out the chemically combined water contained in them. He imagined myriad new opportunities for science. "With men on two heavenly bodies—earth and moon—we'll be in a beautiful position to do triangulations and survey the distances exactly within our own solar system." It was Stapp brewing up a utopia. The idea of a moon colony, he said, was "a beautiful concept."

He received little if any compensation for his public appearances, but upon returning to San Antonio in December, Colonel Stapp found a letter waiting from the great Hungarian mathematician and physicist Dr. Theodore von Kármán announcing that Stapp had been elected to the International Academy of Astronautics. Recognition of his work by other leading figures in the world of science seemed to bring Stapp a satisfaction he didn't get from the various military honors he'd received. He had von Kármán's letter framed.

· · ·

Joe Kittinger, meanwhile, with the two preliminary missions under his belt and working toward the flight of *Excelsior III*, had learned to respect the hazards of space even as he'd marveled at its stark beauty. "Dr. Stapp once described it for me," he said. "He put it all in perspective. Being in space is like

Captain Joe Kittinger in the altitude chamber at Wright-Patterson Air Force Base testing the pressure garments and helmet he would wear on his record-setting balloon flight to the edge of space. Despite rigorous tests, the right pressure glove would fail during the ascent of Excelsior III, *exposing Kittinger's hand to the near-vacuum and jeopardizing the mission. (Photo courtesy of Joe Kittinger)*

being surrounded by arsenic. Outside of you is death." It was like the jungles surrounding Bahia or the endless Mojave. Life was struggle. To survive was to fight. Outside of you is death.

On the morning of August 16, 1960, Kittinger launched in *Excelsior III* from Tularosa, New Mexico, and took his balloon to an altitude of 102,800 feet, despite a failure in one of his pressure gloves that exposed his hand to the near-vacuum. Had Kittinger told his ground crew about the glove, he would have been ordered to abort the flight. He simply endured the painful swelling.

After floating at equilibrium for eleven minutes, he jumped and free-fell to earth. Captain Kittinger reached a top speed of 614 miles per hour in free fall—the fastest any man had ever traveled without a vehicle—and with the help of Excelsior's now-reliable multistage parachute, managed to avoid the flat spin that had nearly killed him on his first attempt. The team had proved the viability not only of their novel parachute system, but also of man's ability to survive near-supersonic free fall in the event of a high-altitude emergency.

It was an epic achievement that would be celebrated with cover stories in *Life* and *National Geographic*, bringing Kittinger his own measure of the celebrity both Stapp and David Simons had tasted before him. Publisher E. P. Dutton would rush out a book on Kittinger and Excelsior titled *The Long, Lonely Leap*, for which Colonel Stapp would write the foreword. Even the Mercury astronauts professed awe. When a reporter asked Alan Shepard, NASA's first spaceman, if he would have been willing to duplicate Kittinger's feat, he said: "Hell no. Absolutely not." But the real lasting value of Project Excelsior, and the vindication of John Paul Stapp's original vision, was the multistage parachute that could now offer high-altitude aircrews the promise of survival following emergency escape from the stratosphere. Within a few years, all high-altitude aircraft of the United States military would be equipped with Francis Beaupre's design. Project Excelsior's glory went to Joe Kittinger, but a large measure was due John Paul Stapp. Kittinger would never tire of reminding the world of that fact.

However, with Project Mercury in ascendancy and his ally Stapp in San Antonio, Kittinger felt the full scrutiny of Systems Command descend on him following the flight and jump from *Excelsior III*. The Excelsior team had planned to make at least one additional jump, but the Air Force had wearied of the risks and General Schriever scotched the idea almost as soon as he got wind of it. Before Stapp's departure from Dayton, Kittinger had already begun planning for a series of astronomy balloon flights that were to be funded and staffed partially by private contractors and academic institutions. After several failed launch attempts, Kittinger and civilian astronomer William White reached 81,500 feet in a balloon equipped with a 12.5-inch telescope and made some unprecedented observations of Mars and the galaxy beyond. Project Stargazer, however, was plagued by design and equipment problems that not even Kittinger's buddies in the Mercury program could help with—though they tried—and the whole effort was mothballed the following spring. "I needed Stapp," he wrote, "but he could no longer help me."

In January of 1961, Stapp was a guest of honor at the Air Force Association's fifteenth anniversary black-tie celebration at the Mayflower Hotel in Washington, DC. He liked these events more than he was ever willing to admit, and though he always felt that he was on a short leash, more and more often now he was being invited to rub shoulders with the headquarters brass. It was during this period that Stapp began to seriously assess his own chances of making general. He had always assumed that he had made too many enemies along the way, but now he began to wonder. They said before you could make general you'd first have to "lose your balls," and he very much intended to keep his, even if he was starting to sense that perhaps the Air Force was going to find it hard to deny him.

At times, as the awards and honors piled up, a promotion came to seem—if not imminent—at least inevitable. Much of the rest of the time, Stapp found himself consumed with niggling administrivia that he imagined generals had ways of avoiding. His superiors were constantly reminding him that his various non-Air Force speaking engagements and appearances were not to be paid for with Air Force dollars, and Stapp spent hours each month negotiating with civilian organizations over advances of funds or reimbursement for his own expenses. Yet even in the face of these hassles, Stapp enjoyed the drama of his exhausting schedule and—it should not be denied—the attention and adoration of his audiences.

He made himself available to comment or consult on a variety of issues. Commercial airliner crashes had become almost common front-page fare in the preceding years. In February 1959, American Airlines Flight 320 plunged into the East River while on final approach to New York's La Guardia airport, killing sixty-five of seventy-five on board. In October 1960, Eastern Airlines Flight 375 crashed shortly after takeoff from Logan International in Boston, killing sixty-two of seventy-two. Most spectacularly and terrifyingly of all, United Airlines Flight 826 collided in midair over New York City with TWA Flight 266, resulting in 128 total dead. While in Chicago, where the deadliest single-aircraft crash to date would occur that September, when TWA Flight 529 would go down shortly after takeoff from Midway Airport, killing all seventy-eight passengers, Stapp invited local reporters to his room at the Palmer House one evening. He urged them to tell their readers what he was telling his colleagues: occupants of airplanes involved in crash landings often die needlessly. He explained that federal regulations required only a 9-g seat and failed to prescribe the seat's orientation. "What is needed," he told them, "is a seat that can take at least 20 g's in ALL directions, not just one direction. It should face the rear of the plane so its occupants will be forced into it in the event of a crash, instead of being thrown forward."

It was a time, Stapp thought, that might be ripe for bold new ideas. John Fitzgerald Kennedy, the youngest American president ever elected, gave a speech to a joint session of Congress on May 25, 1961, in which he confidently committed the United States to landing a man on the moon and returning him safely to earth by the close of the decade. The speech was an audacious gamble; four years after the initial shock of *Sputnik*, public opinion was split on whether federal funds should be spent on rockets and astronauts. John Paul Stapp, however, loved it. Space was now part of the national agenda. There was no turning back.

As the year wore on, Stapp increased his pace. He addressed the Symposium on Biomechanics of Body Restraint and Head Injury in Philadelphia, sponsored by the Office of Naval Research, and chaired a session on acceleration tolerance. He attended the Twelfth International Astronautical Congress in Washington in July, and then flew directly to Wright-Patterson to participate in a top secret planning session on a proposed global satellite surveillance system.

Most satisfying for Stapp, the fifth Car Crash Conference (there had been no meeting in 1960)—now being referred to by participants as the Stapp Automotive Crash Conference—convened in September at the University of

Minnesota. For the first time, the conference featured the presentation of formal papers that were published in a hardbound proceedings volume. "Ever so slowly," Stapp wrote, "our efforts on behalf of automobile safety are beginning to be effective as the vaccines and antibiotics gradually eliminate the more glamorous diseases that have cluttered public attention to the disregard of our vastly more serious traffic plague."

October was a typically breakneck month for the overscheduled colonel. There were conferences, seminars, television appearances, radio and newspaper interviews, and an American Rocket Society convention at which he shared a banquet table with the vice president of the United States. Lyndon Johnson, Stapp observed in his trip report to headquarters, "was dressed in a Confederate gray tuxedo and stood up faster than any Texan present when 'The Eyes of Texas' was played by the orchestra. The undersigned could not help remembering that his grandfather and Vice President Johnson's grandfather chased cattle rustlers together in Johnson and Burnet counties."

Meanwhile, in the United States Congress, Representative Kenneth Roberts was moving forward with his long-running investigations of automobile safety. In a series of detailed articles that year, the *Medical Tribune*, a newspaper-format publication for doctors, had shone a spotlight on the car crash injury problem and had offered a number of potential solutions. In one of those articles, Colonel Stapp had indicted the automakers publicly for the first time: "The automobile industry is the only one whose product can still be sold after killing thousands and injuring millions of customers every year." Despite the stated opposition of Detroit to the notion of federal safety legislation, Representative Roberts had begun to pursue the establishment of a governmental organization that would be chartered with oversight and policy relating to car crashes and safety design. He had introduced House Bill 133: "To Establish a National Accident Prevention Center," before, on February 8, 1962, once again inviting his reliable Air Force witness to testify.

Having learned from his previous congressional appearances that the truly convincing arguments are built on dollars, Stapp came prepared with an armful of statistics demonstrating the economic loss to the United States from traffic injuries and fatalities to military personnel. "Our total Defense

Department loss in the five years from 1955 to 1960," he testified, "is $127 million on motor vehicle accidents." Stapp also, of course, took the opportunity to lament the defunding of his crash research at Holloman, studies he noted cost the government only about $40,000 per year on average, less than the cost of a single missile.

Chairman Roberts: "But the Air Force is no longer doing this particular type work that you carried on?"

Colonel Stapp: "No, sir; we are not."

Roberts: "Because of lack of appropriations?"

Stapp: "That and lack of authority or authorization to do it . . ."

There was a lot of good work being done on the car crash problem, Stapp reported, but America needed, deserved, a central repository of the best data and a commitment to collaborate across company lines.

Representative Paul Rogers of Florida asked Stapp one last question. "Would you think it might be feasible to set up an accident research institute similar to the way we have done it with cancer?"

"Such approaches justify themselves with the results they have produced," Stapp replied, "and I should think that we could extrapolate to the accident situation and say that we could do at least as well."

Congress was not, however, a unanimous cheering section for Roberts's auto safety hearings or, indeed, the auto safety cause, and the opposition came from both sides of the aisle. There were charges that efforts to require even the nation's military services to mandate restraint devices in vehicles they purchased would be too intrusive. Representative Walter Rogers of Texas, a Democrat, charged that mandates for military cars and trucks was nothing more than a promotional stunt that would result in the car companies "selling the government all kinds of safety devices." Representative John Bennett of Michigan, a Democrat, called such regulation "dictatorial."

As the political debate droned on in Washington, Stapp made literally dozens of trips in the middle and latter months of 1962, including stops in Gothenburg, Sweden, where he again met with safety engineers from Volvo, and Varna, Bulgaria, where he consulted with Soviet rocket scientists. He did a ten-minute live interview on NBC TV in which he defended NASA's management of the nation's space program against a recent attack in *Harper's* magazine. In New York City he was treated to a lavish dinner at the Playboy Club by the science editor of *Life* magazine, about which Stapp remarked: "The presence of bunny girls scarcely interrupted a lively discussion on lunar astronomy."

Despite his ongoing struggles with Don Flickinger and others at Systems Command and Air Force Headquarters, Stapp's influence on a whole portfolio of national and world issues had hit a high-water mark.

23

NAZI DOCTORS AND THE NEVER-ENDING ROAD TRIP

> Although the "experiments" were conducted by fewer than two hundred murderous quacks—albeit some of them held eminent posts in the medical world—their criminal work was known to thousands of leading physicians of the Reich, not a single one of whom, so far as the record shows, ever uttered the slightest public protest.
>
> —*William L. Shirer,* **The Rise and Fall of the Third Reich**

THERE WERE TIMES, however brief, when Stapp was able to step off the never-ending road trip and spend a few consecutive days at home with Lili in San Antonio, where they lived in the Terrell Hills neighborhood north of downtown. But the Stapps' domestic life would take an unexpected turn when in September 1962, Lili's younger sister Peggy—one of a contingent Stapp referred to as the "outlaw Lanese sisters"—died suddenly, leaving behind two teenage children with no means of support. While neither John Paul nor Lili had the faintest idea of what would be required, they offered to take in the orphaned children.

Dorian Bates and Douglas Lewis, ages fifteen and thirteen at the time, were moved from New York City to Texas, and into the home of their famously

workaholic uncle, who traveled most of the time, and their taciturn, sometimes haughty, aunt, who valued nothing so much as her privacy. Dorian was
a free spirit who chafed against what must have seemed a solemn suburban
environment, and Douglas had serious developmental problems that became
alarming to the Stapps almost immediately. Diagnosed with schizophrenia,
Douglas had been a ward of the state of New York on and off from the age of
thirteen months, and a resident patient at Rockland State Hospital for Mental and Nervous Diseases since 1959.

Shortly after his arrival in San Antonio, Douglas was examined by a neu-
rologist who detected "definite evidence of birth brain damage." In addition to
his other problems, Douglas had severe dyslexia, which resulted in poor performance at the Northwood School in San Antonio, where he was enrolled. A
psychologist there worked closely with him, as did a reading specialist. But
eventually, due to stalled academic progress and inability to adapt socially,
the Stapps enrolled Douglas in a boarding school for students needing special
attention, tutoring, and medical care. Colonel Stapp's hope was, he said, to
be able to help Douglas become "an honest and independent citizen, able to
make his living and contribute to society in spite of his deficiencies." Yet, not
very surprisingly, the boy would come to resent the Stapps for moving him out
of their home and into a dormitory. Though they meant well, and despite John
Paul's lifelong desire for children, he and Lillian could not have been more ill-
suited to their surrogate parent role. It was confusing for all of them, and Colonel Stapp would refer to Dorian and Douglas alternately as his children, as
his niece and nephew, and as his wards, as if he never quite came to terms
with his own role as guardian or his new responsibility as a father figure.

At the same time he and Lili were struggling with sudden and unexpected parental duties, Colonel Stapp was facing new challenges at Brooks,
one of which was having to share the stage with Hubertus Strughold. The
German had come to Brooks with the School of Aviation Medicine when it
had been relocated from Randolph in 1959. Strughold's reputation as the
U.S. Air Force's top space medicine expert had solidified in the intervening
years. He'd been named chief scientist of NASA's Aerospace Medical Division
in 1961, and—partly on the basis of his own self-promotion—had become
known to the public as the father of space medicine. Strughold liked to hand
out 8 x 10 glossies of himself on horseback. Then, suspicions began to surface

that he'd actually known more about unethical medical experiments in Germany than he'd ever admitted.

Hermann Becker-Freyseng, a researcher at Strughold's aeromedical institute in Berlin who was sentenced to twenty years in prison on the basis of Nuremberg testimony, had witnessed experiments conducted at Dachau by the infamous Nazi doctor Sigmund Rascher. As part of Rascher's agenda, prisoners—including Jews, Russians, gypsies, and even Catholic priests—had been forced to drink massive quantities of putrid seawater. Some of them died of heart attacks. Aeromedical experiments involved locking prisoners into decompression chambers without pressure garments or pressurized oxygen, and exposing them to near-vacuum conditions that "gave men such pressure in their heads that they would go mad and pull out their hair in an effort to relieve such pressure." Dachau doctors slowly froze prisoners to death, exposing them to malaria and typhoid, sterilizing them with extended exposure to x-rays, subjecting them to bone transplants without anesthesia, and dosing them with powerful psychoactive drugs. Rascher personally asphyxiated a number of the prisoners prior to conducting autopsies.

Hubertus Strughold, Nuremberg testimony suggests, likely neither participated in nor observed the Dachau experiments firsthand. Yet Becker-Freyseng and others told interrogators that Strughold had not only been aware of those experiments as, or shortly after, they occurred, but that he had had the unique authority to stop them at any time. Despite his denials, it was confirmed that Strughold had attended conferences in Berlin and Nuremberg where Rascher presented the results of his Dachau work. In the words of one American historian, "nearly all of the [aeromedical] work that occurred in Germany can be linked to Hubertus Strughold." Hitler personally promoted Strughold to full colonel shortly before the collapse of the Third Reich.

At Brooks, Stapp couldn't easily avoid Strughold—on several occasions he was put in the position of serving on committees Strughold chaired, or addressing students or conference audiences alongside Strughold—and much of his work with NASA had to be coordinated with Strughold. Stapp tolerated Strughold in public, and even exchanged Christmas cards with him, but he was always uneasy living in his orbit.

. . .

Hubertus Strughold, a hugely controversial figure in the history of aviation and space medicine. Before he was brought to the U.S. following World War II, Strughold had been the top aeromedical man for the German Luftwaffe. (Photo courtesy of Wilford Stapp)

During Stapp's time at Brooks, car crash protection had been steadily gaining traction as a topic of national discussion. While the automakers in Detroit had continued to press the argument that the public simply didn't value safety design enough to pay for it, insurance companies and law enforcement had begun to weigh in. A California doctor who'd worked as a consultant for both car insurance providers and police departments declared: "We have spent too damn much time worrying about the cause of accidents. It's time we started worrying about the cause of injuries." The American Medical Association had issued calls for Detroit to get serious about seatbelts. The American College of Surgeons had approved formal recommendations that car companies emphasize the safety of their customers as a foundational principle of auto design, and that they include seatbelts capable of withstanding 20 g's.

In Washington, DC, Representative Kenneth Roberts, who had been in attendance at the Stapp conference the previous year, had continued his quest to convince his colleagues to fund a federal organization to oversee and coordinate automotive safety developments, despite continued pressure from the auto industry to drop the idea. American traffic fatalities were still on the rise—the death total for 1962 was an all-time high of 38,980, up 6 percent from the previous year—and the automakers knew now that they were in for a fight. Just that February, Benson Ford, vice president of Ford, revealed the concerns of the industry when he told a meeting of car dealers in Miami Beach: "A large part of the blame for any further increase in traffic accidents will rub off on us, no matter who or what is responsible. And when it does, we will pay heavily. We run a very real risk of being subjected to restrictive legislation, passed in an atmosphere of hostility and misunderstanding, legislation that will impair our flexibility, raise our costs and limit our sales . . ."

So once again, in the spring of 1963, Roberts invited Colonel Stapp to appear before his committee in support of H.R. 133, which would amend the Public Health Service Act to establish a National Accident Prevention Center.

Stapp echoed points made in his previous appearances, but this time—perhaps with some coaching from Roberts, with whom he'd lunched in the congressional dining room prior to the hearing—he focused more specifically on the House bill at hand. He also noted that the Air Force's efforts to improve ground vehicle safety for its own personnel had already resulted in a 50 percent reduction in car crash fatalities, which equated to an annual monetary savings of about $12 million. Stapp made a point of extolling the superior record of Western European nations in the arena of crash protection and singled out Sweden as a worthy example of what could be accomplished.

Representative Leo O'Brien of New York asked Stapp if he could explain how Sweden had achieved its remarkable record.

"Yes sir," Stapp answered. "Eighty percent of the automobiles in Sweden are equipped with seat belts, and in addition most of them have a diagonal body strap going over the outside shoulder and attached to the side post of the car." It was one of the first descriptions of the three-point, bandolier-style seatbelt in federal policy debate over automobile restraint systems. Stapp had first seen the three-point belt in Hugh DeHaven's living room years earlier. DeHaven had built his prototype back in 1937.

A month after the hearings, in May 1963, Stapp got a call from Dr. William Crook, the president of San Marcos Academy, inviting him to be the school's commencement speaker for that year's graduation ceremony, even though Stapp had disparaged the academy in the 1955 *Time* cover story as a "school for displaced hellions." While Stapp reminded Crook that his memories of San Marcos were not completely delightful and that "More than once I had to intervene physically on behalf of Latin-American students who were being insulted and abused by some of the Hellions," he accepted. His commencement speech centered on Mahatma Gandhi. "Resist not evil," he said, "but overcome evil with good."

Stapp's rhetoric in his speeches was tinged with an almost biblical fervor by this time, aligning his concerns seemingly more than ever with those of his late father. Yet this never bred in him any particular affection for religious organizations, doctrine, or churches, especially not for his parents' church. According to Wilford Stapp, a deeply thoughtful and religious man: "I think Paul was somewhat appalled by the Baptist Church. He thought a lot of Baptists were great people, but he didn't want to be one of them. When Paul heard somebody say they know what God's will is—well, he just thought they were nuts."

Meanwhile, much of Stapp's daily life at Brooks was now spent at a typewriter. He entered into a contract with Northrop in mid-1963 to write three reports and a publishable manuscript documenting his years of deceleration and windblast tests with the company. His goal was not only to compile and devise ways to effectively present the results of his work, but to put it into an appropriate context for the benefit of future researchers. He knew, as he commented to Red Lombard in a letter, that it represented "a unique group of experiments not likely to be repeated even by those who might criticize." He added: "After we have done our best job on this data, I don't think we will owe apologies to anyone."

Though it took him away from the typewriter, he also found himself spending more and more time consulting with NASA in Houston. When Major Ellis Taylor, a project officer at Holloman who'd been working with NASA, got into a dispute with the commander of the First Medical Division, resulting in Taylor's unceremonious transfer to an Air Force unit in Izmir, Turkey, Stapp agreed to assume his responsibilities on a temporary basis. "I wasn't going to let NASA down," he wrote.

Stapp picked up a field program at Holloman begun by Major Taylor, an omnidirectional landing impact study for Project Apollo using the Daisy Track and volunteer subjects, studying body orientation and trying to determine the safest touchdown configuration for returning astronauts. The researchers needed to make detailed observations, and capture subjective impressions from their volunteers, in eight different diagonal axes of impact to which the Apollo cabin might be exposed. Stapp organized several cohorts of human volunteers. The subjects were exposed to impacts up to 20 g's with a 1,000-g-per-second rate of onset. Some of the volunteers were given barium to drink prior to the experiment to facilitate x-ray exposures of their organs prior to and immediately following the impact. The results led to some crucial modifications to the Apollo astronaut couch. "At the cost of a few stiff backs, kinked necks, bruised elbows and occasional profanity," Stapp told an interviewer, with his trademark chuckle, "the Apollo capsule has been made safe for the three astronauts who will have perils enough and left over in the unknown hazards of the first flight to the moon."

Just as he had during his Muroc days, Stapp used Mondays to prepare for the runs to be conducted Tuesday through Thursday, and then devoted Fridays to inspecting the equipment and summarizing the week's data. "I think we saved time because we were able to work with great precision for three days, have complete security and safety, and that sure beat having the kind of incidents that, for instance, NASA had on Apollo I."

Stapp would be extremely critical of the lapses in discipline and misguided planning that led to the launch pad fire that killed three of the Apollo astronauts, including Gus Grissom, whom Stapp had selected as one of the original Mercury crew. Not only had NASA put flammable Velcro in the oxygen-rich atmosphere of the overpressurized capsule, but they'd conducted the test on a Saturday. "The attitudes of people on overtime on Saturday, when they are usually at home doing things with their children and the whole lackadaisical, dragged-out way of operating—that invites disaster. And they got it."

His fieldwork at Holloman complete, Stapp left almost immediately for Europe and the 1963 meeting of the International Astronautical Federation, which was held in Paris. He was one of that year's honorees, and he looked forward to an encounter with Soviet cosmonaut and first man in space

Yuri Gagarin. Unfortunately, the occasion turned out to be an awkward and unsatisfying one for both men. Stapp wrote to Lili about the scene at the Palais du Dauphine: "[Gagarin] looks just a little like Peter Lorre without the makeup. He is very small—about 5 ft. 2 and about 134 lbs. A Frenchman walked ahead of him saying, 'Make way for the Russians!' but nobody moved. I heard Gagarin yell out, 'Henk, henk, henk, Democracee, Democracee!' It seems he was burned up because I upstaged him and got the attention and ignored him ... I said I would pull a switch and give Gagarin *my* autograph. I took a personal card and wrote on it, 'My compliments to the first Space Cosmonaut,' and signed it. I shook the little man's hand and gave him my card. He looked blank, slowly put my card in his pocket, and then stared down like a guilty little boy ... What a childish little man!"

From Paris, Stapp took a train to Rome for the annual School of Aviation Medicine meeting—and quickly fell in love with Italy. He was made a guest of honor at the Castel Sant'Angelo, the Pope's castle. The next day, in a delegation to the Vatican, he was seated in the front row for an address by Pope Paul VI. "He made a marvelous speech in French," Stapp wrote to Lili. "The pope then came forward. When he came to me, General Lomonaco [Lieutenant General Tommaso Lomonaco, chief air surgeon of the Italian Air Force] lifted my right arm, pulled back the sleeve, showing him my broken wrist. He then told him I was a great hero of science who suffered injuries in experiments for the good of mankind. This set the pope off. In French, he blessed my actions, exhorted me to continue my good work and ended by saying he would pray for me ... he is absolutely sure he speaks for God." Stapp later told a reporter back in Texas: "As I listened to the Pope commend space research, I could not help but think of how things had changed since the first space scientist Galileo got *his* reception."

The only downside to the entire trip was that Stapp had been forced to make most of it in the company of Hubertus Strughold. The Air Force had given them identical itineraries.

. . .

By the summer of 1963, Joe Kittinger—along with hundreds of his fellow fighter and bomber pilots—was already flying B-26 missions for the Air Commandos in South Vietnam as part of an operation code-named Farm Gate.

Ostensibly designed to train South Vietnamese pilots how to defend their country against aggression from North Vietnam, it was in reality the start of an American troop buildup for another proxy war with the Soviets.

Then, in November of 1963, President Kennedy—who had authorized the deployment of the Air Commandos in Vietnam and who had paid a visit to the researchers at Holloman only a few weeks earlier—was murdered in Dallas. Though a Texan subsequently ascended for the first time to the presidency—and not just a Texan, but a Hill Country Texan, whose grandfather had consorted with John Paul Stapp's grandfather—Stapp had become almost desperate to get out of the Lone Star State. In spite of proximity to his brother, and despite his affection for the city of San Antonio—Stapp served on the board of its Chamber Music Society—he felt that his useful time there had come to an end. Some at Brooks had even begun to suggest that the always combative Stapp was no longer of completely sound mind.

These insinuations began as whispers, but all the officers were aware of them. The charge, as it came back to Stapp, was "that I was losing my marbles from too many sled rides, and should be examined and let out of the AF." A formal complaint was even filed with the Deputy Surgeon General of the Air Force, alleging that Stapp was collecting his USAF paycheck for doing little more than making public appearances for the purpose of burnishing his own reputation. In response, Stapp issued a lengthy report chronicling his activities during just the two previous months, including the five major papers he'd published, the courses he'd taught, and the Air Force and NASA working groups to which he'd contributed. "The undersigned," he wrote, "is not unhappy—when the STAMP OUT STAPP CLUB is inactive, it is either nursing plantar bruises or Stapp has not been producing at a level above the Club's threshold for criticism." When the Navy decided to rename one of its units of measurement—the force of one g acting on an object for one second had previously been referred to as a "jerk"—an anonymous member of the SOS club suggested that perhaps it should be called a "stapp." The term stuck, providing yet another bit of curious immortality to its namesake.

From any perspective, it was time for a change, and in the spring of 1964, Stapp was informed that he would be transferred later that year to the Armed Forces Institute of Pathology (AFIP) in Washington, DC. This was welcome news, but that September, before he could begin any serious planning for his new role, he made his first trip behind the Iron Curtain to address the International Astronautical Federation Congress in Poland. At the Palace of Science and Culture in Warsaw, Stapp delivered a paper on the impact experiments he'd conducted at Holloman for Project Apollo. He wrote a friend that his talk "was very well received, particularly by the Russians." In spite of the rhetoric about cooperation—in which Stapp truly believed; he was heartened during the Warsaw meetings that "They [the Russians] used to come to these meetings and... suggest the lines they were working on, but they would never come right in the open with it. Now they are spelling out their experiments"—he never enjoyed anything so much as one-upping the Soviets. He seemed inspired by the entire Cold War dynamic and wrote to Lili about the "cloak and dagger" atmosphere, which is not to say he was much intrigued by Poland itself. He described it as "under the grim police state control imposed by the damned Russians! It has to be seen to be realized... The people were tired and frightened, yet still kind and friendly." Stapp took in two operas, visited a nightclub called the Crocodyl, and attended a concert in the parlor of Chopin's home. "I will have some interesting little things for you when I return," he promised Lili, "but not from Poland. I didn't want to contribute dollars to the commies."

Shortly after his return, Stapp found himself on a flight from Washington, DC, to Little Rock, Arkansas, with NASA administrator James Webb, who—as it turned out—was a great admirer. Webb was fascinated by Stapp's report of his Polish trip. Stapp explained that Soviet scientists had told him, at a private meeting at the Hotel Bristol in Warsaw, of extensive experiments they'd conducted on +Gx and –Gx tolerances aimed specifically at perfecting cabin configurations for their cosmonauts. Interestingly, the Soviets had revealed a deep interest in the question of whether certain drugs might be used to help increase tolerance to the punishment of ballistic reentry and landing. Stapp also told Webb he believed the Soviet space program was unlikely to challenge NASA in the race to land a man on the moon. When asked, the secretary of the USSR Academy of Sciences had confided to Stapp that the Soviet goal for a manned landing was "1969, plus X," while declining to speculate on the duration of X.

Webb listened as Stapp went on to offer a suggestion. NASA should consider, Stapp argued, an early circumlunar reconnaissance mission. He recommended a three-astronaut configuration in low lunar orbit as an essential prelude to a manned landing. "It would permit direct observation by experts, as well as a camera and instrument recordings for return to earth instead of relying entirely on unmanned telemetry records . . . Such a flight would also greatly ease the pressure relating to success or failure of subsequent lunar landing missions." This was the full dress rehearsal concept Stapp had employed on Project Manhigh, but it was not a notion that had garnered much support at NASA up to that time. The compressed schedule that governed the frantic drive to get America to the moon had been built on an all-or-nothing approach.

Stapp's consultation may very well have affected the space agency's strategic policy debates that led to Apollo 8 four years later. That flight became the first manned mission to leave earth orbit, and astronauts Borman, Lovell, and Anders became the first human beings to both see the dark side of the moon and photograph the whole earth. The crew orbited the moon ten times before returning to earth, and succeeded, as Stapp had predicted, in taking some of the pressure off NASA.

Spectacular Christmas Eve photos of earthrise following Apollo 8's fourth pass around the moon were immediately iconic and a publicity coup beyond Webb's wildest dreams. One particular image snapped by William Anders with a Hasselblad camera—Anders had glanced out and said to Frank Borman: "Oh my God! Look at that picture over there!"—became one of the most famous and most frequently reproduced photographs in history. The bright, jewel-like planet earth, half topaz and half in shadow, set against a perfect blackness, with the scarred ridges and craters of the moon in the foreground, made its way onto postage stamps, T-shirts, and millions upon millions of color posters. Telegrams came pouring into Houston from Americans who found inspiration in NASA's flexing of the country's engineering muscle and in this undeniable evidence that the humiliating days of *Sputnik* were over for good.

PART IV

THE GHOSTS THAT NEVER HAPPENED

24

WASHINGTON

But I tell you, my lord fool, out of this nettle, danger, we pluck
this flower, safety.

—*William Shakespeare,* **Henry IV, Part I**

AS NASA MOVED CONFIDENTLY toward its exotic objective of a manned
moon landing, Stapp busied himself with plans for the Eighth Stapp
Car Crash Conference, which would convene in the fall of 1964 in Detroit. The
location was convenient because Colonel Stapp had already agreed to deliver
a series of lectures to medical students at Wayne State University at that
same time. Emblematic of the auto industry's increasingly direct involve-
ment in the conference, the first day's session was held at Ford Motor Company laboratories, the second day on the campus of Wayne State, and the
third at the General Motors Proving Grounds. Safety presentations offered
by GM personnel focused on everything except cars: driving techniques for
handling blowouts and skids, highway construction methods, and improved
guardrails and bridge abutments. More than 200 researchers, doctors, and
engineers attended. Governor George Romney opened the proceedings and
the vice president of Chrysler was the banquet speaker. The conference had
by this point clearly moved beyond the control of John Paul Stapp and into
the realm of the Society of Automotive Engineers and its members' employ-
ers. This was also the year that a law student from Harvard named Ralph
Nader made his first appearance at the conference. Nader—neither doctor,

scientist, nor engineer—failed to make much of an impression on Stapp or anyone else.

Following the final night of the conference, Stapp went to a local movie house with some of the Wayne State neurosurgery faculty to see the Detroit premiere of the Hollywood epic comedy *It's a Mad, Mad, Mad, Mad World*. Stapp loved the film and noted its relevance to the conference in his trip report: "The chief item of interest in this picture was the use of automobiles driven by stunt drivers in slapstick comedy demonstrations of every degree and kind of automobile recklessness and crashes that could be dreamed of by the producers. More than 50 cars were demolished to make this picture. The picture opened with a fatal one-man crash presented as a hilarious farce. With his last gasp, the victim kicks a tin bucket over a hillside!" It was, Stapp told one of his colleagues, an antidote to taking life too seriously.

Planning for the following year's conference would begin that fall, and there were suggestions that the organization ought to draft a mission statement and publicly declare its objectives. Stapp hated the idea and voiced his objections to Professor Merrill Cragun at the University of Minnesota, where the 1965 conference would be held. "The Stapp Car Crash Conferences," he insisted, "are a non-partisan forum . . . We have never taken sides on anything. There is no political clank; there is only a sincere opportunity for discussion by all participants, with privacy and respect for their views, whatever they might be . . . What each one gets out of the meetings is what he has learned and no party lines or propaganda need be subscribed to. We only seek the truth . . . We try to love everybody at the Car Crash Conferences. So much for the sermon."

Time would reveal the extent to which Stapp's sermonizing would be capable of influencing the direction of the conference he had created and whether he could retain the leverage he needed to preserve the "happy, amorphous anarchy" he preferred, or whether the "fanatical partisans" would prevail. Either way, Stapp still loved a battle.

In January of 1965, Colonel Stapp accepted an invitation to participate in a two-day medical evaluation, part of Navy doctor Ashton Graybiel's study of individuals who'd been subjected to massive g-forces. It was called a vestibular function test, and Stapp believed it was important to make himself available to Graybiel. He also hoped it would settle, once and for all, the question of whether he had suffered some sort of permanent brain damage as a result of his rocket sled runs. Graybiel didn't report back to Stapp until August, but he had good news. Though Stapp was a little overweight and his glucose tolerance test was marginally abnormal—Stapp would develop diabetes in his later years—his health was, in all other respects including cognitive capacity, within "the typically normal range."

Four months after the test was conducted, Stapp made the move from San Antonio to Washington. Even as he told some of his colleagues that he felt the transfer was "an attempt to suppress me," he seemed genuinely excited about the possibilities before him. He was to be given control of the Aerospace Pathology Division of the Armed Forces Institute of Pathology, an organization he'd once referred to as "the international conscience of the medical profession." It should have been a good spot for him, especially considering there were those at the Department of Defense who thought the venerable institute had slipped a few notches. In fact, two brigadier generals, Deputy Surgeon General Kenneth Pletcher and AFIP commander Joseph Blumberg, had hand-picked Colonel Stapp to head the aerospace division and establish state-of-the-art impact pathology research at AFIP. Pletcher and Blumberg wanted more focus on military crash investigation and they needed Stapp's credibility, his experience, and his energy. Stapp believed he was being given carte blanche to overhaul Aerospace Pathology. He wrote a friend: "Stapp is being put on a white charger and told to rescue the department in distress . . . and the Surgeon General's Office is betting on me!"

Stapp found quarters at Bolling Air Force Base, where he would stay until Lili could complete the sale of their house in Texas and join him. Shortly after his arrival, much as he had during his last stint at Wright-Patterson, he began the process of reinventing his new organization. He announced plans for a comprehensive impact program that would use large anesthetized animal subjects. He reached out to the best impact researchers he knew, offering to help arrange reassignment if they would agree to join him at AFIP. To Major Ellis Taylor, with whom he'd worked at Holloman, he wrote: "I will need your help badly and as soon as you can be transferred. We have a lot of hard, worthwhile work to do and you are just the man that can do it!" One controversial series of impact experiments Stapp was proposing to fund would involve tossing live hogs from helicopters and moving trucks.

That idea may or may not have affected Stapp's standing at AFIP, but prospects of once again running his own operation by his own rules, the way he'd done at Edwards and at Holloman, were dashed almost before planning had begun. Despite all his expectations, Stapp's transfer from Brooks amounted, he felt, to a full bait-and-switch. With one of his sponsors retiring unexpectedly, his immediate superior at AFIP was an Army officer who proceeded to make Stapp's life miserable by strangling his reorganization efforts with webs of red tape. "I arrived to find myself assigned a one-man branch with no funds or personnel," Stapp wrote to a colleague, "and a secretary for only two months, with no replacement." His role was quickly whittled down to Chief of Environmental Pathology. By late August, Stapp was already so frustrated that he began cautioning researchers to stay away. He wrote to Taylor: "The Army and the Communist system have one thing in common—irrational dedication to bureaucratic regulationitis to the exclusion of common sense and achievement of any goals. But, as I used to say at Edwards, thank God my enemies are stupid—where would I be if they were smart?"

Wilford Stapp believes his brother's tenure at AFIP was the low point of his career: "A frustrating job. He was treated in a menial way, and almost regarded as some kind of a nut, like a mad scientist."

Regardless of the circumstances, Stapp resolved to accomplish three things while at AFIP. He wanted to complete a multivolume work on biodynamics, a project he'd toyed with at Brooks but to which he had now rededicated himself; to establish a permanent federal accident research laboratory; and to establish a registry of accident pathology that would catalog detailed data on impact accidents. It was an ambitious list.

In the second week of June 1965, Stapp received a letter from W. W. Norton and Company in New York inquiring about his unfinished autobiography. Editor Carol Houck, having reviewed a chapter Stapp had written for a Norton publication titled *Space: Its Impact on Man and Society*, encouraged him to forward the 200-some pages of *Nothing Ordinary* that had been completed. Stapp was tempted, but was determined to finish his study of biodynamics first. "I hope you will bear with me," he wrote, "while I dispose of my debts of conscience to the government."

Later that same month, even as the house in San Antonio remained on the market—he would eventually be forced to sell it at a loss—Stapp brought

Lili and Douglas to the Capital and they all moved into a rented house in the suburban neighborhood of North Chevy Chase, Maryland. Though the Stapps were still supporting her financially, Dorian Bates was out of high school by this time. She was breaking away from her aunt and uncle, with whom she'd never established much of a bond, and she refused to make the move east with her brother. The Stapps, as a result, lost touch with her.

Meanwhile, Colonel Stapp and Lili were determined to help Douglas achieve some success in his new environment. This resolve, coupled with Douglas's own progress, seemed to bode well. Douglas got a paper route delivering the *Washington Post*. He woke at five each morning before school and built the route up from 83 to 132 customers within just a few months. With the money he earned, he purchased a tape recorder which he used to listen to audio recordings of books his dyslexia made impossible for him to read. His uncle got him admitted to classes with a series of reading specialists, and, not wanting to put him back into boarding school, enrolled him at Rock Terrace, a public high school in Rockville, Maryland, that specialized in children with cognitive disabilities. It didn't always go well, however, and not long after starting at Rock Terrace, older kids began stealing Douglas's lunch money and gym clothes.

As he struggled to find the best way to help Douglas, Stapp got a letter from his brother Robert, whom he had not heard from in some time and who required a different kind of help. Robert was now remarried with two more children, and wrote that he had just secured a job teaching art in Arkadelphia, Alabama. Still, he was in debt, he said, and needed money. He apologized for having failed to repay an earlier loan, and wrote that John Paul was "the only brother who had taken concrete steps to show me that I am of value." John Paul sent Robert a check, as he had always done before, but told Lili they should not count on it being repaid.

That same fall, a letter addressed to Dorian arrived at the house in Chevy Chase. Not having heard from her in weeks, Stapp opened the letter to find an invoice for payment owed on an expensive man's watch purchased on credit from a jewelry store in San Marcos, Texas. Stapp sent payment to cover the debt, but chastised the store's owner for extending credit to "an unemployed student dependent without consulting her parents." The watch was a gift from Dorian to the man she would marry and who would become the father of her first child. It would be the last interaction, directly or indirectly, that the Stapps would have with Dorian Bates.

Within weeks of Lili and Douglas's arrival, Stapp was on the road again. Late in the summer of 1965, he flew to Greece for the Sixteenth Congress of the International Astronautical Federation and the International Academy of Astronautics. He submitted a formal proposal to the latter organization for the establishment of a committee to form a cooperative orbital laboratory program. The Americans and Soviets were already discussing the concept of joint lunar and orbital labs. Stapp's proposal was in support of facilities that could accommodate "member scientists from nations having no space vehicle projects," effectively inviting the rest of the world's research communities into the practical application of space science. He proposed that Professor Leonid Sedov of the Soviet Union and Dr. Wernher von Braun of the United States cochair a committee to evaluate projects submitted by non-space nation scientists. Von Braun loved the idea and, with his support, a committee was approved to work on it.

When Stapp returned to a hot, muggy Washington that September, he found a mundane world of military affairs lying in wait. While getting anything done at AFIP was, in his estimation, "like trying to drag race a Galapagos tortoise on a glacier," he kept at it. One of his successes was convincing the Army to adopt a new helmet design of Red Lombard's for its helicopter pilots. The aviator helmets in use in Vietnam left much to be desired. Investigations of some 800 plane crashes had shown that the majority of head injuries were to the front half of the head, with a high percentage of damage to the lower half of the face. Lombard's design featured a hinged plastic lower faceplate to protect the chin and mouth.

The thing about safety equipment like Lombard's helmets was that you never knew when a real-world demonstration would prove—or disprove—the integrity of the design. Researchers could test all they wanted, but at some point their work would be judged by performance under fire. So it was with the legacy of Project Excelsior. On the morning of January 25, 1966, Lockheed test pilot Bill Weaver and navigator Jim Zwayer took off from Edwards Air Force Base in an SR-71 Blackbird, a long-range reconnaissance aircraft capable of tremendous speed. At an altitude of 78,800 feet, doing Mach 3.18 over northeastern New Mexico, the Blackbird lost thrust in the right engine and entered a violent yaw. Three seconds later, the entire titanium structure of the plane shuddered and came apart. There hadn't even been time to fire the ejection seats. Bill Weaver later recalled regaining consciousness in midair, but thinking that he must be dead. His pressure suit was inflated and he was free-falling feet first through the frozen stratosphere with a small stabilizer chute tugging at the airstream just enough to prevent him from spinning. At 15,000 feet, the main chute was released automatically and Weaver touched down a few minutes later, nearly landing on the back of a startled antelope. He was the first survivor of a disabled aircraft deep in the stratosphere, courtesy of the Beaupre multistage parachute and a lot of luck. Jim Zwayer's neck was broken during the Blackbird breakup, killing him instantly, but he too was returned to earth beneath the Excelsior chute. Kittinger and Stapp both took a lot of pride in Bill Weaver's survival. It was the sweet fruit of their labors in safety research.

Later that spring, Stapp reached out to one of his close Stateside colleagues and confidants in the business of safety research, Dr. Richard G. "Jerry" Snyder. Snyder was the nation's leading expert on impact injury resulting from falls: airplane crashes, accidental falls from tall buildings, suicide attempts, failed parachute openings. He had attended Air Force flight school with Joe Kittinger at Nellis Air Force Base in 1949 and gone on to fighter pilot duty in the Korean War. On February 4, 1953, Snyder departed Goose Bay in Labrador in a three-plane formation of F-84Gs bound for Europe. Shortly after takeoff, bad weather forced them to return to Goose Bay. In the meantime, the airport had become enveloped in a thick, swirling fog. The planes circled the runway, waiting for a chance to break through the murk. With fuel getting low, the lead pilot flamed out and crashed. Snyder and the number three pilot decided to attempt a desperate blind approach, with the fog still clinging to the ice below.

"Our planes exploded," Snyder reported. "Just disintegrated. I woke up and my engine was burning a short distance away. There was nothing left but my cockpit." Snyder spent eighteen hours in temperatures that reached 16 degrees below zero Fahrenheit before he was rescued. He had two crushed ankles, five vertebral fractures, and multiple head injuries. He spent a total of a year and a half in multiple hospitals. It was during his recuperation that he learned about the biodynamics research of John Paul Stapp and his efforts to improve pilot seats and restraint systems. Snyder credited the reinforced cockpit of the F-84G and ultimately his survival in Labrador to the direct results of that work. With Stapp as a role model, Snyder went on to become a forensic anthropologist and an independent authority on specific kinds of impact.

Snyder conducted his own impact studies using animals and human cadavers for the National Advisory Committee for Aeronautics (NACA, NASA's precursor) and worked for a time at the Federal Aviation Administration in Oklahoma City. In 1965, as part of a program Colonel Stapp had agreed to set up for the Bureau of Standards, he'd carried out experiments with chimpanzees and pregnant baboons using the Holloman Daisy Track. Snyder carved a niche for himself pioneering the study of restraint systems for pregnant women and children, a tricky area that had mostly been ignored by other deceleration and impact testers. As a crash investigator, one who owned his own P-51—the model he'd flown during flight school at Nellis in the forties—he was always on standby to fly off to the latest site of a crash or fall casualty. If a commuter plane went down in upstate New York or somebody jumped off the Golden Gate Bridge, Snyder would be on the scene within hours. "I didn't investigate the causes of crashes," he said, echoing Stapp, "but the causes of injury and death. I was focused on how to protect the human."

The Bureau of Standards work Stapp conducted at Holloman represented the last live crash testing he would be personally involved in. The purpose of these tests was to establish standards for restraint devices and other safety measures that would soon be required for any cars purchased and operated by members of the United States government. Stapp tested seatbelt and seat anchorage configurations, padded instrument panels and visors, impact-absorbing steering wheels, safety latches and hinges, brake systems, bumpers, backup lights, and rearview mirrors. He also conducted some of the first tests of airbag systems. It was the closest thing to a holistic evaluation of automobile safety design as had been attempted outside of the auto industry labs.

Not long after Colonel Stapp arrived at AFIP, Jerry Snyder left the FAA and accepted a position with the Ford Motor Company in Detroit—"in the Lion's mouth," as he characterized it in a letter to Stapp—as chief research scientist with a charter to develop an impact research group. "So far," he wrote, "I am very encouraged with my reception at Ford . . . We will have new laboratories and they have promised me 17 positions. So far I have had excellent cooperation and no real constraints on what and how I want to run a research program."

At Ford, Snyder quietly introduced the practice of using animals for impact testing. While the research was done very discreetly—in secret, often at night—the word eventually got out and the American Society for the Prevention of Cruelty to Animals began targeting Ford for protests. In Snyder's estimation, the Ford organization had a genuine concern for safety, but an even greater concern for the company's reputation. The live animal tests were restricted and eventually eliminated.

The most important advances in the auto safety world, Stapp believed, had to do with restraint systems, and his own fifteen-year campaign for seatbelts in cars had always been based in large part on a personal response to tragedy. Then, on January 16, 1965, Stapp got word that his first wife, Nylah, had died of head injuries when the car she was riding in—a car unequipped with seatbelts—hit a bridge abutment in Tampa, Florida, and flipped into a ditch. But now he knew how to change the outcome, and an event just the previous November, in El Paso, had showcased—in a very personal way for the Stapp family—what simple restraints could do.

It was 5:20 p.m. and already dark when Mary Louise and Mague Stapp, Celso's daughters, left the family house on Country Club Drive for town in a new Ford Galaxie that had been specially equipped with after-market seatbelts by a local dealer. Mary Louise was doing 45 miles per hour when an oncoming driver made a U-turn and swung into the Stapp girls' lane. Mary Louise braked hard and pulled to the right to avoid a collision, but the other car struck the Galaxie just as it hit the curb. The Galaxie rolled over three or four times as it pitched down a 35-foot embankment into an arroyo, landing upright on its wheels. Though the Galaxie was demolished—the front windshield shattered and the back window was completely ejected—both girls were buckled in and came away with nothing more serious than bruised backs and shoulders. They missed only a single day of school.

Celso wrote to his brother: "Your seat belts saved their lives. Ten days before, a girl went over the side about a half-mile from where Mary Louise

and Mague went over and was killed because she was sitting on her seat belt
and was going fairly fast. Thanks for the seat belts. Have a nice Thanksgiving.
We have much to be thankful for."

Given some of the brewing political developments in Washington, it
looked as if 1966 might be a watershed year in John Paul Stapp's fight to have
seatbelts installed as standard equipment in every new American car. "Stapp
Urges Safer Autos: Abrupt Stop Expert Calls for U.S. Controls" was the headline in the *Dayton Daily News* on March 25. Notwithstanding his problems
at AFIP, Stapp's reputation and recognition of his campaign for auto safety
had only grown since his move to the capital. In addition, that January, the
assistant director of astronautics at the Smithsonian Institution contacted
Stapp with a formal request to have the *Sonic Wind* rocket sled donated to
the National Collection.

. . .

Kenneth Roberts had been defeated for reelection two years earlier, but by
January of 1966 Representative James MacKay of Georgia was prepared to
carry the torch. MacKay had introduced a new bill in the House that aimed
to establish a national traffic safety agency and, having heard Stapp's previous testimony, was eager to have him back for another—and, MacKay and
Stapp both hoped, final—round of hearings. A press release from MacKay's
office set the tone: "The handwriting is unmistakably on the wall. We either
live together more safely … or we die separately in traffic accidents which
indiscriminately kill nearly fifty thousand Americans, injure three million,
and destroy eight billion dollars of our national wealth annually."

On the evening of May 3, Colonel Stapp was called as the last witness of
the day when only four of the committee's thirty-three members remained in
the chamber. Regardless, it was a newly impassioned Stapp who addressed
what remained of the committee about what he, as a doctor, characterized
as nothing less than an epidemic. The nation, he proclaimed, had lost three
times as many servicemen worldwide in car crashes as had died in hostile
action in Vietnam during the same period. His former reserve and equanim-
ity were gone. Colonel Stapp opened with a sober but blistering indictment of
the American auto industry, accusing it of utter failure with regard to safety
design. He called out an industry spokesman who just the day before had

been quoted in the *Washington Post* denigrating the idea of federal supervision of automotive safety, citing a supposed ineffectiveness of the FAA on air safety. "It would seem," Stapp told the committee, "that self-regulation of safety by industry becomes like self-surgery. No matter how well-meant, both the objectivity and the sponge count become difficult." Stapp was ready with facts and well-chosen analogies.

He went to the defense of the FAA. Back in 1959, he said, the U.S. had one fatal accident for every 85,000 hours flown, yet by 1964 the nation had one fatal accident for every 700,000 hours flown. "Accident investigations under Federal regulation," Stapp asserted, "have accomplished relentless pursuit of all mistakes and corrections where they are due." The auto industry spokesman, he suggested to the committee, had no idea what he was talking about.

At that point, an irritated Torbert MacDonald of Massachusetts objected to Colonel Stapp's comments.

"Everything you say is probably very true," MacDonald said, "but it doesn't really have anything to do with the bill . . ."

Stapp ignored MacDonald and continued presenting the latest data on auto accidents among U.S. armed forces personnel.

MacDonald interrupted. "Mr. Chairman, how many minutes will this go on?"

After a brief rules squabble between the remaining committee members over a motion from MacDonald to have the hearing concluded and adjourned, Stapp continued with his presentation on worldwide armed forces crash fatalities. Once again, MacDonald interrupted him.

"What does this have to do with death on the American highway? We have people in Okinawa . . . we have military people all over the world," MacDonald badgered Stapp. "Is this about Okinawa?" As MacDonald's volume rose, Stapp modulated downward—to MacDonald's intense annoyance.

"I can't hear you, sir," MacDonald complained.

Stapp continued: "With adequate research on the best documented population in the world, the Armed Forces, where we can get the most accurate figures, we can arrive at ways of reducing accident rates."

"In Turkey?" MacDonald fumed. "In France? In South Vietnam? In North Dakota or in Georgia, where tempers flare high, I am told?" This was a dig at MacKay, but also probably at Stapp and his own Southern twang.

"These are consolidated figures for the entire world," Stapp said.

MacDonald was heating up. "Are Georgia and Alabama in the world?" he snapped. "Sometimes I doubt it, but I guess they are."

"All over the world for the Armed Forces," Stapp replied. "That is our beat."

"In Iran? In Egypt? You know, camel drivers come into the accident rate."

This continued until it was agreed that Stapp would return the following day to complete his testimony.

By the next morning, Representative MacDonald was nowhere to be found and Stapp was allowed to resume his testimony. "A man," he said, "can walk away from a 60 miles per hour crash against a stone wall and not necessarily be injured." Furthermore, half of all vehicle fatalities on the streets and highways of the nation "occur under survivable conditions." They had proved this at Edwards and again at Holloman. The best crash safety data in the world, he insisted, was already in the hands of the military, and even while the Armed Forces Appropriations Committee in the House had forbidden Stapp and others to pursue research on "off-base accidents"—which was, Stapp suggested, like doing research on malaria but being confined to studying "on-base mosquitoes"—they'd managed to get a lot of valuable crash work done. The problem was that there was no way to share the data other than through the SAE and the Stapp Car Crash Conference, and they risked losing the data altogether unless a federal organization could be established to secure it, house it, and make it accessible to the next generation of researchers.

Simultaneous with MacKay's hearings, the issue of auto safety was heating up in the other house of Congress. Stapp had consulted with Senator Gaylord Nelson of Wisconsin on the auto safety bill Nelson was proposing to introduce in the Senate, and Stapp pledged to do anything he could to help. Meanwhile, a Senate auto safety committee under the chairmanship of Senator Abraham Ribicoff of Connecticut was investigating death on America's highways. In the course of the committee's work, a bizarre drama that no one could have foreseen was playing out, one that would ultimately change the game for everyone involved.

The previous November, a formerly obscure attorney and congressional staffer named Ralph Nader, the son of Lebanese immigrants, had published a blockbuster book titled *Unsafe at Any Speed: The Designed-In Dangers of the American Automobile*. It hit number five on the best-seller list and quickly sold half a million copies. The book made the case that many American car models were fundamentally unsafe, and it focused particular attention on the Chevrolet Corvair. The Corvair was a sporty compact that had been involved in an alarming number of one-car accidents, mostly rollovers and spinouts. As Nader had researched his book, more than 100 lawsuits were pending alleging General Motors' negligence with regard to the Corvair. Colonel Stapp was aware of Nader's work, and began sharing data and ideas with him during this period.

Shortly after the book's appearance, Nader began to suspect he was being followed. His phone would ring in the middle of the night with anonymous, vaguely ominous threats. "Why don't you go back to Connecticut, buddy boy?" Attractive women he'd never seen before sidled up to him in the grocery store and tried to seduce him. Nader began making telephone calls only on pay phones on the assumption that his own phone was bugged. When a pair of suspicious men falsely identifying themselves tried to tail Nader into the Senate Office Building, a security guard apprehended them. The *Washington Post* and the *New Republic* ran stories about the incident.

"The surveillance became so amateurish in the end," Nader said, "that it was almost like slapstick comedy." Ford and Chrysler both issued immediate, unequivocal denials that they were behind any surveillance of Nader, but General Motors released only a "no comment." Eventually, however, as reporters learned the identities of the people involved and traced their associations to the General Motors legal department, GM was forced to admit that it had in fact approved and conducted a months-long undercover "investigation" of Mr. Nader. GM had left few stones unturned in an effort to find incriminating evidence that would show Nader to be an anti-Semite or a Communist or a homosexual or on the payroll of some unsavory organization—a pathetic effort that completely failed to dig up anything that could be used against him.

Once GM went public with its mea culpa, an angry Ribicoff convened a hearing to look into the harassment of one of his committee's witnesses by one of the world's largest and most powerful corporations. Reporters and photographers crowded into the hearing room. The first witness called was James Roche, president of General Motors, and the senators went after him.

Robert Kennedy, the bare-knuckled former attorney general of the United States and now senator from New York, vilified Roche. Fred Harris of Oklahoma and Henry Jackson of Washington State did likewise. Roche tried to argue that his company had been involved in nothing more than the due diligence that was required by GM's legal teams fighting the flood of lawsuits against its products, but he came off as a villain.

The bungled efforts of General Motors to develop compromising information about Ralph Nader transformed the young lawyer and author into an overnight folk hero. Nader's photo ran above the fold on the front pages of the nation's newspapers, in many cases alongside the grim image of James Roche not-quite-atoning for his company's sins. "To the extent that General Motors bears responsibility," Roche had told the senators in a carefully parsed statement drawn up by former presidential aide Theodore Sorensen, now Roche's personal attorney, "I want to apologize here and now to members of this committee and Mr. Nader." The incident turned the tide of the American public's estimation of the auto companies. When Senator Kennedy offered Nader a chance to explain his motivations in uncovering and publicizing the industry's record of safety design, Nader replied: "If I was engaged in activities for the prevention of cruelty to animals, nobody would ever ask me that question. Because I happen to have a scale of my priorities which leads me to engage in activities for the prevention of cruelty to humans, my motivations are constantly inquired into."

A couple of weeks after his Senate testimony, Nader was a guest on NBC's *Meet The Press*. "Mr. Nader," challenged Richard Wilson of Cowles Publications, "it is still hard for me to imagine how you save a man's life going at 50 miles an hour when he runs into an immovable object or even a moving object."

"Well, let me cite you a statement," Nader offered in reply, "by Colonel John Stapp of the U.S. Air Force who is a vehicle safety specialist and has subjected his own body to punishing deceleration rates. He made a very good analogy. He said the auto industry has been ingenious in protecting shocks vertically through pneumatic tires, shock absorbers, springs and seats. Now, why don't they apply these principles to protecting shock horizontally when crashes come in on the side of the automobile? This is perfectly possible with a minimum of inconvenience to the driver."

With momentum building, sympathetic congressmen began to pile on the car companies. Senator Vance Hartke of Indiana called the MacKay bill "a powerful coordinated national effort to stop 'murder by motor'." Bobby Kennedy badgered GM chairman Frederic Donner into admitting that his company devoted less than one tenth of one percent of its product design investment to safety. A barrage of auto safety publicity aimed at state agencies came from a new lobbying effort called Women's Crusade for Seat Belts. A *New York Times* editorial declared that the auto industry had a moral responsibility to the public: "The industry has an obligation both to talk and act on safety, in order to bring the wanton slaughter on America's highways under control."

"All of this," a *Washington Post* review of Ralph Nader's book concluded, "has helped create a situation in which Federal intervention is now probably unavoidable." The smart money in the capital was on passage of some version of a seatbelt bill by the end of the session.

Only a few weeks before the Ribicoff hearings, President Lyndon Johnson had made known his own feelings on the issue of auto safety. "We can hardly tolerate such anarchy on wheels," he said gravely. Thanks to the unexpected gift of the GM–Nader fiasco, Johnson now saw that he had the political cover he needed to sign an auto safety bill if Congress could manage to get one to his desk that summer. Which is not to suggest that the car companies and the car lobby on Capitol Hill had thrown in the towel on regulation. The chairman of the American Business Council, W. B. Murphy, dismissed the whole thing as an "auto kick . . . on the same order as the hula hoop—a fad. Six months from now, we'll probably be on another kick."

With the political winds at their backs, however, Congress moved quickly, and on September 9, 1966, at a White House Rose Garden ceremony, President Johnson signed two bills into law: the National Traffic and Motor Vehicle Safety Act, and the Highway Safety Act. From that moment forward—though the legislation failed to include a key criminal provision favored by both Nader and Stapp that would have made corporate executives personally liable for the results of egregiously negligent safety design—the federal government became responsible for establishing and enforcing safety standards for vehicles and the roads they travel on.

In Johnson's address accompanying the signings, he explained the laws' significance to the nation: "In this century, more than 1,500,000 of our fellow citizens have died on our streets and highways; nearly three times as many Americans as we have lost in all our wars." In words intended for the automakers in Detroit, he added, "Safety is no luxury item, no optional extra; it must be a normal cost of doing business."

Colonel John Paul Stapp was standing behind the president as he spoke and was one of the first to shake his hand following the address. This was Stapp's moment of triumph. His fight had finally come full circle, from the Battle of Muroc to the White House, and no one else could appreciate the years of sand, smoke, and broken bones that had gone into it—even if Ralph Nader would get the lion's share of the acclaim that day. Stapp later remarked, of Nader's dramatic appearance at the signing ceremony: "Mr. Ralph Nader, flanked by two body guards, arrived just as the ceremony was about to begin, and immediately drew all the press away from the front benches, leaving President Johnson chagrined." But beyond any publicity generated by the event, this was a measure of vindication for both Nader and Stapp. GM eventually settled a lawsuit filed by Nader for $425,000, while refusing to admit any wrongdoing. Nader used the money to set up a public interest organization focused on consumer protection.

The Corvair was discontinued in 1969.

25

SLUGGING AWAY AT THE STANDARDS

> You might think that before they denounce unwelcome
> research findings, major corporations would devote their
> considerable resources to checking out the safety of the
> products they propose to manufacture. And if they missed
> something, if independent scientists suggest a hazard, why
> would the companies protest?
>
> —*Carl Sagan*, **The Demon-Haunted World**

THE BILLS SIGNED INTO LAW that September established a National Highway
Safety Bureau that was charged with establishing safety standards for
all new cars and trucks—and related equipment such as tires—by January
31, 1967, and issuing revised standards one year later. The bills authorized
appropriations of more than $380 million for the period 1967–69. One effect
of the new legislation was that the auto industry and its legal staffs immediately geared up to fight potential mandates: seatbelts, shoulder straps,
headrests, shatterproof windshields, energy-absorbing steering wheels and
columns, flashing hazard lights, dual braking systems—they were prepared
to resist all of it. Ford alone may have spent as much as $40 million fight-
ing mandatory seatbelts. Industrial safety research was temporarily frozen.
Within two years of the signing of the auto safety bills, the car companies—
even Ford, which had at one time been a leader in safety research—had
scaled back their safety research groups.

Two weeks after the triumphant ceremony at the White House, Stapp was off on a groundbreaking journey he'd been trying to arrange for years. He'd first met Leonid Ivanovitch Sedov during the International Astronautical Federation meeting in Paris in 1963. Sedov had served as president of the IAF and had been the first chairman of the Soviet space exploration program. While both men were passionate advocates and spokesmen for their own nation's approach to space, Stapp and Sedov shared a desire to foster communication and even cooperation between the superpowers on matters of extraterrestrial research. Sedov could be provocative, as in a statement to an American delegate to a conference in Barcelona shortly after the launch of *Sputnik*: "You Americans have a better standard of living than we have. But the American loves his car, his refrigerator, his house. He does not, as the Russians do, love his country." But in correspondence with Colonel Stapp he had shown himself to be genuinely committed to the sharing of knowledge between the Soviet Union and the United States. Sedov had been quoted in the *New York Times* as saying that he envisioned cooperation with the U.S. in the development of space technology.

So, on September 24, 1966, with Sedov helping to arrange visas, Stapp traveled from Washington to Prague, and then secretly on to Moscow and Leningrad, where he spent two nights in each city. Stapp was able to meet with researchers in Leningrad at the Pavlov Institute of Physiology, and with Sedov and his colleagues at the Commission for the Study and Utilization of Space in Moscow, where he was presented with an engraved aluminum casting of himself on *Sonic Wind*. He did some sightseeing and attended two ballets. It was a whirlwind trip of mostly cursory meetings and discussions, but Stapp was hopeful that it might serve as an icebreaker.

It did. One outcome of the trip was a reciprocal arrangement for the following spring, engineered by Stapp, that resulted in a U.S. State Department invitation to Soviet Professor Arnold Barer, whom Stapp knew from the IAF, and Dr. Victor Costin to present papers at the Aerospace Medical Association meeting in Washington. It was at least a ceremonial thawing. In a thank-you letter to his Intourist hostess in Moscow, Stapp would write that his time there had been "the most exciting and instructive week of my life." Part of Stapp's excitement had nothing to do with space technology or science, but came courtesy of the hostess herself, Irina Novikova. They talked music and history and literature. "You are," he wrote to her immediately upon his return to Washington, "beautiful, charming, and gracious, and always so kind hearted." He added: "I would like to keep up a correspondence with you until we can speak again—incidentally, I bought a book of Russian Lessons and will start to work trying the learn the language."

A few weeks later, Stapp wrote again to ask whether she might be able to accompany Professor Barer and Dr. Costin as a translator on their visit to the States. Failing that, Stapp suggested that they might meet at the World's Fair in Montreal in June of 1967. Stapp and Novikova eventually saw each other again in New York City that summer. They had dinner and talked about their mutual love of Haydn.

According to what he told a close friend years later, he soon got cold feet and backed away from what had clearly been building toward a love affair. "I could never do that to Lili," he said. Besides, Stapp confided, he had begun to wonder whether Novikova had been encouraged by the Soviet government to court a relationship with him in order to try and gain access to top secret information about the U.S. space program. He wondered if she'd been sent to him as a spy. Nevertheless, Stapp offered to help Novikova immigrate to the United States. She eventually settled in New York. Shortly after her arrival, Stapp broke off his correspondence and never saw the Russian woman again.

Stapp stayed busy in the summer of 1966, working to finish his multi-volume study of biodynamics amid the hoopla surrounding the auto safety bills. He packaged up his manuscript and forwarded it to a McGraw-Hill editor who had expressed interest. The response was discouraging. It read, in part: "I very much fear that it will not have a commercial market strong enough to support publication. There is no question of the great importance of your studies – nor even of the increased interest in vehicle safety that is now current – but I do not think that these factors spell commercial success." Stapp explained to McGraw-Hill that he'd never envisioned a commercial market for this work, and that profit had never been any part of his motive. This was a technical, not a popular work. "I had no intention of taking even an author's royalties, but to contribute my services free," he wrote the editor. He would continue to search for a publisher during the remaining months of that year, without success.

Meanwhile, an alphabet soup of new federal agencies, administrations, and departments was waiting to be staffed by scientists and engineers who understood vehicle safety. From the National Traffic and Motor Vehicle Safety Act came the National Highway Safety Bureau (NHSB); the new bureaucracy's first director was William Haddon, a New York public health doctor who had been an early attendee of the Stapp Car Crash Conferences. When the Department of Transportation was created in October of 1966, the NHSB—which would shortly become the National Highway Traffic Safety Administration (NHTSA)—was incorporated into the new cabinet department along with the Federal Highway Administration (FHA). The FHA's administrator, Lowell Bridwell, with the enthusiastic backing of Dr. Haddon, quickly arranged with the Air Force for Colonel Stapp to be detailed to the executive branch as chief medical scientist first of FHA and then of NHSB, and eventually of NHTSA. A press release announced: "The world's outstanding first-hand authority on how much crash force a human being can stand is going to work for the National Highway Safety Bureau to reduce death and injury on the Nation's highways." Stapp would join NHSB as the number three man in the organization and would enjoy an authority and access to resources he had expected, but never received, at AFIP.

By the summer of 1967, a reenergized Stapp had already secured a $307,000 contract to work with universities to support the Registry of Accident Pathology he had founded but not been able to fully staff or fund at AFIP—it would eventually catalog more than a million autopsies of accident victims—and a $92,000 contract to conduct live automotive restraint tests at Holloman for the purpose of establishing detailed requirements for federal mandates. To Stapp, it was both liberation and vindication. He was finally in a position to formulate and—through the auspices of the federal government—help dictate national auto safety standards with which Detroit would be obliged to comply.

What became a long, complex wrangle over the government's initial twenty-three proposed mandatory federal automobile safety standards began almost as soon as the NHSB was formed. The car companies vigorously opposed the implementation plan for Safety Standard 201, which dealt with car interiors and was intended to eliminate protruding handles and knobs that were known to cause disfiguring injuries even in low-speed crashes. Instead of the eleven-month period stipulated to retrofit interiors of new car models, the industry insisted it would need a minimum of three years to implement the required changes.

While Stapp had always believed the auto industry would eventually come to see the wisdom of federal safety standards, he was never naïve about how long that might take. It was revealed during a Senate investigation by the Ribicoff Committee just a few months later that GM, Ford, and Chrysler had conspired to effectively boycott selling their 1968 cars to the United States government as a way of protesting new safety requirements. If it had not been for the cooperation of American Motors, the General Services Administration might have been unable to purchase new domestic vehicles at all that year. It was noted during the Senate hearing that despite their pointed refusal to bid on government civilian vehicle contracts, the Big Three had continued to receive lucrative military contracts courtesy of the American taxpayer. In the previous year, GM alone had received some 625 million dollars' worth of business from the Defense Department.

Industry spokesmen and lobbyists applied relentless pressure on the various committees charged with authoring the standards. In Stapp's words: "They started slugging away at the standards . . . and they managed to knock out several standards." But Stapp and his colleagues fought back attempts to revise or discredit critical Standards 208, 209, and 210, which prescribed upper torso restraint systems for car occupants. With the latest Volvo data in hand, coupled with Stapp's own research at Holloman and Edwards, NHSB was able to successfully defend its work. According to Stapp, who pushed particularly hard on Standard 208: "They [auto industry interests] quit fighting that one . . . and began to cooperate."

One of the outcomes of the standards work was a law requiring that new American cars, beginning with 1968 models, be equipped with the three-point lap/shoulder belt configuration for driver and front-seat passenger that had already been standard equipment in Volvos for five years. European automakers had long led the United States in safety design and implementation, and back in 1965 Colonel Stapp had convinced Volvo engineers—who already made three-point seatbelt harnesses available to its customers as factory-installed options—to conduct a study. Volvo collected data on 27,000 accidents involving its cars; 9,200 of the individuals involved had worn seatbelts and the rest had not. For the belted drivers and passengers, even at crash speeds up to 70 miles per hour, not a single accident had been fatal. For the unbelted, fatalities were reported at speeds as low as 12 miles per hour.

Nils Bohlin was the Swedish engineer who'd invented Volvo's three-point system in 1958—based on a design documented by Hugh DeHaven twenty years earlier—in what Bohlin described as a "sudden flow of inspiration." By the time Detroit was begrudgingly equipping its latest models with Bohlin's restraint design for outboard front seats, Volvo was already equipping rear seats of its cars with the devices. "GM and Ford didn't like it," Bohlin recalled. "They said negative things about it. They said the belt was not only not good enough, but that it could be dangerous." Bohlin had made his first presentation on the three-point restraint system to the Stapp Car Crash Conference back in 1964. The statistical study requested by Stapp and presented by Bohlin and his Volvo colleagues at the Eleventh Stapp Car Crash Conference proved to be extremely influential on the NHSB. Stapp suggested loudly and on more than one occasion that Bohlin ought to be shortlisted for a Nobel Prize.

Ralph Nader attended many of the standards meetings and continued to be a major irritant to the car companies. Nader's increasingly confrontational style, however, was coming to strike Stapp as counterproductive. "I did not call names," Stapp said. Stapp's principle was: don't create dragons with placards and loud shouting; instead, be ready with your solution and sell the hell out of it. Though he had great respect for the young attorney's intellect and drive, Stapp began to regard Nader as too much flamboyant self-promoter and too little constructive consultant—just the sort of charge, ironically, that had been leveled at Stapp himself repeatedly over the years.

The agendas of these early standards meetings were crowded. Restraint systems weren't the only mandates established by NHSB and NHTSA committees. Standards on safety glass called for plastic laminate glazing that would provide a pocket of finely crushed and mostly harmless surface around a head impact at 30 g's. Standards for door locks required them to hold tight up to a pressure of 2,500 pounds. In time, air bag standards would create a walking legion of car crash survivors Stapp referred to as "the ghosts that never happened."

Stapp would spend much of the next two years in seemingly endless and often frustrating committee and task force meetings as the federal government (largely the nascent Department of Transportation and the Department of Health, Education, and Welfare)—in consultation with a wide variety of subject-matter experts, industry lobbyists, and federal agencies— attempted to develop its regulatory portfolio governing vehicle safety. When all was said and done, NHTSA, during Stapp's tenure, would complete and implement eighteen crashworthiness standards for passenger cars, most of them intended to be temporary minimums that would be regularly reviewed and updated as safety technology improved and investigative tools evolved.

It was at times frustrating work for Stapp, complicated by the small army of stakeholders and political interests hovering on the periphery. Nevertheless, even as he was buried in the standards work, Colonel Stapp did his best to keep an eye on the race to the moon. At NASA's invitation, he traveled to the Kennedy Space Center in Florida to witness the launch of Apollo 11 on July 16, 1969. He wrote to a friend about the experience: "While I was watching it, I was saying to myself, 'This is better than being at the dock watching Columbus leave the harbor in 1492!'" On a Sunday, four days later, Stapp, along with half a billion people around the world, witnessed the conclusion of the decade-long space race play itself out in grainy, glorious imagery. People everywhere huddled around television sets, many of them teary, barely believing their eyes. They were hushed as the lunar module, guided at first by computers and finally by the steady hand of the mission commander, Neil Armstrong, swooped down to land in a lumpy boulder field alongside a giant crater.

Stapp and Lili sat together in their home in suburban Washington, glued to the television. They watched CBS anchor Walter Cronkite, seemingly stunned, struggle to find words to describe what they were all seeing. Cronkite finally gave up. "Oh, boy!" he said, removing his glasses, and sighing. "Whew!"

At 10:56 Eastern Daylight Time that evening, July 20, Armstrong hopped down into the lunar dust—and in that instant it was done. The United States of America had met the most audacious engineering challenge in all of history. It is not known if Stapp, who'd been a part of it from the beginning, raised a cheer, hugged Lili, or whether he perhaps allowed himself to contemplate what it all meant and what it had taken to make it a reality. For Stapp, of course, success was always about the round trip, and he would have wanted to see America's astronauts returned safely to earth before joining any celebration. Triumph had never been even close to a guarantee. President Richard Nixon's speechwriter, William Safire, had secretly prepared a contingency statement for his boss titled "In the Event of Moon Disaster." There's no question that Stapp felt slighted that night not to have been included more prominently in the endgame. In the years to come he would lament that his labors and the labors of many of his aeromedical colleagues had not been fairly represented when the histories of the space program were recorded. Historian Maura Mackowski has written that Stapp and others in the military services who had made valuable, sometimes crucial, contributions to the race to the moon collectively became "the Moses of the space age: they were not allowed to enter the promised land."

Yet Stapp persevered in his continuing effort to exert his influence on the direction of the space program. One breezy, late-summer week at Cloudcroft, New Mexico, in the mountains above Alamogordo, only two months after the moon landing, Stapp chaired an International Academy of Astronautics conference on what had come to be called the International Orbiting Laboratory. This was the concept Stapp had first raised at the IAF conference in Athens four years earlier, which would become the basis for the International Space Station. The idea was to create a working orbital venue for scientists of all nations, and a Soviet delegation that Stapp had invited to Cloudcroft brought an enthusiasm for cooperation from the other side of the Iron Curtain, despite the disapproval of the Air Force brass, which Stapp did his best to ignore. While there was a suitable dose of well-earned cynicism among Americans regarding the Russians' sudden post-Apollo kumbaya, it was a validation of Stapp's efforts to claim space as an arena of science apart from military matters. "The whole business of the conquest of space doesn't belong to any one country," Russian space medicine expert Dr. Oleg Gazenko told reporters at the Cloudcroft conference. "It belongs to all mankind."

In his opening remarks to the conference, Colonel Stapp embraced Gazenko's vision. "The message in Armstrong's words," Stapp proclaimed, "is that earth men will take the giant step of exploring the solar system and the Universe together." In the self-appointed role of extraterrestrial sheriff, he warned that Americans and Russians were going to have to learn to "check their badges and guns on Earth."

Though Stapp continued to try to shape the future of space policy in the months after Apollo 11, his was a lone voice in the political wilderness. Despite his past achievements and his recent success with vehicle safety, what had been clear for some time was that Stapp would not be receiving the star he coveted from the Air Force. There would be no General Stapp. Perhaps it had never been in the cards at all. He admitted to friends that he was disappointed, but mostly because he'd wanted to leave Lili with a general's pension someday. On July 31, 1970, two weeks after his sixtieth birthday, Colonel John Paul Stapp retired from the United States Air Force—the only full-time employer he'd had since his days at Decatur College. It was a monumental moment in Stapp's life, but one that was attended by a total absence of fanfare. He filed some paperwork with headquarters and made a few calls to colleagues, had a quiet dinner at home with Lili, hung up his uniforms, and simply got on with his business. He was immediately placed on the payroll of the Department of Transportation on a two-year civilian contract, which allowed him to continue his work at NHTSA—as Dr. Stapp, now—with little interruption.

Secretary of Transportation John Volpe, his new boss, presented Stapp with the Distinguished Service Medal. "Colonel Stapp," Volpe reminded reporters at a brief ceremony, "wrote the first proposal for a set of federal seat belt standards . . . I am sure that he can take a large part of the credit for the fact that preliminary figures for 1970 highway deaths show a marked decrease—perhaps as many as 1400 fewer highway deaths last year than in 1969." Yet amid the kudos, the Air Force's Stamp Out Stapp club took one last parting shot. An anonymous officer somewhere at headquarters started a nasty rumor that, now retired, Stapp was planning to file a massive lawsuit against the Department of Defense to collect damages for the injuries he'd suffered on the rocket sleds.

As a civilian, Dr. Stapp continued to respond to requests for his appearance at conferences and symposia all over the world. Of particular interest was his participation at a domestic event in the fall of 1970 in Dayton sponsored jointly by the Aerospace Medical Research Laboratory and the National Academy of Sciences. The Symposium on Biodynamic Models and Their Applications represented an increasingly influential component of crash studies. Given that the management of human volunteers and anesthetized animals as research subjects was becoming ever more controversial, the research community was accelerating efforts to replace live subjects in its testing. The Dayton meeting dealt both with human surrogates (sophisticated anthropometric dummies) and mathematical models that sought to recreate crash conditions using computer simulations. The auto industry was highly motivated to find ways of doing product design that would not require biological specimens of any sort.

In his closing remarks to the symposium, composed on the notes pages of the event's agenda as he listened to the speakers, Dr. Stapp acknowledged and honored the hope shared by most of the attendees: "What boundless optimism spurs your intrepid assaults on this most obstinate and irregular object you have chosen for mathematical analysis and modeling—MAN! This fifty liter rawhide bag of gas, juices, jellies, gristle and threads moveably suspended on more than 200 bones presided over by a cranium, seldom predictable and worst of all *living*, presents a challenge to discourage a computer into incoherence. I salute your courage in seeking to make logic out of this seeming chaos through biodynamic mathematical modeling."

Nevertheless, Stapp was not and had never been very sanguine about the ultimate validity of these models, and he had been concerned for years about what he saw in some of his colleagues—especially those connected with the auto industry and the university programs they funded—as a rush to replace live testing with purely mathematical simulations. As early as the mid-1960s he'd complained that models were being elevated to presumed infallibility much too rapidly. There would never and could never be, in the opinion of Stapp and other influential researchers such as Dr. Richard Snyder, a completely reliable substitute for biological test subjects. Live tests were difficult, logistically messy, sometimes unpleasant, and occasionally downright brutal, but they were essential in order to validate the new mathematical models.

These became defining arguments in the crash research community—but there were others. In January 1971, Stapp went to Detroit for the annual SAE Congress and Exhibition. During a roundtable discussion he chaired, representatives from Ford and Chrysler rose to lambaste the work of DOT and its standards committees—and, by implication, Stapp himself. The speakers complained about proliferating government standards being created by "uninformed civil servants." Taking the high road, Stapp praised the cooperation of the auto industry and singled out some good work that had begun in Detroit on passive restraint systems, particularly air bags.

Air bags were one of the most promising areas of innovation and NHTSA estimated their use could save as many as 9,000 American lives annually and prevent 700,000 disabling injuries. Six months earlier, DOT Secretary Volpe had announced a departmental objective of pushing for installation of air bags in American cars with "utmost speed." But, as Stapp noted in his trip report in reference to his own complimentary words, "There was no applause for these remarks." The battle lines, it seemed, had not only been drawn, but reinforced. In fact, the war was raging even between the automakers. Perhaps surprisingly, General Motors had taken a fresh lead not only in developing and promoting marketable safety systems, but in the area of air pollution control as well. Following the meeting, Stapp privately characterized the attitudes of Ford, Chrysler, and American Motors as "reactionary."

Ford fought back against rival GM and the federal government with a series of ads later that year unveiling the company's anti-air bag stance. "We believe," one ad said, "in cars that are engineered, not legislated." Allstate Insurance Company threw in with the government and took out its own high-profile counter-ads quoting NHTSA's administrator to the effect that the United States might cut traffic deaths in half if all cars were equipped with air bags.

Yet despite DOT's efforts and growing congressional activism, the federal government would prove to be an unreliable supporter of Stapp's and NHTSA's auto safety regulations. As administrations changed, so would attitudes toward auto safety. On the morning of April 27, 1971, President Richard Nixon met privately in the Oval Office with the top two men at the Ford Motor Company: chairman Henry Ford II and president Lee Iacocca. Ford told Nixon that his company's major concern was "this total safety problem." The company, he said, was particularly worried about implementation of the air bag standards NHTSA had agreed on. Iacocca was more direct. He told the president, with stunning contempt for his company's customers, that the new safety regulations were nothing more than an attempt to protect the American people from "their own idiocy." And it wasn't just air bags, in Iacocca's view. "The shoulder harnesses, the headrests are a complete waste of money," he said. It was all about competitiveness. How could the American auto industry compete if it had to keep burning money installing all this safety equipment? "The Japs," he told Nixon, "are in the wings ready to eat us up alive."

Nixon was quick to assure his guests that he did not share his own Secretary of Transportation's urgency about safety devices. "Maybe there are safety problems," Nixon mused. "I think they're greatly exaggerated." He went on to blast "Naderism" and insult all those who had served the cause of reducing death and injury from car crashes. "They're a group of people that aren't one really damn bit interested in safety or clean air. They're interested in destroying the system. They're enemies of the system." The president made it clear to the Ford executives, as if there had been any doubt, that he would always be pro-business. He promised to look into the air bag issue.

Three days after the meeting with Ford and Iacocca, Nixon's assistant John Ehrlichman called Secretary Volpe to tell him that the president wanted the air bag implementation regulations delayed. Volpe was adamant that any delay would damage the administration's credibility. "There will be an avalanche, and I mean an avalanche, of protest in every newspaper in this country," he warned. But Ehrlichman was unmoved, and maneuvered the discussion into a tactical consideration of how the regulations could be delayed—and perhaps eventually defeated—without bringing unnecessary fallout on the White House.

Against his own instincts, Volpe issued a revised ruling that summer that delayed the new regulations from taking effect. Shortly thereafter, Ehrlichman assigned White House aide Chuck Colson to contact the automakers to let them know that it was the president himself who had pushed for the revised ruling. It was important to the administration that the industry knew who its friends were. All of this meant that John Paul Stapp was now fighting not only the recalcitrance of the auto industry on one side and activists such as Nader accusing NHTSA of dragging its feet on the other, but intra-administration sabotage personally perpetrated by the most powerful man in the country.

Stapp was enough of a realist to understand that working on behalf of the federal government on safety legislation would always mean living with compromise. The Air Force bureaucracy had at least prepared him for this.

As his time in Washington wound down, facing the fact that he would not be able to achieve everything he'd once hoped for, he counseled himself to celebrate the successes and to remain hopeful about the rest. He told a reporter before that year's Stapp Car Crash Conference, "I don't think we are through inventing answers ... We by no means are at the ultimate state of development in either the air bag or the restraint system."

Throughout his life, Stapp had remained mostly disdainful of political parties and politicians—"the political animal sitting on a fence with his ear to the ground, tuned to the sensitivities of his pious patriotic constituents"— though he had been a faithfully registered Democrat since casting his first vote for Franklin Roosevelt in the presidential election of 1932. According to Wilford Stapp: "He knew where he stood on all the issues, so he was never a neutral person. He was interested in policy, but not the politicians. As far as human rights and having respect for mankind, he was a total liberal. As far as money spending, he was a conservative." In his last decades Stapp would complain about taxes but support both the legalization of marijuana and government-funded health care. With regard to the latter, he insisted, as a medical doctor, that a healthy society required healthy citizens, and he wanted assurance that the waiter who brought him his steak was not also bringing him tuberculosis.

As Stapp continued to fight the administration and the auto industry, he was heartened by developments in the world of deceleration and impact research. A new generation of scientists who'd grown up with the examples of Stapp and Snyder and Charles Lombard and Hugh DeHaven were carrying the work forward with a new level of sophistication and accuracy. At their well-appointed lab on the grounds of a heavily guarded NASA assembly facility in Michoud, on the eastern edge of New Orleans, Dr. Channing Ewing, Dr. Daniel Thomas, Dr. George Beeler, Bill Muzzy, and others were designing and conducting experiments intended to reveal the extraordinarily subtle mechanics of head and spine injury. Stapp had given up attempting to get good data on head movement in deceleration or impact work way back in 1951 "because [of] angular motion of the head and instability of the helmet on the head." This was very difficult stuff.

The origins of the Naval Biodynamics Laboratory (NBDL) are found in a 1965 proposal for Army funding authored by Dr. Ewing and supported by

Army General Robert Cutting who arranged for the approvals required to use human volunteer test subjects. Ewing had used the acceleration track at Wayne State University in Michigan, where he and his team ran a small group of Army enlisted men in the first series of experiments. The project was moved to Michoud in 1971, where a 700-foot test track was constructed, and everything Stapp heard about the work going on there was encouraging. He made it a point to follow the progress of the NBDL studies and resolved to do what he could to support the doctors and engineers involved.

Stapp was distracted during this period, however, by events at home. At the time of his transfer to the Department of Transportation, he and Lili moved first into a rented house in Bethesda, Maryland, and eventually into Washington, DC, Northwest. Douglas Lewis's behavior had become increasingly erratic. He'd been caught stealing on several occasions. He had continued to need the assistance of behavioral and psychological counseling. Lili told her husband that she was sometimes afraid of Douglas and was uncomfortable staying alone with him during Stapp's frequent travels.

At some point in the early summer of 1971, the day-to-day environment in the Stapp household became untenable. The Stapps' ability to cope with Douglas, who had been hauled in by the DC cops for a variety of minor transgressions—Dr. Stapp was called to post bail for Douglas following an arrest for marijuana possession, though the charge was later dismissed by prosecutors—had continued to deteriorate. The final straw came when Douglas physically threatened Lili and made death threats against both his aunt and uncle. Lili called her husband at his office in tears, reporting that Douglas had returned home after being out all night and that he was out of control, yelling and smashing things. She said she was afraid for her life. Stapp rushed home and confronted Douglas, now a grown man of twenty-one, and ordered him to leave the house—permanently. There was a shouting match, a bad one. Stapp promised Douglas that if he ever harmed Lili he would kill him. In a rage, Stapp grabbed Douglas and pushed him out the door. While his reaction seems out of character for the generally even-tempered doctor, it illustrates the degree to which he'd become frustrated in dealing with Douglas, and perhaps also the frustration with his own impotence to treat or even to properly understand Douglas's deep-seated problems.

On June 17, in response to a complaint from the Stapps, the U.S. District Court for the District of Columbia ordered Douglas remanded against his will to St. Elizabeth's Hospital for psychiatric observation and evaluation. About a week after Douglas's commitment to St. Elizabeth's, Stapp and Lili moved into a three-bedroom house on a half-acre of land they'd purchased in Fairfax, Virginia. Stapp wrote to a friend who knew about the situation: "Our unlisted phone number is 280-2710. Please bear in mind it is unlisted—we moved while Douglas is at St. Elizabeth's for observation after twice attacking Lillian with death threats. We don't know whether he will be released or not, and he is dangerous, with a fixation of all the hostilities toward his mother, directed at Lillian. It is out of the question to have him live with us any more, or even to know where we live, if we can help it." Stapp had rarely spoken to those he knew about Douglas. His colleagues and even some of his longtime acquaintances had been completely unaware that the Stapps had taken in the teenagers in the first place. It was almost as if Douglas and Dorian had been family secrets that were not to be divulged.

What happened to Douglas once he was released from St. Elizabeth's is unknown. The mental health resources of the day, even for someone with Stapp's connections and insight, were hit-and-miss. Neither the Stapps nor anyone else in the family ever saw or heard from Douglas Lewis again. It was a sad failure for the Stapps and for John Paul personally—the otherwise generous family father figure whom Joe Kittinger had once called "the bravest man in the United States Air Force" all but hiding from his own nephew. According to Wilford Stapp: "Paul was always afraid of that boy."

Then, over Christmas 1971, the Stapps received word that Dorian Bates had died of lupus. They had not heard from her in six years, and had not even been aware of her illness, but the news hit them hard. The Stapps had liked Dorian and regretted that they'd not been able to do more for her. John Paul and Lili drove to Austin just before New Year's to meet Dorian's twenty-five-year-old widower and their two children, a son and a daughter. In February, Stapp wrote to his friend James Ryan in Minnesota about the tragedy, still clearly distraught.

Sad times in the Stapp household got worse in mid-May of 1972, when word came from the Air Force that Joe Kittinger's F-4 Phantom had been shot down over North Vietnam. Both the pilot and copilot had been declared missing in action. The forty-four-year-old Kittinger had been on his third tour in Southeast Asia, serving in Thailand as commander of the most famous fighter squadron in the war, the 555th, or Triple Nickel. Stapp was shaken. Lili, who of all her husband's colleagues had been fondest of Joe, wept. It is no exaggeration to say Joe was like a son to both of them, the son that Douglas had never had the chance to be.

True to form, Stapp found distraction in his work. A civilian now and freed from the encumbrances of the Air Force, he agreed to serve for the first time as a consultant and expert witness for the plaintiff in a personal injury lawsuit against a car company. The case involved a seventeen-year-old high school student named Vivian Lee Hobbs, who had been in the passenger seat of a yellow Volkswagen Beetle traveling north at about 45 miles per hour on Route 1 near Beltsville, Virginia, a suburb of Washington, DC. It was a Saturday, June 17, 1972, the same day that Richard Nixon's "plumbers" burglarized the Watergate Hotel complex just a few miles away. Hobbs's fiancé, Greg Berzinski, swerved to avoid a drunk driver in a black Ford Thunderbird who'd turned in front of them. The Thunderbird hit the VW on its left side, spun it around counterclockwise, and sent it skidding backward for 100 feet into an embankment, where it went partially airborne and struck a telephone pole. Vivian's seat failed under the impact and she was hurled rearward. Her head smashed into the metal rim of the rear windshield and then whipped forward, snapping three vertebrae in her neck. Greg's seat held and he stumbled from the VW in shock.

It was the kind of case that would naturally have attracted John Paul Stapp even if he had somehow managed to forget the young woman from Baylor and the drunk driver at Hollywood and Vine fifty years earlier. Contacted by Vivian's legal team, he readily agreed to assist, pouring himself into an exhaustive investigation of the crash. He inspected the wreck of the VW, spent hours at the crash site getting his own measurements, and he conducted detailed interviews with both Hobbs and Berzinski. His narrative was a meticulous recreation of the event: "At the instant of collision with the telephone pole, if not before, the right front seat back failed backward until it was against the rear seat back and the left side of the right front seat, which was off the seat track, rotated backward toward the front of the rear seat cushion leaving an opening between the two front seats through which the passenger slid backward and toward the left rear corner, head first, face down ... The top and right side of her head rammed into the left upper rear corner of the rebounding car top with convergence of stress on the right sides of the third and fourth neck vertebrae producing explosive fractures and shearing pressures on the spinal cord."

According to Dr. Nick Perrone, who, along with Stapp, served as an expert witness for the plaintiff in *Berzinski v. Volkswagen*, the prosecution was outgunned in both resources and legal experience. Volkswagen brought in Dr. Derwyn Severy, a former colleague of Stapp's who had once pushed for increased seat-back strength, to testify that the seats in the yellow Beetle were in fact perfectly adequate. Perrone makes the point that defense witnesses for car companies in cases like this can earn huge fees for their services.

Volkswagen won the case and Vivian Hobbs, a quadriplegic, paralyzed for life, got nothing—although she did defy the predictions of her doctors, who thought she'd only survive a few days at most. Breathing with the aid of a respirator, she married Greg as planned and the couple eventually had three children. Vivian entered the University of Maryland and graduated in 1978 with a degree in biochemistry, and three years later got her law degree from Georgetown. She would later go on to work with Senator Robert Dole to help write and pass a landmark civil rights law: the Americans with Disabilities Act of 1990.

The experience of working on the Hobbs case soured Stapp on the legal process. He felt defiled by the defense's efforts to discredit his research and his reputation. The tactics may have been nothing new, but Stapp had expected better. He came away from the experience believing that the courts willfully ignored valid scientific evidence and allowed skilled, unscrupulous attorneys and their witnesses to misrepresent the facts through sophistry and deceptive judicial theater. The trial embodied everything Stapp had come to hate about Washington, DC.

He wanted out.

26

WHO WANTS A RESURRECTION, ANYWAY?

Sometimes I feel beaten to death by a steady procession of
Decembers.

—*John Paul Stapp, to the journalist Ed Rees*

THE STAPPS' TIME in Washington had overlain the tortured administrations of Lyndon Johnson and Richard Nixon, massive antiwar protests
in the streets, rising crime rates, and, following the assassination of Dr.
Martin Luther King, Jr., aggrieved citizens trying to burn it to the ground.
Despite the friends they'd made and the myriad honors Stapp had received
there, even the curious ones, such as his enshrinement as a life-size figure
in the National Historical Wax Museum (alongside Charles Lindbergh and
Alan Shepard), they would be glad to escape the bitter winters and muggy
summers that had once caused Stapp to muse: "The Capital should be at Colorado Springs, where the elevation would be too much for 80 year old Senators, which we can do without."

The sixty-two-year-old Dr. Stapp had spent seven years in Washington.
Since his childhood in Bahia, he had never lived in one place longer. Having
completed the term of his contract with the Department of Transportation,
and with health concerns beginning to weigh on him—he had lately been
experiencing periods of blurry vision and labored breathing after climbing
just a few stairs—he decided to make a change. In January 1973, the Stapps

put their Fairfax house on the market and left "the District of Corruption"—as they liked to call it—for good.

They might well have returned to Texas, either to the Stapp family environs in the Hill Country, perhaps to Austin—that "most interesting place"—or more likely to the familiarity of San Antonio and proximity to Wilford and Margaret. Instead, they chose the high desert of Alamogordo, where so much of Stapp's best work had been accomplished and where he was still beloved. A place where, as he said, the highways were toll-less and the streets were free of muggers and parking meters. Where the Sacramento Mountains towered over the Tularosa Basin and the sun shone almost every day of the year. Yet retirement for a man of constant motion would take some practice. One early morning Lili found Stapp out busily weeding the garden, on his hands and knees in a sport coat, well-cinched necktie, and wingtips.

Shortly after settling into a new ranch-style house in a prosperous Alamogordo neighborhood, the Stapps got a call from an information officer Stapp knew at Holloman with the news that Joe Kittinger had been confirmed alive and was believed held prisoner in the Hoa Lo prison camp, known to the public as the Hanoi Hilton. National Security Advisor Henry Kissinger and Vietnamese Communist Party chief Le Duc Tho had just signed the Paris Peace Accords, and there was some reason to believe, the information said, that release of at least some of the prisoners might be imminent. Had Stapp been a praying man, he said, he would have prayed a whopper.

In late February, the first group of the Hanoi POWs was released, and on March 23, John Paul and Lili got a message saying that a second group had come out, and that Kittinger was one of them. The prisoners would go first to the Philippines for medical tests and debriefing, and would be there for a week before coming back to the United States. Stapp made calls to those who knew Joe to spread the news. Once Kittinger was back on U.S. soil, in Florida where his parents lived, Stapp caught up with him by phone. Joe told him the story. He'd been hit by an air-to-air missile and had ejected in excess of Mach 1 at about 18,000 feet. The ejection seat had worked perfectly, as had the chute, and Kittinger had landed in a rice farm with a gash in his leg but otherwise in good shape. He'd been captured immediately, then imprisoned and tortured. It had been a very hard ten months, he told Stapp, but said he

was alive at all only because of the escape system work Stapp and his colleagues had done at the Aero Med Lab, and at Muroc and Holloman. He'd
been waiting all this time, he said, to say thanks.

Despite the peace accords and prisoner releases, fighting continued in
Vietnam. Stapp kept up with events through friends and former colleagues,
and occasionally drove out to the base with Lili for dinner at the officers'
club, where he could catch up on the news and be fawned over by the staff
and the regulars. For almost the first time in his life, John Paul Stapp had
the luxury of spare time, although only a little considering he hardly cut
back his travel. While the time between trips was now mostly his own, he
still racked up a quarter million air miles a year and continued to respond to
nearly every invitation to address conventions; local, national and international organizations; classrooms and science fairs; award ceremonies; media
interviews—few of which paid more than just a stipend.

The Stapps had not gotten wealthy in Washington and, unlike some
of his colleagues, Stapp had never filed for a patent or taken a share in
any of the enterprises he'd worked on during his time with the Air Force.
He claimed to have turned down any number of lucrative private sector
positions and had rejected appeals to endorse products and companies,
including—according to Lili, who kept close track of such things—both
General Motors and Ford. The Liberty Mutual Insurance Company, a pro-
gressive voice for auto safety regulation over the years, also made him a very
tempting offer to serve as a consultant. Stapp never seriously considered it.
Regardless of the snarky suspicions of the SOS crowd, he'd never had any
intention of seeking restitution for injuries suffered in the line of duty. Sue
the Air Force for allowing him to run his projects? The idea was offensive
to him. He had never, in fact, collected a single dollar of hazardous-duty
pay. His base salary the final year of his service at DOT was $35,612—good
wages, considering that the median household income for Americans at
that time was only $8,500—but much less than what he might have com-
manded in private medical practice. His wants, however, were always modest and cash had never been much of a motivator. "Money just didn't enter
into Paul's life at all," said Margaret Stapp.

Not that he didn't intend to keep earning it. Shortly after arriving back
in New Mexico, Stapp accepted a position as an adjunct professor on the fac-

ulty of the University of Southern California beginning that same fall. The
position required Stapp to be on campus in Los Angeles one evening a week
for a three-hour session. The schedule suited him. He could drive to El Paso,
catch a plane for L.A., teach his classes, and return to Alamogordo the following morning. He felt very comfortable in the classroom and his courses
quickly became popular. Stapp continued to teach at USC through the 1976–
77 school year.

In 1979, Dr. Stapp was inducted into the International Space Hall of Fame,
the brainchild of a former local mayor who reasoned that if Cooperstown, New
York, could put itself on the map with the Baseball Hall of Fame, Alamogordo
surely ought to be able to claim its own measure of cultural relevance with
a world-class aerospace museum. The striking bronze-cube architecture was
set prominently in the rugged foothills on the eastern outskirts of town and
a five-minute drive from the Stapp house on suburban Rockwood Street. Dr.
Stapp would become the acting chairman of the Space Center Commission
and a familiar face around the museum. An outdoor exhibit was christened
the John P. Stapp Air and Space Park, and it featured the original red and
white *Sonic Wind* rocket sled.

After his years of battles within the Air Force and with the auto industry,
Stapp was finally able to bask a bit in the respect of his chosen community.
"It was all *his* out there," his brother Wilford said. "They made a lot of fuss
over him and made him feel at home." The Stapps' closest friends in town
were Monroe and Marion Curtis. Monroe Curtis had been a range officer
at Holloman during Stapp's tenure as head of the Aero Medical Field Lab,
and later worked at Holloman on NASA research projects. The couples spent
many evenings together and the Curtises came to understand John Paul and
Lili as well as anyone ever had, and as time went on would come to see themselves as guardians of the Stapps' privacy.

When he was in town, Stapp could often be found in the late afternoon
having a snack at Opal's Bar, sitting in the corner scribbling puns and poems
and swapping stories with friends who would stop by on their way home
from work. Tourists were always showing up unannounced at his house. To
Lili's great annoyance, Stapp would invite them inside and sit at the kitchen
table talking and laughing for as long as they cared to stay.

One subject his visitors regularly asked Stapp to address was the growing controversy about whether the debris discovered near Roswell just across the mountains to the east more than thirty years earlier had really been the remnants of a Project Mogul balloon or the crash site of a flying saucer. In 1980, Charles Berlitz and William Moore published a book titled *The Roswell Incident* claiming that the Air Force had taken custody of alien beings and gone to great lengths to keep the story hushed up. Charges of conspiracy became fare for the tabloids: "ROSWELL: WHAT REALLY HAPPENED! FLYING SAUCER MYSTERY! UFOS ARE REAL AND THE GOVERNMENT KNOWS IT!" Stapp rarely offered much in the way of comment about the hubbub that eventually turned sleepy little Roswell into Alien Central, complete with tacky gift shops and a UFO museum. The attention was a bonanza for the tourist trade in New Mexico, one of America's poorest states, and Stapp refused to make any public statements that might undercut it. He even turned down requests to be interviewed on the record by the Air Force as part of its attempt to corral the conspiracy theories. After all, he told a friend, the work they'd been doing with dummies and balloons back in the 1940s and 1950s had been part of secret Air Force projects, and he saw no reason to discuss any of it. Besides, he admitted, all the silliness gave him a certain perverse pleasure. "Error," he'd once jotted on the back of a report he'd been sent for review, "survives on the insatiable human craving for unreality."

While there may have been topics Stapp didn't want to discuss, one thing he never minded talking about was the Big Run. It was the high point of his field research career, though he was always quick to point out that the land speed record had never been his goal. Then, on October 4, 1983, Scotsman Richard Noble knocked Stapp out of the record books. Noble's high-tech Thrust2 car, powered by a single Rolls-Royce jet engine, hit 650.88 miles per hour on the great playa of the Black Rock Desert north of Reno, Nevada, giving the world a new fastest man—though it had taken nearly thirty years to surpass *Sonic Wind*. Stapp remarked with a chuckle that he wondered how Noble would have felt about going from his top speed to zero in less than a second and a half.

Stapp's own speed days, of course, were long past. His brother Robert had just gone through open heart surgery, and John Paul's own health had begun to worry him. He had put his body through a punishing regimen during the years from 1947 through 1954, and now he made a game of attributing spe-

When his tenure with the Department of Transportation was up, Stapp
moved back to New Mexico. He loved out-of-the-way Alamogordo and
the majesty of the White Sands. As Wilford Stapp observed, "It was all
his out there." (Photo courtesy of Wilford Stapp)

cific aches and pains to particular sled runs. To make matters worse, he was
overweight and out of shape. He bought a stationary bicycle and did his best
to follow a regular exercise routine, but it was a challenge. He wrote to his
longtime friend Red Lombard: "Old age is just plain brutal and all the idiotic
consolation nonsense is just flowers for the funeral—who wants a resurrec-
tion, anyway?"

Stapp remained energized during this period, however, by discussions
with neuroscientists and doctors who were making advances in the field of
brain impact research. Friends, such as Dr. Keith Jamieson, an eminent neu-
rosurgeon who had agreed to store some of Stapp's controversial animal data
at his own facility in Los Angeles, kept him up to date. Particularly exciting
were reports from Michoud, Louisiana, where the Naval Biodynamics Lab
continued its work on impact and deceleration tolerances. Funding from
the Office of Naval Research had made it possible for the NBDL researchers
to procure a cohort of chimpanzees, baboons, and rhesus monkeys—along
with some 300 human volunteers—for an exhaustive battery of impact
experiments. The test subjects, whether human or primate, were thoroughly
instrumented and data was collected from the latest in accelerometer arrays,
physiological recording devices, and high-speed motion picture cameras.
The lab was able to run both horizontal and vertical accelerations, and to

expose subjects to frontal, lateral, oblique, and axial impact orientations. The goal, at a relatively modest cost of between 4.5 and 5 million dollars a year, was to amass a definitive data set of head and spine impact research for use by those designing vehicles, restraints, helmets, and other protective devices for the purpose of reducing human injury and death—specifically injury and death on the battlefield.

However, auto safety test labs such as Dynamic Science in Phoenix, where much of the original testing conducted during Stapp's tenure at NHTSA had been carried out, were already under heavy political pressure by this point. The Ronald Reagan administration, which took control of the DOT in 1981, was very friendly with the auto industry. According to Mark Pozzi, a principal researcher at Dynamic Science: "The Reagan people gutted all the independent test labs." When considering the history of federal auto safety efforts, there is Before Reagan and After Reagan.

Then, an incident 1,500 miles away from New Orleans changed the landscape of what would be possible for head and spine impact research in the United States from that point forward. In May of 1984, a group calling itself the Animal Liberation Front staged a midnight raid on the University of Pennsylvania's Head Injury Clinic in Philadelphia, trashing the lab and stealing some sixty hours of film documenting the clinic's research using primate subjects. The films were handed over to People for the Ethical Treatment of Animals (PETA), which edited them to produce a twenty-six-minute documentary called *Unnecessary Fuss*. The title came from the clinic's director, Dr. Thomas Gennarelli, who claimed that what he saw as hysterical reaction to legitimate research was really much ado about nothing. PETA distributed its film to sympathetic members of Congress and to news outlets. The ensuing public uproar resulted in withdrawal of National Institute of Health funding for the research, and the university disbanding the project and shuttering the lab.

The incident served as a flashpoint for organized resistance to live animal testing of all kinds. In the months following, more than a dozen lab breakins occurred nationwide. It became more and more difficult for researchers to obtain funding for or to conduct live testing—on human volunteers or on animals. Not long after the outcry from the Pennsylvania incident, many key scientific personnel were transferred from or voluntarily left the Naval test lab.

No matter how circumspect Stapp and the Air Force had been through the years with regard to their animal research, it had been a constant public relations battle. The drumbeat of opposition affected Stapp, particularly in his later years. Animal welfare groups had hectored him and his colleagues from the earliest days—certainly well before courts and some national governments began to assert that animals might properly and legally be regarded as persons, with all the attendant rights that implies—and especially heavy pressure had been directed at the animal facilities at Holloman.

Complaints directed at his animal work had always struck Stapp as sentimentally naïve at best. His attitude had been shaped during his childhood in the equatorial jungles of Brazil, where both flora and fauna—"I was taught the value," he said, "of human survival against our plant and animal antagonists"—had represented tangible threats. Stapp believed in the tradition of American philosopher and social reformer John Dewey, that animal testing was not only morally defensible but a moral requirement. Dewey wrote: "Scientific men are under definite obligation to experiment upon animals so far as that is the alternative to random and possibly harmful experimentation upon human beings ... Such experimentation is more than a right; it is a duty."

Stapp maintained that all the major medical discoveries of the twentieth century had been made possible by animal testing, and one of his favorite examples was the polio vaccine. More than 9,000 monkeys and 150 chimp subjects, along with 133 human volunteers, had participated in medical trials. Each of those test subjects, Stapp argued, had been critical to the process. The result? Near-eradication of a crippling disease.

Through the years, Stapp and his colleagues had received everything from concerned inquiries from humane societies and the ASPCA, to harassment and even threats from more activist groups such as PETA and In Defense of Animals, and finally from militant organizations such as the Animal Liberation Front. According to friend and colleague Mark Pozzi, Stapp was "deathly afraid of these threats from the animal rights crowd." Lili confirmed that they'd received harassing phone calls and even, on occasion, taunts and heckling from strangers when they appeared in public. The attention paid to animal testing only increased during Stapp's final years, as the Air Force attempted to transition its 10-million-dollar state-of-the-art primate research facility at Holloman, along with some 140 chimps, to private ownership.

Like the auto companies, the U.S. military wanted out of the animal testing business. The scrutiny wasn't worth it. International chimpanzee expert Jane Goodall went to Alamogordo to examine the animal facilities and draw attention to them, but was denied access by the Air Force. A California-based group calling itself Last Chance for Animals erected a billboard in Alamogordo offering a $100,000 reward "leading to the conviction of animal researchers for animal cruelty." Stapp couldn't help feeling that the attacks were aimed squarely at him. He accused his detractors of "zoophilia, the human race's worst perversion."

. . .

In 1984, with his vision deteriorating, Stapp underwent cataract surgery on his right eye and had a plastic lens implanted. It was the same eye he'd been worried about losing following his final sled run at Edwards. Though it was not a life-threatening situation, the operation caused Stapp to contemplate his own mortality. Lili recalled him talking for the first time about the benefits she could expect from the Air Force when he was gone. Then, just a few weeks after the cataract procedure, death claimed the first of the Stapp brothers. John Paul's relationship with Robert Stapp had been strained since their Texas days, brought on in part by Robert's jealousy of his more successful brothers. But when Robert died in Kansas City that April from a massive brain hemorrhage, it came as both shock and warning. In one of Robert's last letters to his older brother, he had complained of the difficulty of making any money selling his artwork. "So I now sit out at the edge of the street with my feet in the ditch," he wrote, "and watch for the postman to bring me my gove'ment check."

Though time was hacking away at his own body, Dr. Stapp's psychic energy seemed hardly to flag at all. Stapp got some of his best ideas in hardware stores. "He had a real passion for hardware stores," Margaret Stapp recalled, almost mystified. "We would go with him and sometimes practically spend the day just because that's where he was. You wanted to be around him." He would wander the aisles and dream. Monorail electric trains; climate-controlled cities; fly-by-wire systems for the nation's airlines;

the employment of unused underground cavities, laterally connected to subway tunnels, to grow vegetables and fruit using artificial ultraviolet lighting systems powered by the energy of the moving trains. In the summer of 1984, Stapp hired Charles Barr, a longtime friend and former head of PR at Northrop, as his publicity agent to help find appropriate outlets for his steady current of ideas.

Despite Barr's attempts to interest the commercial world, most of Stapp's offbeat and often whimsical ideas for film and literary projects came to not much at all, but according to those who knew him at that time he was always working on something. Throughout his life he'd kept up the habit, learned from his mother, of creating lists of new words (zebrule, pinguid, apodictic, xanthrochroid) and attempting to incorporate them into his conversation and writings. One of his favorites was "periclitate"—meaning "to expose to danger"—and he enjoyed confessing that he was a self-periclitator. One of his projects was *Dr. Stapp's Almanac and Rational Calendar*, which collected decades of Stapp ephemera, the epigrams, puns, jingles, and haikus he'd been scribbling on cocktail napkins and sales receipts since his college days, and organized it into a replacement for the—in Stapp's view—irrational Gregorian calendar with its variable-length months and unpredictable day-date relationships.

Dr. Stapp's calendar is divided into thirteen months, with each month composed of four seven-day weeks. The first day of each month is a Saturday, the last day a Friday. In order to get a 365-day year, December 29—or New Year's Eve—has no weekday name, coming between Friday the 28th and Saturday, January 1. The thirteenth month occurs between June and July, and is called Midyear. Birthdays and holidays always fall on the same weekday. His system, Stapp claimed, produces a calendar "as rational as your watch."

Each day of the year is paired with a Stappism—mostly corny, occasionally revealing:

- Wednesday, February 19: "A pat on the back frequently is but searching for a place to stab."
- Monday, March 3: "Talent: Consolation prize for all the things you are not."
- Thursday, May 13: "Corruption is the manure of prosperity."

- Wednesday, Midyear 19: "Aging: For a woman, waking with that morning stiffness; for a man, waking without it!"
- Thursday, August 6: "Dignity is a fat old man too lazy to walk fast."
- Wednesday, November 26: "If you are being run out of town, get in front of the crowd and make it look like a parade."

It is almost as if Stapp was in a rush to get his ideas down on paper before it was too late. He alternated between short spells of depression and longer periods of intense creativity. In addition to occasional consulting work on car crash investigations, he maintained some contacts at NASA. By the early 1980s, the U.S. space program, with its emphasis on international cooperation and research, had already realized the gist of Stapp's vision for it, though the shuttle flights had by that point already come to seem almost routine to the American public. Launches on live TV played to dwindling audiences. Interest picked up in 1983 when Sally Ride became the first American woman in space, and again in the summer of 1985 when a biochemist from Oklahoma, Dr. Shannon Wells Lucid, made her first space flight aboard the shuttle *Discovery*. Dr. Stapp paid special attention and felt a special pride, and not just because he had helped pave the way for human space missions. He and Lucid were cousins. Her parents had been Baptist missionaries in China, where she was born. The name Shannon came from Stapp's mother's side of the family. Lucid went on to participate in five space flights, and Stapp—who liked her tremendously—made an effort to be at Edwards Air Force Base for the landings of each of the shuttles that brought her back. "Had I but one of my nine lives remaining," he wrote her with a flourish, "I would be honored to render it to you as a spare to take with you on your next perilous journey to advance the civilization and science of Planet Earth."

Stapp and family suffered another blow when, on September 25, 1985, his brother Celso died in El Paso at the age of seventy-one. The brothers from Bahia who had both become doctors as well as colonels in the United States Air Force had been each other's biggest supporters as often as they'd been rivals and tormenters. Those who were close to him could see it was a gut punch for John Paul to lose a second sibling in the space of eighteen months. His blood connection to the family roots in Brazil now rested solely with Wilford.

Feeling unmoored, Stapp busied himself that fall on a paper for the Stapp Car Crash Conference. He used case studies to illustrate the relationship between vehicle seats and restraint systems, and offered specific recommendations for NHTSA and the auto industry that, the paper claimed, had the potential to save 10,000 lives a year. As permanent chairman of the Car Crash organization, Stapp continued to attend advisory committee meetings and the annual conference, scribbling notes and asking anyone who'd listen why they were continuing to debate issues that in his mind had been resolved back in the 1940s and 1950s.

On January 28, 1986, the perception of space flight as routine was erased in an instant when the shuttle *Challenger* exploded seventy-three seconds after takeoff. One of those on board, Sharon Christa McAuliffe, a New Hampshire educator, had been part of NASA's Teacher in Space Program and had attracted the attention of hundreds of thousands of students worldwide, who watched the agonizing drama on live television. Reports estimated that some 85 percent of Americans had either seen the broadcast or heard about it within an hour of the event. Stapp had been privately critical of NASA's decision not to include an escape system for shuttle astronauts, and spoke bitterly about the tragedy to friends when it was learned that at least some of the crew had survived the spectacular fireball and were alive when the hardened crew compartment hit the Atlantic Ocean.

Engineers working for contractor Morton Thiokol had expressed concern in the days leading up to the launch that forecasted low temperatures for the central Florida coast could affect the integrity of rubber O-rings meant to seal the joints of the rocket boosters, and recommended delaying the launch. They were ignored by NASA managers. Joe Kittinger, who had become one of the leading advocates for aircrew escape technology, his own life having been saved twice by ejection seat systems, was outspoken: "The *Challenger* tragedy was the result of a lot of selfish political people who just didn't give a shit." Like Stapp, Kittinger wanted to know why an escape hatch and the Beaupre multistage parachute had not been provided to the crew.

As an odd footnote to this phase of American space exploration, Hubertus Strughold died in September 1986, just eight months after the *Challenger* disaster and just as a U.S. Justice Department's Office of Special Investigations examination of his wartime activities was ramping up. Previously suppressed Nuremberg documents had surfaced, showing Strughold on a list of individuals who had been sought for war crimes. When Strughold died, the government dropped the investigation, but his reputation never recovered, and recognition of his contribution to the space effort would be increasingly downplayed in the official histories.

Meanwhile, Shannon Lucid's final shuttle mission in 1989, which helped NASA with its own reputation problem in the post-*Challenger* period, included a grueling 179-day stay aboard *Mir*, the cramped Russian space station, in which, at age fifty-three, she carried out hundreds of experiments of her own design, definitively revealing John Paul Stapp's one-time objection to female astronauts for what it was: simple prejudice. Lili Stapp enjoyed reminding her husband of this fact, and brought it up for discussion on the occasion of each of Lucid's historic missions.

. . .

Stapp had just turned eighty-one in 1991 when he was invited to the White House to receive the National Medal of Technology from President George H. W. Bush in a Rose Garden ceremony. Lili clipped newspaper and magazine articles about the event, made copies at the Alamogordo library, and mailed them to acquaintances. Stapp considered the technology medal his greatest honor, though he groused about the Desert Storm invasion of Iraq that had begun that January, and about the fact that the federal government had been unwilling to cover his travel expenses to and from Washington for the ceremony. To defray costs, he had carted along a case of his Rational Calendars and wandered the halls of the Capitol selling them at ten bucks a pop.

He had more projects in the works. He titled his next volume, completed in 1992, *For Your Moments of Inertia—From Levity to Gravity: A Treatise Celebrating Your Right to Laugh*. It was another collection of Stapp puns, doggerel, limericks, and haikus, and was a case in point of the Stapp contradiction: deadly serious man of science vs. the playful—even frivolous, perhaps even ridiculous—wordsmith. To most, it was evidence of the breadth of Stapp's interests and the offbeat sense of humor that all who knew him understood to be a core component of his character. To others, those who encountered him as an old man at one of the Car Crash Conferences sitting alone beside a table with stacks of his self-published calendars and books for sale, it was sad—one of the great minds of his time reduced to hawking haikus. Wilford Stapp relished these contradictions in his brother. "Paul was a *very* complex person. And yet, he had almost a simplistic side of him that came out in his humor. He felt profoundly about a lot of things that he didn't talk a whole lot about. That's why he made jokes and all that. He didn't always want to talk about the serious part."

"In Literature," Stapp wrote in the preface to *For Your Moments of Inertia*, "contrary to the laws of physics, levity outweighs gravity." Laughter was therapeutic, Stapp had always told patients of the Curbstone Clinic. Life was already top-heavy with tragedy and folly. Stapp once cracked up an Albuquerque restaurant when his waitress asked if a bloody beefsteak was perhaps *too* rare. "Tie off the bleeders and prep the O.R.," Stapp had shrieked. "By God, we can save this steer!"

In September 1995, Stapp learned that the Hubertus Strughold Aeromedical Library at Brooks Air Force Base was going to be rechristened and that Strughold's name was to be removed from the facade of the building. The Anti-Defamation League issued a press release: "Paying tribute to Dr. Strughold was an obscene mockery of the pain and death suffered by his victims." Shortly after the Air Force took his name off their library, his likeness was removed from a stained glass mural, *The World History of Medicine*, on the Ohio State University campus; Strughold's image had been etched into the glass alongside such luminaries as Hippocrates and Marie Curie. A German historical committee would later uncover proof of unethical experiments conducted on-site at Strughold's Berlin institute during the war. Epileptic children had been placed in vacuum chambers for the study of the effects of high-altitude sickness and hypoxia. Stapp had long spoken of Strughold with contempt, but in his later years he would spit the name out bitterly whenever he felt required to say it.

· · ·

Stapp had always pushed hard to keep his eponymous auto safety conference operating on, as he liked to say, a "free will basis," encouraging uninhibited discussion and staying strictly independent of special interests, especially those of the auto industry. That independence was, in fact, the whole point. "We have quietly discouraged resolutions, manifestos, constitutions, or any other commitments to any position pro or con," he had written back in 1964. "No party lines or propaganda need be subscribed to. We only seek the truth about vehicle occupant safety and protection... The Industry loves cars but appears to hate the people that buy them!" But the SAE had officially assumed administration of the conference by 1966, as the gathering continued to set new attendance records each year, and the truth was that the conference's advisory committee was, by the 1990s, almost totally dominated by Detroit interests.

In order to support the efforts of researchers around the country who were doing what Stapp judged to be good vehicle safety research, he had founded a public interest organization he called the New Mexico Research Institute (NMRI) in the early nineties. One of the first projects the NMRI took on was to help the research community gain access to the impact acceleration data collection built at the Naval Biodynamics Lab. The NBDL was finally closed for good in 1996; experimentation ended and the facility was dismantled. Still, Stapp publicly endorsed the research and pronounced the data "outstanding, statistically valid results applicable to military and civilian ground, sea, air and space transportation requirements for impact injury protection" and insisted it was "vital to Legal, medical and insurance issues." Though only a small fraction of the data had ever been analyzed, Stapp judged the promise of future advances based on it to be great, and he wanted to do everything within his power to support it, which is why he agreed to write the introduction to the NBDL's final report on its work at the Michoud facility.

Even as he lent his credibility to the NBDL researchers, Stapp found himself fighting a rearguard action against his own safety conference. In 1997, whether because Stapp's aging and poor health had reduced his effectiveness or because his voice was no longer welcome, the Advisory Committee of the Stapp Car Crash Conference ousted Stapp, effectively removing the last bit of influence he retained over the organization that bore his name. A few years earlier, the executive advisory committee had begun to feud with SAE over operational issues and by 1999, the Stapp Foundation, which is associated with the University of Michigan Transportation Research Institute (UMTRI), had wrested control from the SAE. Those interests would always be indifferent—if not hostile—to independent voices, eventually even those of such eminent researchers as Dr. Richard Snyder, who would resign from the advisory committee in protest in 2003, and even, in the end, to Dr. John Paul Stapp himself, who was expressing in private his disappointment at what the conference had become.

"Couldn't they have waited?" he asked a friend after receiving the news of his excommunication. "I'm an old man. I'll be dead soon, and then they can do whatever they want."

To Lili, Stapp seemed defeated, but he rallied later that summer when he received an invitation to sponsor Joe Kittinger at Kittinger's induction into the Aviation Hall of Fame in Dayton. The two men had enjoyed a long-time mutual admiration society. Kittinger had not only been Stapp's most trusted pilot, he was also the man who'd agreed without complaint to forgo a potential high-profile roster spot on the Mercury astronaut team in order to complete Project Excelsior. Joe's admiration for Stapp was unbounded, and his personal affection for the man just old enough to be his father ran deep,

Following their retirement from the U.S. Air Force, John Paul Stapp and Joe Kittinger, charter members of the Survivors Club, recall the hero days at Holloman when they regularly pursued activities "not conducive to longevity." (Photo courtesy of Joe Kittinger)

despite their fundamental differences: Stapp the intellectual, connoisseur
of the ballet and opera, worldly man of science; Kittinger the beer-drinking,
country music-loving, good ol' boy fighter pilot who christened his inner
circle the Redneck Squadron. In his induction speech on behalf of Kittinger,
Stapp hit on the one thing that may have most effectively connected them:
they had both, Stapp reminded the assembly in Dayton that night, achieved
advanced age (they were eighty-six and sixty-nine, respectively) despite lives
pursuing activities "not conducive to longevity." Charter members of the Survivors Club, brothers in man-made-miracle.

One of the things they talked—and laughed—about in Dayton that night
was the conclusion of a bizarre, episodic drama they'd both been a part of:
the so-called Roswell Incident(s). The Air Force had finally wrapped up its
internal investigation three years earlier, and, in 1997, shortly before the Hall
of Fame ceremony, published *The Roswell Report: Case Closed*. With affidavits
by Kittinger, Duke Gildenberg, and Dan Fulgham—the captain injured in the
Project Excelsior balloon training flight crash landing who'd been mistaken
for an extraterrestrial visitor—the Air Force made its case that the incident
had resulted from simple misunderstandings surrounding two secret balloon projects, Mogul and Excelsior. Kittinger's sworn statement, included in
the report, concluded with the words: "I am not part of any conspiracy to
withhold or provide misleading information to the United States Government or the American public." Though Stapp had resolutely refused to make
any statement of his own on the grounds that the Roswell story was good for
New Mexico, he never ceased to enjoy the spectacle.

By the following year, 1998, Stapp's health had deteriorated further—
he sometimes gasped just to catch his breath—but he continued to travel.
In September of that year, while he was in Los Angeles for the Society of
Experimental Test Pilots' annual symposium, he paid an impromptu visit
to the Simon Wiesenthal Center, headquarters of the Jewish human rights
organization dedicated to preserving the evidence and teaching the lessons
of the Nazi Holocaust. The untranslated German aeromedical documents
Stapp had secretly—and, he assumed, illegally—kept since 1946 had been
on his mind as he contemplated his own end, and he'd been trying to decide
their fate. He did not want the documents to be discovered among his possessions after his death, fearing that they might be used by the Air Force as

an excuse to deny Lili full military benefits, but he was unwilling to simply destroy them.

He had no contacts there, but Stapp inquired at the Wiesenthal Center whether he could talk to someone about his knowledge of what he believed were the war crimes of Hubertus Strughold and others. No one there, however, knew who Stapp was and the officials did not take him seriously, which might have had something to do with his overall disheveled appearance, thanks to a bad limp and some unseemly weight gain. For whatever reason, he was politely rebuffed.

When Stapp returned to Alamogordo that fall, he sent several boxes containing the Nazi documents to his friend Dana Kilanowski with instructions to donate their contents to the Wiesenthal Center upon his death. Stapp was confident he could trust Kilanowski, and yet the historian did not want the documents in her home. She had leafed through them briefly and been sufficiently horrified that she quickly boxed them back up and sealed them. The next day, she drove the boxes to a military documents archive and arranged to have them stored.

By that point, John Paul Stapp had been diagnosed with prostate cancer. It was not an especially unusual diagnosis for a man of eighty-eight, but it was one more piece of evidence that the days remaining to him would not be easy ones. Despite his health worries—limping as he lugged his satchel bulging with files and technical reports, diabetic and struggling just to breathe—Dr. Stapp gamely traveled to Detroit in March of 1999 to attend the SAE International Congress and Exposition. Some of the young automotive engineers snickered behind his back. "Too many sled rides," was what they'd tell one another, having no real conception of what Stapp had put himself through for the cause of safety before they were born.

Following one of the conference sessions at Cobo Center, Stapp lost his balance and stumbled on an escalator. Later that same day, he slipped and took a bad fall on an icy sidewalk on his way out. Friends who heard about the incidents were appalled to learn that Stapp had been offered no personal assistance—by the SAE or anyone else—while he was in Detroit. The American automobile industry had left John Paul Stapp and what he represented on the roadside, as if it tolerated his presence, but little more. He ate alone and often sat in the back of the room by himself as he listened to the conference speakers. When he returned to Alamogordo, Lili was shocked to see him covered with bruises. Stapp's close friend Mark Pozzi could hardly believe it. "He looked like he'd fallen off a motorcycle at 60 miles per hour." "I've been through worse," Stapp told them. "I'm fine."

Not much later, Stapp was hit by the first of a series of heart attacks while stacking boxes of files in his garage. He was rushed to a hospital in Las Cruces, accompanied by Monroe Curtis, who rode with him in the ambulance, and by Dr. Keith Jamieson, who flew in from L.A. Doctors discovered blood clots that had formed as a result of his Detroit bruises. Stapp's respiratory problems were also getting worse—he was diagnosed with emphysema at this same time and was told he had only 50 percent of normal lung capacity—exacerbated by the 4,300-foot altitude of the Tularosa Valley. He required supplemental oxygen even after being released.

By June, Stapp had decided to seek treatment in San Antonio, where it was easier for him to breathe. Lili accompanied him to Texas, and got him settled in a spare bedroom at Wilford and Margaret's house on Bluehill Road. After a few days, Lili returned to Alamogordo to attend to affairs there and Wilford drove his brother to medical appointments at the hospital at Lackland Air Force Base. In October, with Wilford accompanying him, John Paul attended the Forty-Third Stapp Car Crash Conference in San Diego. In a wheelchair and breathing from an oxygen cylinder, he sat through as many of the presentations as he could endure. But it was clear to everyone that he was in very bad shape.

One day, just weeks after returning to San Antonio, according to Wilford, John Paul woke up and announced that it was time to go home. No further explanation, but it chilled Wilford. He knew immediately what it meant, and knew it would be the last time he'd see his brother, the man he admired more than any other in the world. Stapp flew to El Paso, where he was met by Monroe Curtis and driven the 90 bleak miles to Alamogordo, to home, to Lili.

A celebrated soloist on stage, Lili had been the ultimate loner in life. At times, toward the end, she seemed not even to acknowledge who Stapp was or what he'd accomplished. The world of science and technology, and the byzantine traditions of military culture, bored her. Her obsessions were always the fine arts. In her later years, she would spend hours leafing through her books of Japanese paintings, chain-smoking in the dark, watching and rewatching videotapes of the world's great ballets and operas.

John Paul and Lili, while content for most of the years of their long marriage, had soured toward each other in old age. Stapp sometimes complained that he felt "a great loneliness" in his personal life. He had confided in Marion Curtis that the two great loves of his life were the young woman at Baylor who'd died in the Hollywood car crash—whose name he still refused to say, even to the Curtises—and Celia Richards, the beautiful redhead from Kerosene Flats. To Lili's great annoyance, Stapp had kept photos of Richards in the top drawer of his desk at home. He told the Curtises he'd never stopped loving her.

Early one morning a few days after his return from San Antonio, on November 13, 1999—it was still dark—Lili walked into the living room to find her husband pitched forward out of his favorite chair and lying face down on the rug. She stepped over him on her way to the kitchen and made herself breakfast. Friends had suspected for some time that Lili was suffering the early stages of dementia. She said later she thought he was asleep, which was why it had taken her nearly two more hours to call Monroe Curtis to say that John Paul Stapp was dead.

27

HONOR AND VIOLATION

> Our only lasting reward is a better world we leave behind us,
> since people's praise is as meaningless and fickle as their
> scorn.
>
> —*John Paul Stapp, Brooks Air Force Base oral history*

MONROE CURTIS RUSHED to the house and immediately dialed Keith Jamieson in L.A. for advice. Dr. Jamieson, who years earlier had coined the term "Stapp straps" for seatbelts, called the Holloman base medical center to arrange for an ambulance to pick up the body and deliver it to the hospital in Las Cruces, where Stapp could be legally declared dead. Jamieson also reported to the commander at Holloman that Stapp's house and garage contained sensitive material, including films of Air Force experiments using animal subjects. He knew because Stapp had shown the boxes to him only months earlier. Jamieson also alerted an FBI field agent in Albuquerque that Stapp had been in possession of confidential material that needed to be secured. Later that afternoon, Monroe Curtis phoned Jamieson to tell him that people were "all over Stapp's house" and that they were "taking things." It wasn't clear to Curtis who the people were.

Stapp had kept a small office on Canal Street on the north end of town where he stored his library of technical reports, along with some of the data still in his possession from his Muroc and Holloman experiments. The office also contained the official documents and the seal for the New

Mexico Research Institute. Mark Pozzi had been working with Stapp on an NMRI project that was to have relocated the Daisy Track to land Pozzi had purchased outside of Albuquerque as a component of a new impact research facility—they'd recruited George Nichols as a consultant—and happened to be in Alamogordo when he got word of Stapp's death. Pozzi tried to contact Lili by phone, but he couldn't get through. He wanted to make sure everyone remembered her husband's insistence that his body be autopsied. It had been important to Stapp. What he intended as his parting gift to the Cause would be his own tissues and bones, his organs and especially his brain. He had compiled his own detailed injury inventory. "I want them to take me apart," he'd told Pozzi. "Every nut and bolt." Per Stapp's instruction, each trauma site on his body was to be studied. According to Duke Gildenberg, Stapp suspected that he might have suffered long-term lung damage as a result of his deceleration tests, despite an extensive examination in 1984 that had detected no pulmonary abnormalities. Keith Jamieson, who—along with Monroe Curtis—had visited Stapp earlier in the year at the hospital in Las Cruces, recalls the attending physician there saying "Stapp's lungs were like nothing he'd ever seen," though he'd failed to elaborate on the damage he'd observed. According to Jamieson, the doctor wondered aloud whether the sled runs could have been responsible. Stapp himself had told a writer from the *Alamogordo Daily News* shortly before leaving for Texas in June that his diminished lung capacity was due to "the results from 29 rocket sled rides."

The morning after Stapp's death, as Pozzi drove over to see Lili, his pickup passed by the Canal Street office. What he saw shocked him. A large truck was backed up to the front door of the office. A few people were standing and watching as a crew wrapped whole file cabinets in sheets of Visqueen and dollied them up a ramp and onto the truck.

Pozzi, an officer of NMRI, asked the man who appeared to be in charge for identification. The man refused, declaring only that he had "authorization to take it all." Stapp's last will and testament, and the family trust documents, at least, contained no such authorization, bequeathing his estate to Lili, family members, and to the NMRI. Pozzi demanded that the men allow him to retrieve some of his own personal files that had been stored at the office, but was denied. "It was bizarre. They were basically stealing everything. But I was kind of in shock. I'd just lost my best friend." The office was cleaned out and the contents were hauled away.

Wayne Mattson, a retired military researcher who had been working on a history of Holloman Air Force Base from a desk at the New Mexico Museum of Space History, was another eyewitness. But Mattson's protests, like Pozzi's, were met with indifference. Pozzi remembers somebody, possibly Mattson, wondering aloud whether Stapp might have arranged to turn his papers over to the Car Crash Conference organization and speculating that the movers were probably going to "haul it all off to Michigan."

Curtis would later send a box of Stapp materials that had somehow escaped attention to an ex-Air Force officer friend, suggesting that they might be forwarded to the Stapp Foundation at the University of Michigan so that "it will not get into the hands of Animal Activists nor cause trouble for the Air Force or the Automobile Industry." There's no question the auto industry would have been interested in keeping Stapp's documentation on auto safety failures from falling into the hands of those who might publicize them or make them available to attorneys involved in litigation against automakers. As Pozzi observed: "John Stapp's integrity made him a dangerous man from the perspective of those companies." Nobody really knew for sure who the men with the truck were.

For fifteen years, the destination of those files remained a mystery. Inquiries made by crash researchers at UMTRI, where the Stapp Advisory Committee and Stapp Car Crash Conference were headquartered, met with denials that any of the Stapp files had ended up there. In fact, the fifty-four file cabinets taken from Stapp's Canal Street office had been in Ann Arbor all along. None of the members of the committee contacted by the author could say how the Stapp files got there, or under whose authorization, but Richard Chandler, a longtime member of the Advisory Committee who did a lot of very good work over the years compiling the definitive Stapp bibliography, confirmed their location. "The University of Michigan got what was nominally called Stapp's library," he said. "They put it into storage. It's all stacked in a room in Michigan. There's a lot of material. Big files. Technical reports."

Shortly before this book went to print, the Stapp Advisory Committee donated much of the Stapp material in its organization's possession to

Embry-Riddle Aeronautical University. Twenty-six large boxes were trans-ferred from the University of Michigan to Embry-Riddle's Prescott, Arizona, campus. An archivist at Embry-Riddle was told that the Stapp papers had been culled and that some materials had been discarded prior to shipping. It will remain for those who were aware of the contents of Stapp's office prior to his death to verify the integrity of this collection and to identify any absent items. The surviving material includes published papers and presentations, SAE committee correspondence, technical reports, and the extended works on biodynamics. Most importantly, it is also believed to include much of Stapp's original test data from both Edwards and Holloman: deceleration data, records of both human and animal experiments on rocket sleds and on the Daisy Track, and summaries of the data Stapp amassed in cooperation with Northrop. The surviving Stapp library is now accessible to the safety research community. Its ultimate value is unknown, but, like the head and neck data from the NBDL, its potential is multiplied by the fact that accelera-tion and impact research using human volunteers and animal subjects is no longer possible to fund or conduct.

It is John Paul Stapp's final contribution.

· · ·

Stapp's body was flown to San Antonio on Sunday, November 14, 1999. A USAF pathologist conducted an autopsy on Monday at Lackland Air Force Base. The cause of death was listed as arteriosclerotic cardiovascular dis-ease, the mechanism of death being "lethal arrhythmia," or heart attack.

In spite of the suspicions of Stapp and others that his sled rides had caused lung damage, the pathologist reported finding "no demonstrable evidence of sequelae of G-induced injuries." The rocket sled rides had done no long-term damage, at least none that could be conclusively attributed to them. The investigation uncovered "no evidence of acute or remote trauma." The brain and spinal cord were forwarded on to AFIP in Wash-ington, where multiple cross-sections of the brain were examined. Stapp's brain was and is still considered to be a very valuable specimen. Dr. Daniel Thomas refers to Stapp as "the index case for human impact experiments." Not everyone accepted the conclusions documented in the autopsy report. Keith Jamieson, for one, declared the report "very suspicious." He said:

"That they found nothing just seems incredible to me. Stapp had crappy lungs, and they found nothing?"

An obituary in the *New York Herald-Tribune*—one of hundreds in newspapers and magazines around the world—referred to Dr. Stapp as "a gentleman who can stop on a dime and give you 10 cents change." One news service headlined its obituary with mentions of the two phenomena Stapp would have least hoped to be associated with: "'FASTEST MAN ON EARTH,' COL. JOHN PAUL STAPP, DIES AT 89; HAD CONNECTIONS TO ROSWELL INCIDENT, MURPHY'S LAW." His favorite might well have been a simple homage that appeared in the English paper the *Guardian*: "HE PUT THE SEATBELT IN AMERICAN CARS."

A memorial service was held at First Presbyterian Church in San Antonio on November 22. There were—surprisingly, given the proximity to multiple Air Force bases—no uniformed military officers present, which was evidence to some of Stapp's friends and former colleagues of the disrespect with which the Air Force had sometimes treated one of its own heroes. Joe Kittinger attended, and delivered his own tribute to the fastest man on earth. Dr. Jamieson had taken the initiative to phone the Secretary of the Air Force to get an official statement that could be read aloud at the ceremony.

John Paul Stapp's remains were cremated, as he had wished, and interred at Fort Bliss National Cemetery in El Paso near those of his brother Celso.

28

THE STAPP YEARS

> It is for man to tame the chaos; on every side, whilst he lives, to
> scatter the seeds of science and of song.
>
> —*Ralph Waldo Emerson, "Uses of Great Men"*

O N THE MONDAY FOLLOWING the death of her husband of forty-one years, Lili
Stapp drove herself to the space history museum up the hill and placed a
dozen fresh-cut red roses on the seat of *Sonic Wind*. It was the only time she'd
ever been there alone, and her private tribute surprised and touched friends
who heard about it. Though she clearly reveled in the status that came with
the role of John Paul Stapp's wife, Lili had never seemed to take much real
interest in Stapp's work, and what interest she may have once held had faded
over the years.

Nevertheless, Lillian Lanese had represented the highest ideal of art to
Stapp, and that ideal was something to which he'd aspired all his life. She
provided a classical gravity that seems to have helped hold Stapp in his orbit
for four decades crowded with battle and temptation. Celso had once advised
his brother that what he needed, for the sake of his career, was a suitable
spouse on his arm. "You'll never make colonel without one," he'd insisted,
"and you'll never make general without a really good one." Though Lili was
miscast as an officer's wife, she'd been good for Stapp, and John Paul and Lili
had mostly been good for each other. Without him, she was lost. Only weeks

after leaving the flowers on the rocket sled, Lili was admitted to a rest home in Alamogordo, where she would wander the halls at night stealing cigarettes from the other residents. She then lived for a while with her niece in San Antonio, and was eventually moved again to a care facility there, where she died at the age of eighty-four in 2009.

On the same Monday that Lili left the roses, Dana Kilanowski retrieved the Fed Ex boxes containing the Nazi aeromedical documents and shipped them overnight to the Wiesenthal Center as Stapp had asked her to do. She was glad to be done with them. The following morning at precisely 10 a.m., the time the boxes with her return address and telephone number affixed

The centerpiece of John P. Stapp Air and Space Park at the New Mexico Museum of Space History. This is where Lili laid the roses on November 15, 1999. (Photo courtesy of New Mexico Museum of Space History)

were scheduled to arrive in Los Angeles, Kilanowski's phone rang. When she answered, she heard only dead air . . . and then a click and a dial tone. She wasn't sure what it meant, if anything. She never received any formal notification from anyone at the center that the documents had been received.

· · ·

Today the Holloman High Speed Test Track is 10 miles long and the test vehicles run at hypersonic speeds, the sleds levitating on invisible helium clouds to keep the steel rails from melting. In 2003, a Holloman team ran a 200-pound payload at 6,453 miles per hour. Experiments now routinely achieve velocities beyond Mach 4, and a number have exceeded Mach 10. But, notably, without living test subjects. The days of using the track to study human capacities are now the stuff of history, or of what some of the Air Force researchers at Holloman still refer to as "the Stapp years."

"The only thing I'll allow myself to boast about," Stapp once told a documentary filmmaker, "is getting this kind of a research program done, for as many years and with as many human subjects, without a single fatality or a single disabling injury. Or a single lawsuit!" He burst into a belly laugh, as if he'd put one over on the lawyers.

John Paul Stapp's final distinction, rare in the history of medical science, is his success in moving from applied research in aircraft, spacecraft, and ground vehicle safety to policy formation that drove both federal legislation and the implementation of mandates that helped remake a nation's approach to its own survival. His most far-reaching and profound contributions are surely his work on ground vehicle safety and his war against what he called the kinetic plague of the steel age.

He didn't win every battle. Following Stapp's exit from NHTSA, the agency lost much of its effectiveness. In the view of former Ford executive Dennis Gioia and other industry insiders, NHTSA had become, by the time of John Paul Stapp's death, little more than a consulting and advisory group. In Stapp's time, Gioia said, "they struck fear in the heart of the auto industry. They had the power to dictate, and they did." But due to inadequate funding, poor management, and industry-friendly policies, it didn't last. Over time, the role of driving auto safety improvements fell increasingly to the courts. The result is a dangerously unfinished patchwork of safety regulations. The most glaring safety defects in American cars nearly half a century after Stapp's time at NHTSA include latches (on seatbelts, doors, and hatchbacks) that fail in crash situations, air bags that fail to deploy properly or at all, cars that continue to be designed—decades after the Ford Pinto disasters—with fuel tanks positioned aft of the rear axle in the "crush zone," and vehicles with a propensity for rollover during sudden turns, such as occur when a driver swerves to avoid a hazard. Some of these vehicles have weak roofs that give way in rollover crashes and crush the occupants.

One defect, however, may be the most egregious of all, partly because the fix would, according to experts, be simple and inexpensive. Today, despite the strength of our seatbelts, we remain pathetically vulnerable in even a low-speed rear-end collision. Rear-end impacts represent, according to NHTSA, a full quarter of all car crashes and are the number one cause of quadriplegia in America. Even worse, if small children have been strapped into their protective cocoons in the rear seat, as automakers urge and laws require parents to make sure of, they are in danger of being crushed as the front seats give way and heavier occupants rocket backward. "The safety standards for seat backs are very, very low," says safety engineer Bill Muzzy, "even lower than those you have for your office chair. I duplicated the tests, and I had an aluminum lawn chair pass that test with a 10 percent safety factor."

Finally, the NBDL data—according to Dr. Daniel Thomas, who helped collect and organize it, "the only set of three-dimensional head-neck response data on volunteers, rhesus monkeys, and chimpanzees ever developed,"—remains sequestered and unavailable. It could have important implications for concussion research.

Despite all that remains undone, the Stapp years represent a demonstrable measure of victory in the fight for auto safety. In the year 1955, when Stapp hosted the first car crash conference at Holloman, American drivers logged about 606 billion annual vehicle miles. Total auto fatalities for that year were 36,688. By 2012, thanks to the efforts of Stapp and others, with tens of millions more in population and millions of additional cars and trucks on the roads, and with miles traveled increased by a factor of five to nearly three trillion, fatalities had actually *decreased* to 33,780. It's been suggested that Stapp was indirectly responsible for saving more lives than anyone in history. They were all the ghosts that never happened.

John Paul Stapp's delights and his curiosity were unbounded, but his professional passion, his Cause—as a doctor, as a scientist, as a man—was narrow and straight as a rail: the physical salvation of the living, breathing, ridiculous, unpredictable, and forever imperfect human being.

APPENDIX: HOW FAST?

FOUR DAYS AFTER John Paul Stapp's historic final rocket sled ride on *Sonic Wind*, he filed his weekly status report with his commander at Holloman Air Force Base. In that report he wrote: "At the time of burnout the velocity was roughly estimated at 549 knots (632 mph)." That estimate was quickly released to the press, and subsequent accounts have repeated that value as the top speed achieved by the fastest man on earth. But actual project measurements were later shown to have clocked the sled at 937 feet per second. Using the standard conversion factor, the calculation is as follows:

$$937 \text{ feet per second} \times .681818 = 638.863466 \text{ miles per hour}$$

This book uses the rounded value of 639 miles per hour (Mach 0.84).

Then, what about the claim that the Big Run exceeded the speed of a .45 caliber bullet? It depends on the type of gun and the type of bullet, but the muzzle velocity of a .45 ACP (Automatic Colt Pistol) round—at approximately 860 feet per second, or 586 miles per hour (Mach 0.77)—is significantly less than the top speed of *Sonic Wind*.

The fastest man on earth didn't break the speed of sound, but he *was* faster than a speeding bullet.

NOTES

John Paul Stapp's surviving professional library is housed at Embry-Riddle Aeronautical University's campus in Prescott, Arizona. Cataloguing and indexing of that material is in progress at the time of this writing. Stapp's personal papers, on loan from Wilford Stapp, are currently in the possession of the author; their permanent location is still to be determined. All references not otherwise sourced are contained in this latter collection.

Full source notes for all quotations, along with a complete bibliography, can be found at http://books.wwnorton.com/books/sonic-wind/.

CHAPTER 1: WHERE ARE MY CHILDREN TO-NIGHT?

Material on JPS's childhood in Brazil is based on his unfinished, unpublished memoir; the letters of JPS, Charles Stapp, and Mary Louise Stapp; interviews with JPS and Wilford Stapp; Wayne Mattson, "Fastest Man on Earth, Pioneer of Aerospace Medicine and Maverick of the Tularosa Basin," *New Mexico Space Journal*, Winter 2004, 3–15; E. Stanton, "John Paul Stapp," *Del Sol New Mexico*, December 1982; and an interview for an oral history of the Aerospace Medical Division, Brooks AFB (1984, uncompiled). Material on Baptist missionaries in Brazil benefits from various online sources, including "The Bagbys: Missionaries from Texas to South America," radio script, KWBU-FM, January 9, 2010. Material on the Stapp family comes from JPS's memoir and "1994 Stapp U.S. Directory." Material on JPS's days at Brownwood School, San Marcos Academy, Baylor University, Decatur College, the University of Texas, and the University of Minnesota comes primarily from JPS's memoir and letters.

CHAPTER 2: THE MOST INTERESTING PLACE

The material on JPS's early days in Texas and his time in Minneapolis is based on his memoir and on the letters of JPS, Charles Stapp, Mary Louise Stapp, and others; the author's interviews with JPS and Wilford Stapp; an interview for an oral history of the Aerospace Medical Division, Brooks AFB; and Dana Kilanowski's interview with Wilford Stapp.

CHAPTER 3: COLONEL TANK AND COLONEL GAS

Material on JPS's early military career is based on his memoir; letters to and from JPS; an interview for an oral history of the Aerospace Medical Division, Brooks AFB; and military and personal records in JPS's files.

CHAPTER 4: SCUM JOBS

The material in this chapter is based on JPS's memoir; letters to and from JPS; an interview for an oral history of the Aerospace Medical Division, Brooks AFB; an interview of JPS by John Bullard and T. A. Glasgow, June 19, 1978; and military and personal records in JPS's files.

CHAPTER 5: THE CITADEL OF AEROMEDICAL RESEARCH

Material on JPS's first months at Wright Field is based on his memoir; letters to and from JPS; Richard Chandler, "Project MX-981," 2001 John Paul Stapp Memorial Lecture (*Stapp Car Crash Journal* 45, November 2001); an interview for an oral history of the Aerospace Medical Division, Brooks AFB; and military and personal records in JPS's files.

CHAPTER 6: THE BATTLE OF MUROC

The material on JPS's years at Muroc/Edwards AFB is based on his memoir; letters to and from JPS; the author's interviews with Dana Kilanowski and Celia Richards Wilson; Dana Kilanowski's interview with George Nichols; Richard Wessel's interview with JPS; JPS, "Measurement for Survival"; an interview for an oral history of the Aerospace Medical Division, Brooks AFB; Nick Spark, "The Fastest Man on Earth," *Annals of Improbable Research,* September–October 2003, 4–24, and "Gee Whiz! How Colonel John Paul Stapp Set the Land Speed Record, Discovered Murphy's Law, and Might Have Saved Your Life," *Journal of the American Aviation Historical Society,* Fall 2003, 162–74; Chandler, "Project MX-981"; Chuck Yeager, *Yeager* (New York: Bantam, 1985); the author's conversations with individuals at Edwards AFB; Peter W. Merlin and Tony Moore, *X-Plane Crashes* (North Branch, MN: Specialty Press, 2008); Lauren Kessler, *The Happy Bottom Riding Club: The Life and Times of Pancho Barnes* (New York: Random House, 2000); the daily journal of Project MX-981; U.S. Air Force Flight Test Center History Office, *Edwards AFB . . . Then and Now* (California: Edwards Air Force Base, 2001); and military and personal records in JPS's files.

CHAPTER 7: OSCAR EIGHTBALL AND THE CURBSTONE CLINIC

In addition to the sources listed for chapter 6, the material in this chapter is based on Dana Kilanowski's interview with Wilford Stapp; and on "Fastest Man Alive: A Legend for his Pioneer Work," *Alamogordo Daily News,* December 7, 1987.

CHAPTER 8: STAPP'S FIRST RIDES

In addition to the sources listed for chapters 6 and 7, the material in this chapter is based on "RAAF Captures Flying Saucer on Ranch in Roswell Region," *Roswell Daily Record,* July 8, 1947.

CHAPTER 9: ANTICIPATING STRICT COMPLIANCE

In addition to the sources listed for chapters 7 and 8, the material in this chapter is based on an interview of JPS by Bullard and Glasgow; Shirley Thomas, *Men of Space,* vol. 1 (Philadelphia: Chilton, 1960); and JPS, "Engineers and the English Language."

CHAPTER 10: G-FORCE CHIMPS

In addition to the sources listed for chapters 6, 7, and 9, the material in this chapter is based on Daniel Ford, *Glen Edwards: The Diary of a Bomber Pilot* (Washington, DC: Smithsonian Institution Press, 1998), and Richard Chandler, "John Paul Stapp and Deceleration Research, Part III: Project 7850 and Other Research at Holloman Air Force Base," 2003 John Paul Stapp Memorial Lecture (*Stapp Car Crash Journal* 47, October 2003).

CHAPTER 11: BROTHER, WE ARE THE GOATS

In addition to the sources listed for chapters 7, 8, and 10, the material in this chapter is based on Spark, "The Fastest Man on Earth"; and various newspaper articles.

CHAPTER 12: EVICTED

In addition to the sources listed for chapters 7, 8, 10, and 11, the material in this chapter is based on the author's interview with Daniel Thomas; JPS's May 1966 congressional testimony; and official USAF Aero Medical Laboratory memoranda.

CHAPTER 13: SOLO ON LOVERS LANE

Material on JPS's second assignment at Wright-Patterson AFB and his early years at Holloman AFB is based on his memoir; letters to and from JPS; the author's interview with JPS; David Bushnell, *History of Research in Space Biology and Biodynamics at the Air Force Missile Development Center, Holloman Air Force Base, New Mexico, 1946–1958* (New Mexico: Holloman Air Force Base, 1959); Thomas, *Men of Space,* vol. 1; George F. Meeter, *The Holloman Story* (Albuquerque: University of New Mexico Press, 1967); "Biography of Colonel John Paul Stapp, USAF MC" (uncompiled USAF biography in JPS's personal papers); Jon Franklin and John Sutherland, *Guinea Pig Doctors: The Drama of Medical Research through Self-Experimentation* (New York: Morrow, 1984); Charles Coombs, *Survival in the Sky* (New York: Morrow, 1956); "The Fastest Man on Earth," *Time*, September 12, 1955; Ed Rees's dispatches for *Time* (in JPS's personal papers); Hugh DeHaven, "Mechanical Analysis of Survival in Falls from Heights of 50–100 Feet," *War Medicine* 2 (1942) (reprinted in William Haddon, Jr., Edward A. Suchman, and David Klein, *Accident Research: Methods and Approaches,* New York: Harper and Row, 1964, 546); an interview for an oral history of the Aerospace Medical Division, Brooks AFB; Dana Kilanowski's interview with George Nichols; the author's interview with Celia Richards Wilson; Kennedy, *Touching Space*; military and personal records in JPS's files; and various magazine and newspaper articles.

CHAPTER 14: WAITING FOR A TIGER TO SPRING

In addition to the sources listed for chapter 13, the material in this chapter is based on James J. Haggerty, "Fastest Man on Earth," *Collier's*, June 25, 1954, and Chandler, "Project MX-981."

CHAPTER 15: SUBJECT HAD CONSIDERABLE APPREHENSION

In addition to the sources listed for chapters 13 and 14, the material in this chapter is based on Maura Phillips Mackowski, *Testing the Limits: Aviation Medicine and the Origins of Manned Space Flight* (College Station, TX: Texas A & M University Press, 2006); Lyle, Stapp, and Button, "Ophthalmological Hydrostatic Pressure Syndrome," *Transactions of the American Ophthalmological Society* 54 (1956): 121–8; "Health in the Heavens," *Journal of the American Medical Association* 164, no. 7 (June 15, 1957); JPS's status report, December 14, 1954; and various newspaper articles.

CHAPTER 16: THE CONFERENCE

In addition to the sources listed for chapters 13, 14, and 15, the material in this chapter is based on an anonymous interview of JPS (Alamogordo, February 6, 1986); and JPS's abstract, "Twenty-Five Years of Stapp Car Crash Conferences."

CHAPTER 17: SPACE SCIENTISTS IN HOLLYWOOD

The material in this chapter is based on letters to and from JPS; Bushnell, *History of Research in Space Biology and Biodynamics*; Mattson, "Fastest Man on Earth"; Snyder, "Human Impact Tolerance" (SAE 700398, Society of Automotive Engineers, 1970); the author's interviews with JPS and Dana Kilanowski; JPS, "Justification for Car Crash Project"; JPS's abstract, "Twenty-Five Years of Stapp Car Crash Conferences"; JPS, "Science as a Social Force"; Joel W. Eastman, *Styling vs. Safety: The American Automobile Industry and the Development of Automotive Safety, 1900–1966* (Lanham, MD: University Press of America, 1984); military and personal records in JPS's files; and various magazine and newspaper articles.

CHAPTER 18: **SPACE SCIENTISTS IN SPACE**

In addition to the sources listed for chapter 17, the material in this chapter is based on Joseph W. Kittinger with Martin Caidin, *The Long, Lonely Leap* (New York: Dutton, 1961); David G. Simons with Don A. Schanche, *Man High* (Garden City, NY: Doubleday, 1960); the author's interviews with Clifton McClure and Joe Kittinger; anonymous interview of JPS (Alamogordo); "Prelude to Space," *Time*, June 17, 1957; Gregory P. Kennedy, *Touching Space: The Story of Project Manhigh* (Atglen, PA: Schiffer, 2007); official USAF Aero Medical Laboratory memoranda and historical documents; Lillian Levy, "Suit May Protect Astronaut Ejected From Space Capsule," *Science Service*, April 2, 1962; and JPS's 1957 congressional testimony.

CHAPTER 19: **THE MOST SIGNIFICANT APPARATUS**

In addition to the sources listed for chapters 17 and 18, the material in this chapter is based on letters to and from JPS; Project Manhigh technical reports and transcripts; "Ringside in Heaven," *Newsweek*, September 2, 1957; Bushnell, *History of Research in Space Biology and Biodynamics*; Caidin, *Overture to Space* (New York: Duell, Sloan, and Pearce, 1963); the author's interviews with Dana Kilanowski, David Simons, and Duke Gildenberg; Dickson, *Sputnik: The Shock of the Century* (New York: Walker and Co., 2001); Paul G. Hoffman with Neil M. Clark, *Seven Roads to Safety: A Program to Reduce Automobile Accidents* (New York: Harper and Brothers, 1939); Arthur W. Stevens, *Highway Safety and Automobile Styling* (Boston: Christopher, 1941); and JPS's 1957 congressional testimony.

CHAPTER 20: **KICKED UPSTAIRS**

The material in this chapter is based on JPS's memoir; letters to and from JPS; JPS's 1958 congressional testimony; Eastman, *Styling vs. Safety*; the author's interviews with Dana Kilanowski, Clifton McClure, David Simons, Joe Kittinger, and Wilford Stapp; Kennedy, *Touching Space*; Meeter, *The Holloman Story*; military and personal records in JPS's files; and various newspaper and magazine articles.

CHAPTER 21: **THE SURVIVORS CLUB**

The material in this chapter is based on letters to and from JPS; Reichhardt, "First Up?" *Air & Space*, August/September 2000; an interview for an oral history of the Aerospace Medical Division, Brooks AFB; JPS, "The 'G' Spectrum in Space Flight Dynamics" (circa 1960); Joe Kittinger and Craig Ryan, *Come Up and Get Me* (Albuquerque: University of New Mexico Press, 2010); James McAndrew, *The Roswell Report: Case Closed* (Washington, DC: Government Printing Office, 1967); the author's interviews with Joe Kittinger and JPS; Dana Kilanowski's interview with Wilford Stapp; military and personal records in JPS's files; and various newspaper and magazine articles.

CHAPTER 22: EXIT FROM VALLEY FORGE

Material in this chapter is based on letters to and from JPS; JPS's trip reports; an interview for an oral history of the Aerospace Medical Division, Brooks AFB; the author's interviews with Joe Kittinger; Kittinger and Ryan, *Come Up and Get Me*; JPS's 1962 congressional testimony; JPS's Remarks at the 33rd Annual Scientific Meeting of the Aerospace Medical Association; K. Clark, "Roll Chuck Yeager and Evel Knievel into One, and You Get an Idea of Joe Kittinger's Thirst for Danger," *Chicago Tribune Magazine*, June 26, 1988; military and personal records in JPS's files; and various newspaper and magazine articles.

CHAPTER 23: NAZI DOCTORS AND THE NEVER-ENDING ROAD TRIP

The material in this chapter is based on letters to and from JPS; an interview for an oral history of the Aerospace Medical Division, Brooks AFB; the author's interviews with Daniel Thomas, Dana Kilanowski, and Wilford Stapp; Linda Hunt, *Secret Agenda* (New York: St. Martin's Press, 1991); Mackowski, *Testing the Limits*; Tom Bower, *The Paperclip Conspiracy* (London: Michael Joseph, 1987); Eastman, *Styling vs. Safety*; JPS's 1963 congressional testimony; reports to USAF Aerospace Medical Division; JPS's commencement address at San Marcos Academy; an Aerospace Medical Division biographical sketch of JPS; military and personal records in JPS's files; and various newspaper and magazine articles.

CHAPTER 24: WASHINGTON

The material in this chapter is based on letters to and from JPS; an interview for an oral history of the Aerospace Medical Division, Brooks AFB; Dana Kilanowski's interview with Wilford Stapp; the author's interview with Dana Kilanowski; Kevin Graham, *Ralph Nader: Battling for Democracy* (Denver: Windom, 2000); Justin Martin, *Nader: Crusader, Spoiler, Icon* (Cambridge, MA: Perseus, 2002); Thomas Whiteside, *The Investigation of Ralph Nader* (New York: Arbor House, 1972); Dickson, *Sputnik*; transcript of *Meet the Press*, May 29, 1966; JPS, "Proposal to the International Academy of Astronautics for an Orbital International Laboratory Committee"; "Lyndon B. Johnson," www.history.com; JPS, "Informal Comments and Suggestions on Integrated Design of Aircrewmen Protective Headgear"; JPS's 1966 congressional testimony; military and personal records in JPS's files; and various newspaper and magazine articles.

CHAPTER 25: SLUGGING AWAY AT THE STANDARDS

The material in this chapter is based on letters to and from JPS; the author's interviews with Wilford Stapp and Joe Kittinger; Oral History Program CWJS OH-1A Dayton, OH, July 14, 1993; JPS, "Remarks—Opening Session I: IAA International Orbiting Laboratory and Space Science Conference," September 28, 1969; President of the United States, "Citation to Accompany the Award of the Distinguished Service Medal"; John Volpe, Remarks at Ceremony Awarding Distinguished Service Medal to Colonel John P. Stapp, February 8, 1971; JPS, "Closing Remarks: The Future," Symposium on Biodynamic Models and Their Applications, October 28, 1970; Mackowski, *Testing the Limits*; JPS's various memoranda to the National Highway Transportation Safety Administration; Ford, Iacocca, and Nixon on *Frontline*, www.pbs.org/wgbh/pages/frontline/shows/rollover/nixon/; military and personal records in JPS's files; and various newspaper and magazine articles.

CHAPTER 26: WHO WANTS A RESURRECTION, ANYWAY?

The material in this chapter is based on letters to and from JPS; the author's interviews with JPS, Margaret Stapp, Wilford Stapp, Mark Pozzi, and Joe Kittinger; Dana Kilanowski's interview with Wilford Stapp; anonymous interview with JPS (Alamogordo); an interview with JPS, Foundation for the National Medals of Science and Technology, circa 1991; JPS, *For Your Moments of Inertia*; JPS, *Dr. Stapp's Almanac and Rational Calendar*; McAndrew, *The Roswell Report*; "Fastest Man on Earth," *Time*; personal records in JPS's files; and various magazine and newspaper articles.

CHAPTER 27: HONOR AND VIOLATION

The material on Stapp's final years is based on letters to and from JPS; the author's interviews with Mark Pozzi, Wilford Stapp, Daniel Thomas, Dana Kilanowski, John Melvin, Richard Chandler, Richard Snyder, and Keith Jamieson; JPS's autopsy report, Wilford Hall Medical Center, November 18, 1999; JPS, "Requiem"; Michael Lennick's film *The Land of Space and Time* (Foolish Earthling Productions, 2009); personal records in JPS's files; and various magazine and newspaper articles.

CHAPTER 28: THE STAPP YEARS

The material in this chapter is based on interviews with a number of crash investigators and impact researchers (cited); Jeff Plungis and Lisa Zagaroli, "Deadly Driving" series, *Detroit News*, March 3–6, 2002; *60 Minutes*, February 16, 1992; Snyder, "Human Impact Tolerance"; Daniel Thomas, "Important Data on Impact Injury Prevention Needs Protection and Further Analysis with a Long Term Strategy for this National Treasure" (privately distributed); Joseph Hanna and Daniel Kain, "The NFL's Shaky Concussion Policy Exposes the League to Potential Liability Headaches," *NYSBA Entertainment, Arts and Sports Law Journal* 21, no. 3 (Fall/Winter 2010); Mark Fainaru-Wada and Steve Fainaru, *League of Denial: The NFL, Concussions and the Battle for Truth* (New York: Crown Archetype, 2013); David Viano, et al., "Concussion in Professional Football: Comparsion with Boxing Head Impacts—Part 10," *Neurosurgery*, December 2005, 1154–72; Bushnell, "Origin and Operation of the First Holloman Track 1949–1956," Vol. 1 of "History of Tracks and Track Testing at the Air Force Missile Development Center," ASTIA 231907, 1956; Alison L. Schmidt et al., *Establishing the Biodynamics Data Resource (BDR): Human Volunteer Impact Acceleration Research Data in the BDR* (U.S. Army Aeromedical Research Laboratory, October 2009); D. M. Severy, "Photographic instrumentation for collision injury research," *Journal of the Society of Motion Picture and Television Engineers* 67, no. 69 (1958); JPS, "Effects of Mechanical Force on Living Tissues"; a press release from the plaintiff's attorneys in the Flax case; and various newspaper and magazine articles.

INTERVIEWS

All interviews were conducted by the author unless otherwise noted.

Franicis Beaupre (March 8, 1992)

Richard Chandler, conducted by Dana Kilanowski (November 16, 2001)

Richard Chandler (July 5, 2013)

Marion Curtis (November 10, 2012)

Bernard Gildenberg, conducted by Dana Kilanowski (January 19, 2001)

Bernard Gildenberg (October 3, 1991)

Keith Jamieson (December 15-16, 2012)

Dana Kilanowski (May 21, 2011 and December 9, 2012)

Joe Kittinger (July 23-26, 2008)

Clifton McClure (March 2 and August 10, 1992)

John Melvin (June 29, 2013)

William Muzzy (January 20, 2012)

George Nichols, conducted by Dana Kilanowski (July 11, 2003)

Nicholas Perrone (June 30, 2013)

Mark Pozzi (November 7, 2012, and January 4, 10, and 16, 2013)

Kenneth Saczalski (February 10, 2013)

David Simons (September 11, 1989)

Vera Winzen Simons (June 15 and 22, 1992, and June 5, 1994)

Richard Snyder (October 30, 2011)

John Paul Stapp (October 3–5, 1991)

Margaret Stapp (October 16, 17, 2010)

Wilford Stapp (October 16, 17, 2010)

Daniel Thomas (January 21, 2012)

Celia Richards Wilson (May 15, 2011)

LIST OF ILLUSTRATIONS

ACKNOWLEDGMENTS

AS A CAREER MILITARY OFFICER and longtime public figure, John Paul Stapp lived a well-documented life: add to the service records and press coverage his prolific correspondence, his hundreds of publications and speeches, dozens of interviews spanning decades, and a sprawling—if never completed—memoir. The inevitable inconsistencies and disagreements among sources are part of the biographer's challenge. I am grateful to those who've done their best to help escort me through some slippery terrain. They've done what they could—on both the details and the larger architecture of the story. Any errors that appear here are mine.

A number of men who knew and worked with Stapp, some of whom are themselves titans in the field of biomechanical or neurological research, shared their memories, conclusions, and speculations. Among these are Dr. Richard Snyder, Mark Pozzi, Dr. Daniel Thomas, William Muzzy, Dr. Kenneth Saczalski, Dr. Nicholas Perrone, Dr. John Melvin, and Dr. Keith Jamieson. A very special thanks goes to Richard Chandler for his superb Stapp bibliography, of which this book has made liberal use. I'm proud to know these men and count myself extraordinarily lucky to have had their company and their cooperation.

Among others who provided documents, photographs, and recollections: George House, Mike Smith, Ron Keller, and Chris Orwoll, all of the New Mexico Museum of Space History; Richard Bark, Tony Moore, and Jeannine Geiger at Edwards Air Force Base; Elizabeth Suckow at NASA; John Couch at Ralph Edwards Productions; Melissa Gottwald at Embry-Riddle Aeronautical University; Nick Spark; Bobbie McClure; Celia Richards Wilson; Sherry Kittinger; Theresa Lopez Lusczek; Erich Streckfuss; and Marion Curtis.

I am truly grateful for the generous assistance I got from Dana Marcotte

Kilanowski, historian for the Society of Experimental Test Pilots. Dana was close to both John Paul and Lili Stapp. She interviewed them multiple times and shared with me her collection of Stapp documents and photographs. She also served as a sounding board as I worked my way through the story, and helped connect me with some of the other key players. She was a valuable ally throughout the process.

Mark Pozzi not only admired Stapp but considered him his best friend. Mark spent many hours talking to and corresponding with me about a thousand different aspects of the story, despite a challenging work schedule and never-ending ranch chores. He patiently attempted to educate me on the science of crash protection and the dizzying politics of auto safety in America. Mark carries on Stapp's fight, and his passion is inspiring.

Joe Kittinger, the man Stapp called "the greatest pilot and gentleman" he ever met, was a monumental help. Outside of family, Joe probably logged more time with Stapp over the years than anyone else. His perspective is unique, and I have been privileged to spend time with Joe talking about the man he credits with changing the arc of his career and his life.

Thanks to Annette Brown and Kathleen Ninette Smith—on Lili Stapp's side of the family—for their help and encouragement. And deep, heartfelt thanks to Wilford Stapp for trusting me and for helping me understand the meaning of the Cause. Margaret Stapp died before this book was completed, but she also provided great insight into the Stapp story. She was one of the loveliest people I've ever met.

I also want to acknowledge the help I received from my wife, Kathy Narramore. She was a great research assistant at a time when I desperately needed one. She read through and organized many thousands of pages of material so I could focus on constructing the story. She also challenged my thinking about people and events, and got me to see some of them from new perspectives.

Finally, thanks to my superb copy editor, Allegra Huston; to my friend and long-time agent, Al Zuckerman at Writers House; and to the smartest editor I've had the pleasure to work with, Philip Marino at Norton/Liveright.

—*Craig Ryan, Portland, Oregon, 2015*

INDEX

Page numbers in *italics* refer to illustrations.